Smart Nanomaterials Technology

Series Editors

Azamal Husen, Wolaita Sodo University, Wolaita, Ethiopia

Mohammad Jawaid, Laboratory of Biocomposite Technology, Universiti Putra Malaysia, INTROP, Serdang, Selangor, Malaysia

Nanotechnology is a rapidly growing scientific field and has attracted a great interest over the last few years because of its abundant applications in different fields like biology, physics and chemistry. This science deals with the production of minute particles called nanomaterials having dimensions between 1 and 100 nm which may serve as building blocks for various physical and biological systems. On the other hand, there is the class of smart materials where the material that can stimuli by external factors and results a new kind of functional properties. The combination of these two classes forms a new class of smart nanomaterials, which produces unique functional material properties and a great opportunity to larger span of application. Smart nanomaterials have been employed by researchers to use it effectively in agricultural production, soil improvement, disease management, energy and environment, medical science, pharmaceuticals, engineering, food, animal husbandry and forestry sectors.

This book series in Smart Nanomaterials Technology aims to comprehensively cover topics in the fabrication, synthesis and application of these materials for applications in the following fields:

- Energy Systems—Renewable energy, energy storage (supercapacitors and electrochemical cells), hydrogen storage, photocatalytic water splitting for hydrogen production
- Biomedical—controlled release of drugs, treatment of various diseases, biosensors,
- Agricultural—agricultural production, soil improvement, disease management, animal feed, egg, milk and meat production/processing,
- Forestry—wood preservation, protection, disease management
- Environment—wastewater treatment, separation of hazardous contaminants from wastewater, indoor air filters.

Azamal Husen
Editor

Secondary Metabolites Based Green Synthesis of Nanomaterials and Their Applications

Springer

Editor
Azamal Husen
Wolaita Sodo University
Wolaita, Ethiopia

Smart Nanomaterials Technology
ISBN 978-981-99-0926-1 ISBN 978-981-99-0927-8 (eBook)
https://doi.org/10.1007/978-981-99-0927-8

This Springer imprint is published by the registered company Springer Nature Singapore Pte Ltd.
The registered company address is: 152 Beach Road, #21-01/04 Gateway East, Singapore 189721, Singapore

To my father, Asgar Ahmed
(July 1, 1929–December 17, 1993)

Preface

Nanotechnology is gaining importance in every field of science and technology. Green synthesis of nanomaterials involves the use of microorganisms such as bacteria, fungi, viruses, and different lower and higher plants. In this connections, green syntheses of nanomaterials from plant extract becoming popular in comparison to synthesis using microorganisms. Plant based-nanomaterials synthesis are easy, have no need to bring back from the culture medium, and are safe. Additionally, plant-based nanomaterials are eco-friendly, in comparison to physical and chemical modes of synthesis. Several lower and higher plants are rich in terms of the presence of secondary metabolites. These metabolites have been used as medicine in crude extract form or with some other formulations. Both lower and higher plants have been also used to isolate the bioactive compounds in modern medicine as well as in herbal medicine systems. Thus, phytochemicals present in the plant and their parts play an important role in nanomaterials synthesis, mainly due to presence of a significant number of secondary metabolites, for instance, alkaloids, flavonoids, saponins, steroids, tannins, etc. Further, essential and aromatic oils have been also explored for nanomaterials synthesis, and they are also equally useful in terms of their various biological applications. These organic ingredients come from a wide range of plant components, such as leaves, stems, roots, shoots, flowers, bark, and seeds. Globally, the presence of different plants has shown a capability to produce huge and diverse groups of secondary metabolites. The functional groups present in the plant extract acts as capping and stabilizing agent. Most of the time, pure isolated bioactive compounds are more biologically active; hence scholars are focusing their research on the synthesis of nanomaterials using some particular class of secondary metabolites. In recent days, scholars are developing new techniques and or formulation that is suitable for plants secondary metabolites to increase their native functions. Investigations have shown that the green synthesized nanomaterials were found to be more biologically active in comparison to chemically synthesized nanomaterials. These nanomaterials and or nanocomposites found different applications especially in drug delivery, in detection and cure of tumor/cancer cells, in diagnosis of a genetic disorder, photoimaging, and angiogenesis detection. They have also shown several applications in agricultural, horticultural as well as forestry sectors.

The book in hand covers a wide range of topics as mentioned above. It incorporates chapters that the authors have skillfully crafted with clarity and precision, reviewing up-to-date literature with lucid illustrations. The book would cater to the need of graduate students as a textbook and simultaneously be useful for both novices and experienced scientists and or researchers working in the discipline of nanotechnology, nanomedicine, medicinal plants, plant science, economic botany, chemistry, biotechnology, pharmacognosy, pharmaceuticals, industrial chemistry, and many other interdisciplinary subjects. It should also inspire industrialists and policy makers associated with plant-based nano products.

I extend my sincere thanks to all contributors for their timely response and excellent contributions. Finally, my special thanks go to Shagufta, Zaara, Mehwish, and Huzaifa for providing their time and overall extended support to put everything together. I shall be happy to receive comments and criticism, if any, from subject experts and general readers of this book.

Wolaita, Ethiopia Azamal Husen

About This Book

Nanotechnology is gaining importance in every field of science and technology. Green synthesis of nanomaterials involves the use of microorganisms such as bacteria, fungi, viruses, and different lower and higher plants. In this connections, green syntheses of nanomaterials from plant extract becoming popular in comparison to synthesis using microorganisms. Plant based-nanomaterials synthesis are easy, have no need to bring back from the culture medium, and are safe. Additionally, plant-based nanomaterials are eco-friendly, in comparison to physical and chemical modes of synthesis. Several lower and higher plants are rich in terms of the presence of secondary metabolites. These metabolites have been used as medicine in crude extract form or with some other formulations. Both lower and higher plants have been also used to isolate the bioactive compounds in modern medicine as well as in herbal medicine systems. Thus, phytochemicals present in the plant and their parts play an important role in nanomaterials synthesis, mainly due to presence of a significant number of secondary metabolites, for instance, alkaloids, flavonoids, saponins, steroids, tannins, etc. Further, essential and aromatic oils have been also explored for nanomaterials synthesis, and they are also equally useful in terms of their various biological applications. These organic ingredients come from a wide range of plant components, such as leaves, stems, roots, shoots, flowers, bark, and seeds. Globally, the presence of different plants has shown a capability to produce huge and diverse groups of secondary metabolites. The functional groups present in the plant extract acts as capping and stabilizing agent. Most of the time, pure isolated bioactive compounds are more biologically active; hence scholars are focusing their research on the synthesis of nanomaterials using some particular class of secondary metabolites. In recent days, scholars are developing new techniques and/or formulation that is suitable for plants secondary metabolites to increase their native functions. Investigations have shown that the green synthesized nanomaterials were found to be more biologically active in comparison to chemically synthesized nanomaterials. These nanomaterials and/or nanocomposites found different applications especially in drug delivery, in detection and cure of tumor/cancer cells, in diagnosis of a genetic disorder, photoimaging, and angiogenesis detection. They have also shown several applications in agricultural, horticultural as well as forestry sectors.

The book in hand covers a wide range of topics as mentioned above. It incorporates chapters that the authors have skillfully crafted with clarity and precision, reviewing up-to-date literature with lucid illustrations. The book would cater to the need of graduate students as a textbook and simultaneously be useful for both novices and experienced scientists and or researchers working in the discipline of nanotechnology, nanomedicine, medicinal plants, plant science, economic botany, chemistry, biotechnology, pharmacognosy, pharmaceuticals, industrial chemistry, and many other interdisciplinary subjects. It should also inspire industrialists and policy makers associated with plant-based nano products.

Contents

About the Editor

Azamal Husen served as Professor & Head, Department of Biology, University of Gondar, Ethiopia, and is a Foreign Delegate at Wolaita Sodo University, Wolaita, Ethiopia. He was also a Visiting Faculty member of the Forest Research Institute, and the Doon College of Agriculture and Forest at Dehra Dun, India. He specializes in biogenic nanomaterial fabrication and application, plant responses to environmental stresses and nanomaterials at the physiological, biochemical, and molecular levels, herbal medicine, and clonal propagation for improvement of tree species, and has published over 200 research articles. He is contributed to R&D projects of World Bank, ICAR, ICFRE, JBIC, etc. He is on the advisory board of Cambridge Scholars Publishing, UK. Husen has been on the Editorial Board and the panel of reviewers of several reputed journals. He is a Fellow of the Plantae group of the American Society of Plant Biologists, and a Member of the International Society of Root Research, Asian Council of Science Editors, and INPST. He is Editor-in-Chief of the *American Journal of Plant Physiology*, and a Series Editor of *Exploring Medicinal Plants* (Taylor & Francis Group, USA); *Plant Biology, Sustainability, and Climate Change* (Elsevier, USA); and *Smart Nanomaterials Technology* (Springer Nature, Singapore).

Plant-Based Metabolites and Their Uses in Nanomaterials Synthesis: An Overview

Kaliyan Barathikannan, Ramachandran Chelliah, Vijayalakshmi Selvakumar, Fazle Elahi, Momna Rubab, Simpy Sanyal, Su-Jung Yeon, and Deog-Hwan Oh

Abstract The primary disciplines in the field of nanotechnology are biology, physics, chemistry, and material sciences. It also involves the creation of novel therapeutic nanomaterials for biomedical and pharmaceutical uses. Various macro-microscopic organisms, including plants, bacteria, fungi, seaweeds, and microalgae, are responsible for the biological synthesis of nanoparticles. The various endemic pathogens were effectively controlled by biosynthesized nanomaterials while causing fewer harmful side effects. In addition to alkaloids, flavonoids, saponins, steroids, and tannins, the plant also contains a variety of additional nutritional components. These natural substances can be found in plants. These organic ingredients come from a wide range of plant components, such as leaves, stems, roots, shoots, flowers, bark, and seeds. Recent research has demonstrated the potential of plant extracts as a non-hazardous precursor for the synthesis of nanomaterials. The plant extract serves as a reducing and stabilizing agent for the bio-reduction reaction to synthesize novel metallic nanoparticles because it contains a variety of secondary metabolites. A complex mixture of various phytochemicals, including phenolics, sugars, flavonoids,

K. Barathikannan · R. Chelliah (✉) · V. Selvakumar · F. Elahi · M. Rubab · S.-J. Yeon · D.-H. Oh
Department of Food Science and Biotechnology, College of Agriculture and Life Science, Kangwon National University, Chuncheon 24341, Korea
e-mail: ramachandran865@kangwon.ac.kr

K. Barathikannan
Agricultural and Life Science Research Institute, Kangwon National University, Chuncheon 24341, Korea

R. Chelliah
Department of Food Science and Biotechnology, Kangwon Institute of Inclusive Technology (KIIT), Kangwon National University, Chuncheon 24341, Korea

Saveetha School of Engineering, (SIMATS) University, Chennai 600124, Tamil Nadu, India

M. Rubab
Department of Food Science and Technology University of Management and Technology (UMT), Lahore, Pakistan

S. Sanyal
Department of Electrical and Computer Engineering, Sungkyunkwan University, Suwon 16419, Republic of Korea

A. Husen (ed.), *Secondary Metabolites Based Green Synthesis of Nanomaterials and Their Applications*, Smart Nanomaterials Technology,
https://doi.org/10.1007/978-981-99-0927-8_1

xanthones, and others, makes up plant extract. Nanoparticles are synthesized using non-biological techniques (chemical and physical), which are extremely dangerous and harmful to living organisms. In addition, the environmentally friendly, low-cost, one-step biological synthesis of metallic nanoparticles. Several nanomaterials that are better for the environment can be effectively manufactured using plants. These nanoparticles include such as cobalt, copper, silver, gold, palladium, platinum, zinc oxide, and magnetite. Furthermore, the plant-mediated nanoparticles show promise as a potential treatment for a variety of diseases, which includes cancer, HIV, hepatitis, malaria, and other acute infections.

Keywords Nanotechnology · Pathogens · Secondary metabolites · Nanomaterials · Plant extracts

1 Introduction

The nanoparticle has exciting possible applications in a wide range of industries, including energy, nutrition, and medicine. A challenge in biomaterial science has been the biogenic synthesis of monodispersed nanoparticles with specific sizes and shapes. Nanotechnology is a fascinating area of science that is growing quickly. It makes materials on the nano-scale (1–100 nm) that can be used in many different ways [80]. Metal oxide NPs are essential for numerous applications in research and technology. Nanomaterials and metal oxide NPs are increasingly being used in medical applications for cancer treatment, antimicrobial therapeutic agent, biosensing, chemotherapy, and imaging purposes [55, 82]. Additionally, it has produced an outstanding economic benefit for the pharmaceutical sector in the treatment of numerous viral and bacterial diseases. Biosynthesis systems have more compensation due to more biological entities and ecologically favorable procedures. It's been studied how to create nanomaterials using the rich variety and easy accessibility of plants [21, 32]. Recent studies show the successful biosynthesis of nanoparticles, wires, flowers, and tubes. Potential applications for these biologically synthesized nanomaterials include treatment, diagnosis, the development of surgical nano-devices, the production of commercial products, and agricultural sectors [9, 29-37, 39, 48, 49, 83, 85]. The application of nanotechnology has a significant impact on the treatment of chronic diseases in the healthcare sector. Accordingly, the eco-friendly synthesis of nanoparticles is viewed as a building block for following generations to overcome multiple diseases. The crude extracted from the plant contains novel secondary metabolites, which include phenolic acid, flavonoids, alkaloids, and terpenoids [2, 38, 84].

As a result, the production of nanoparticles in a way that is beneficial to the environment is seen as an essential component for future generations in the fight against different diseases. These substances are mostly accountable for the change from ionic to nanostructured materials as a result of the reaction. These primary and secondary metabolites are actively participating in the redox process that is taking place to produce environmentally favorable nanoparticles. Numerous prior researches have shown that biosynthesized nanoparticles are able to efficiently regulate oxidative stress, genotoxicity, and apoptosis-related changes. This has been shown to be the case in a variety of different contexts [27]. Additionally, nanoparticles have a wide variety of uses in the fields of agriculture and plant sciences. For instance, the capability of nanoparticle biotechnological processes technology to convert waste from agriculture and food production into fuel as well as other valuable byproducts [60]. On the basis of this information, the review concentrated on metallic nanoparticles that were biosynthesized from plant derivatives and their uses in the medical and industrial fields, including the treatment of wastewater, the production of cosmetic items, and the processing sector.

2 Traditional Approaches to Metals

Gold has been valued throughout history as a representation of both power and prosperity. Since then, several types of gold metal nanoparticles have been utilized in an effort to better the health of humans. Metallic gold nanoparticles (GNPs) benefit human health and cosmetics nowadays. In the eighteenth century, Egyptians utilized water that had gold metal dissolved in it for the purpose of psychological and spiritual purification [51, 76]. In rural areas, where people still hold high regard for the restorative qualities of gold, it is common practice for peasants to cook their rice with a gold pellet in the expectation that this will cause an increase in mineral absorption by the body. Silver has been used for centuries to keep food from going bad and to keep illnesses in the body under control. Silver is used as a gastric treatment and wound healer [19, 66].

3 Different Methodology for the Development of Metallic Nanoparticle Synthesis

Synthesizing nanoparticles (NPs) can be accomplished using a variety of approaches, including physical, chemical, enzymatic, and even biologically derived. Physical processes include plasma arcing, ball milling, thermal evaporation, spray coating, ultra-thin films, laser ablation dissociation, imprinting approaches, sputtering coating, and surface development [11]. Similar to how electro-deposition, the sol–gel process, chemical solution deposition, chemical vapour deposition, the soft chemical approach, the Langmuir Blodgett method, the catalytic pathway, and hydrolysis are all employed in chemical methods to synthesis NPS. Physical and chemical processes have utilized high levels of radiation as well as extremely concentrated reducing agents and stabilizing chemicals that are harmful to the environment

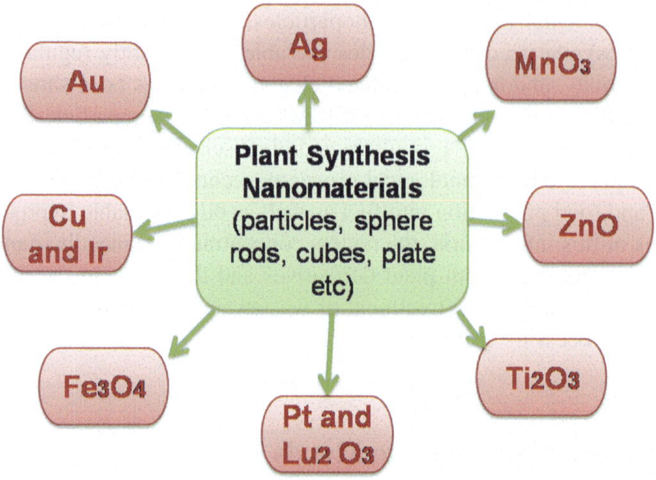

Fig. 1 Different types of green synthesized plant-based metallic nanoparticle

and human health. Therefore, the biological production of nanoparticles is a single-step green synthesis process, and less energy is required to produce eco-friendly NPs [71]. In addition, environmentally friendly materials such as plant extracts, bacteria, fungi, microalgae, cyanobacteria, diatom, seaweed (macroalgae), and enzymes are used in biological processes (Fig. 1).

4 Effect on the Different Parts of Plants Applied to Synthesize Metallic NPs

In recent times, there has been an increase in interest in plant-mediated nanomaterial as a result of the numerous applications it offers in a variety of disciplines as a result of its physic-chemical properties. Synthesized from natural resources, the various metallic nanoparticles, such as gold, silver, platinum, zinc, copper, titanium oxide, magnetite, and nickel, were the focus of the research [13] (Fig. 2). The diverse components of the plant, including the stem, root, fruit, seed, callus, peel, leaves, and flower, are utilized in the biological synthesis of nanomaterials in a wide range of shapes and sizes. These nanomaterials can be used in a variety of applications [73]. The different sizes and characteristics of the nanoparticles made from plant materials. A wide variety of metal concentrations and plant extract concentrations in the reaction medium can affect biosynthesis reactions and change the size and shape of the nanoparticles (Fig. 3).

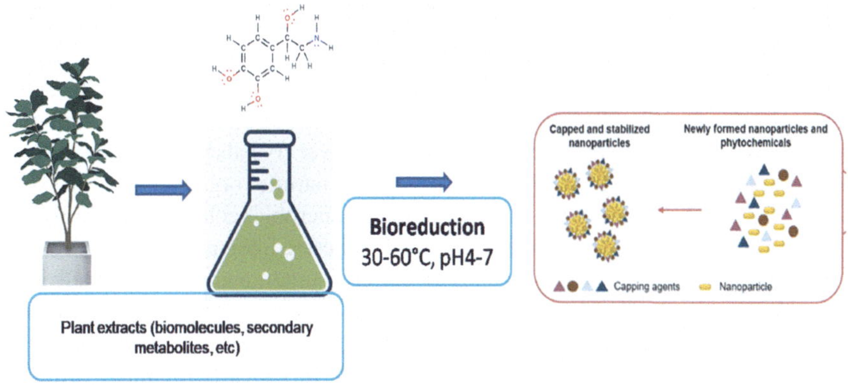

Fig. 2 Basic Methodology in preparation of green synthesized plant based metallic nanoparticle and stabilization

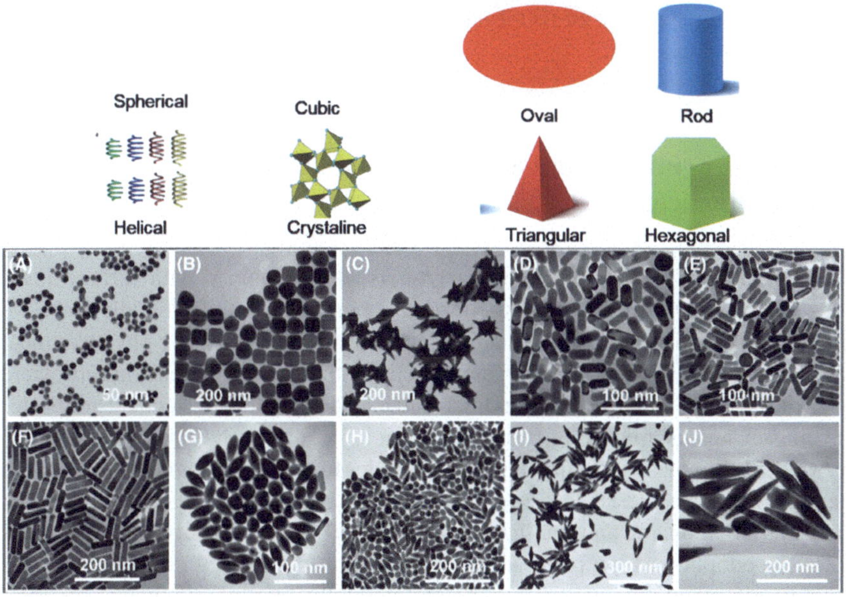

Fig. 3 Different dimensional (size and shape) based biologically synthesized nanoparticle

5 Role of Stem Based Green Synthesized of Nanoparticle (NPs)

The methanol (polar solvent) extracts of the *Callicarpa maingayi* stem was utilized in the manufacture of silver nanoparticles, which resulted in the formation of a [Ag (*Callicarpa maingayi*)] + complex. An aldehyde or ketone is present in the plant extracts, and its primary function is to participate in the process that converts metallic silver nanostructures from silver ions. Polypeptides and amide I are the molecules responsible for encapsulating ionic substances into metallic nanoparticles, as shown by the diverse functional groups AC-0 and C-N. The intricate and incompletely understood molecular study of the biosynthesis of silver crystals [81]. However, some earlier research has offered model methods for how harmful organisms interact with nanoparticles. The lipoproteins that make up the microbial cell wall are destroyed when silver nanoparticles that have been biosynthesized contact with the proteins that are found on the outer membranes of bacteria, fungi, and viruses. Finally, the cell division was stopped, and now the cell leads to death. At room temperature, extracts of *Cissus quadrangularis* are used in the photosynthesis of silver nanoparticles. The stem section of the plant extract reveals very clearly the many different functional groups that are engaged in the oxidation of silver ions. These functional groups include the hydroxyl, aldehyde, and polyphenols. Silver nanoparticles that have been synthesized exhibit greater effectiveness against the pathogenic microorganisms *Bacillus subtilis* and *Klebsiella planticola.* As a direct consequence of this, the biologically synthesized metal nanoparticles functioned exceptionally well as antimicrobial drugs [70].

6 Role of Fruits Mediated Synthesis of Metallic Nanoparticles

The plant fruiting kernels of *Tribulus terrestris* were combined with various quantities of silver nitrate to produce AgNPs that are environmentally friendly and have specific forms. The single-step reduction reaction is caused by active phytochemical components in the extract. Silver nanoparticles with different dimension form were synthesized using the *T. terrestris* extract, and they demonstrated excellent antibacterial efficacy against multidrug-resistant human infections [96]. Similar research has been done on synthesizing palladium nanoparticles from grape polyphenols and using them to combat bacterial infections. Additionally, *Rumex hymenosepalus* extract functions as a reducing and stabilizing agent for the synthesis of silver nanoparticles [72]. The adoption of the pharmacologically solicited chemical conditions for the synthesis of nanomaterials is a major step toward the pharmacological solicitation goal of treating a wide range of endemic diseases. This goal has been significantly supported by this adoption.

7 Effect of Seeds Mediated Green Synthesized of Nanoparticle (NPs)

The fenugreek seed extracts contain a significant amount of polyphenols in addition to other naturally occurring therapeutic elements such as cellulose, tannin, and vitamin supplements. The significant oxidizing agents found in fenugreek seed extract help to minimize the amount of chloroauric acid in the surfactant, which in turn results in enhanced performance. The functional groups COO (carboxylic), CN, and CC can be found in the extract of the seed. The electrochemical durability of gold nanoparticles can be maintained by quercetin, and the functional group of molecules can act as a surfactant for metallic (gold) nanoparticles [7]. The amount of silver ion that could be reduced by an aqueous extract of Macrotyloma uniflorum was increased. It's possible that this is due to the presence of caffeic acid in the extract [6].

8 Effect of Leaves Mediated Green Synthesized of Nanoparticle (NPs)

It was reported that an extract from plant leaves was employed as a mediator in the process of making nanoparticles. Extracts of the leaves of many different plants, including *Centella asiatica*, *Murraya koenigii*, and *Alternanthera sessilis*, have been the subject of research [28]. Moreover, it was discovered that the leaves of P. nigrum contain an essential bioactive ingredient that is engaged in the creation of nanoparticles using an environmentally benign manner [94]. The biological mechanism of produced silver nanoparticles at 100 mg/ml was efficient drug dosage on HEp-2 and HeLa cell lines to influence the essential metabolic function in tumor cell [5]. The AgNPs have potent medications in anticancer drug that can cure a variety of oncology disorders and other terrible conditions. It functions as reducing agents for the creation of silver nanoparticles and may boost the cytotoxic effects of the tumor cells. *P. nigrum* preparations have been shown to include linguine and Piper amide. Earlier studies have described a technique for the environmentally friendly manufacture of silver nanoparticles by employing the extract of the leaves of the *Artemisia nilagirica* plant [98]. It demonstrates a potent instrument that can be utilized by antimicrobial drugs in the here and now as well as in the not too distant future. In a similar fashion, numerous pathogenic disorders in humans can be controlled by silver nanoparticles that are generated from plant resources.

9 Role of Flowers as a Source for NPs Production (NPs)

Environmentally friendly gold nanoparticles (GNPs) are made from rose petals [40]. Sugars and proteins are abundant in extract medium, these functional compounds converts tetrachloroaurate salt into GNPs. Similar to this, *Catharanthus roseus* and *Clitoria ternatea*, two different types of flowers, are employed to create metallic nanoparticles in the desired sizes and shapes [95]. The green chemical process is used to create the medicinally useful gold nanoparticle extract from *Nyctanthes arbor tristis* blossoms, which has been successfully used to suppress hazardous pathogenic bacteria [42]. The water extract (polar solvent) of Mirabilis jalapa flowers reduces gold nanoparticles [58]. The plant metabolites shown in Table 1 are representative of the bioreduction reaction that leads to the creation of metallic nanoparticles and their therapeutic uses.

10 Pharmacological Application of Metallic Nanoparticles

10.1 Anti-Bactericidal Efficiency of Plant Based Green Synthesized Metallic Nanoparticles

The AgNPs were able to effectively damage the polymer components of cell membranes in pathogens. The bilateral response of nanoparticles inevitably results in the breakdown of the cellular membrane and a disruption of the pathway responsible for cellular metabolism in the microbial system. The increased levels of silver nanoparticles have an accelerated membrane penetrability compared to the lower concentration levels, which causes the cellular structure of the bacteria disruption. The optimum absorption coefficient was found in *Rhizophora apiculate* diminished silver nanoparticles, which showed bacteriostatic effect (reduced bacterial number), when compared with AgNO3-treated cells. This is possibly attributed to the smaller size of the particles and greater surface area, both of which result in an increased in membrane permeability and cell wall destruction [1]. The interactions between bacteria and the metallic nanoparticles of silver and gold have resulted in a binding with the active site of the cell membrane, which in turn inhibits the functions of the cell cycle. The biosynthesis of silver nanoparticles was accomplished in a single step using the peel extract of Citrus sinensis as both a capping agent and reducing agent. The effectiveness of the *C. sinensis* peels in reducing silver nanoparticles was demonstrated, and the activity of the extract against *Escherichia coli*, *Pseudomonas aeruginosa* (gram-negative), and *Staphylococcus aureus* (gram-positive) bacteria was demonstrated as well [44]. Earlier studies have shown that silver nanoparticles employed to synthesize from the leaves of the *Acalypha indica* plant can effectively control water-borne pathogens at a concentration as low as 10 lg/ml [47].

Table 1 Different functional activity of green synthesized metallic nanoparticles synthesis

Plants used	Nanoparticles	Parts of plant	Size (nm)	Shapes	Plant metabolites involved in bioreduction	Pharmacological applications	Cited
Artemisia nilagirica	Ag	Leaves	70–90	Spherical	Secondary metabolites	Antimicrobial	[89]
A. mexicana	Ag	Leaves	20–50	Spherical	Protein,	Antimicrobial	[87]
Aloe vera	In_2O_3	Leaf	5–50	Spherical	Biomolecules	Optical properties	[53]
Acalypha indica	Ag, Au	Leaves	20–30	Spherical	Quercetin, plant pigment	Antibacterial	[47]
Alternanthera sessilis	Ag	Whole	40	Spherical	Amine, carboxyl group	Antioxidant, antimicrobial	[61]
Boswellia serrata	Ag	Gum	7–10	Spherical	Protein, enzyme	Antibacterial	[46]
Cinnamon zeylanicum	Ag	Leaves	45	Spherical	Water soluble organics	Antibacterial	[77]
Caria papaya	Ag	Fruit	15	Spherical	Hydroxyl flavones, catechins	Antimicrobial	[41]
Cassia fistula	Au	Stem	55–98	Spherical	Hydroxyl group	Antihypoglycemic	[17]
Citrullus colocynthis	Ag	Calli	5–70	Triangle	Polyphenols	Antioxidant, anticancer	[79]
Dillenia indica	Ag	Fruit	11–24	Spherical	Biomolecules	Antibacterial	[88]
Citrus sinensis	Ag	Peel	35	Spherical	Water soluble compounds	Antibacterial	[44]
Mirabilis jalapa	Au	Flowers	~100	Spherical	Polysaccharides	Antimicrobial	[97]
Euphorbia prostrata Rahuman [100]	Ag	Leaves	52	Rod, spherical	Protein, polyphenols	Antiplasmodial	[100]
Dioscorea bulbifera	Ag	Tuber	8–20	Rod, triangular	Diosgenin, ascorbic acid	Antimicrobial	[25]
Withania somnifera al. [59]	Ag	Leaves	5–40	Irregular, spherical	Methyl 7-oxooctadecanoate	Antimicrobial	[59]

(continued)

Table 1 (continued)

Plants used	Nanoparticles	Parts of plant	Size (nm)	Shapes	Plant metabolites involved in bioreduction	Pharmacological applications	Cited
Tinospora cordifolia	Ag	Leaves	34	Spherical	Phenolic compound	Antilarvicidal	[43]
Melia azedarach	Ag	Leaves	78	Irregular	Tannic acid, polyphenols	Cytotoxicity	[92]
Gelsemium sempervirens	Ag	Whole	112	Spherical	Protein, amide, amine group	Cytotoxicity	[18]
Lippia citriodora	Ag	Leaves	15–30	Spherical,	Isoverbascoside compound	Antimicrobial	[16]
Iresine herbstii	Ag	Leaves	44–64	Cubic	Biomolecules phenolic compound	Biological activities	[20]
Melia azedarach	Ag	Leaves	78	Irregular	Tannic acid, polyphenols	Cytotoxicity	[92]
Trigonella-foenum graecum	Au	Seed	15–25	Spherical	Flavonoids	Catalytic	[8]

10.2 Anti-Fungicidal Efficacy of Plant Based Green Synthesized Nanoparticles

Biologically synthesized metal oxide nanoparticles have more antifungal and antibacterial potential than fluconazole and amphotericin [64]. The membrane damage in Candida species was very clearly demonstrated by the plant-derived Ag nanoparticles. Along with disruption of fungal based cellular functions and injury to their intercellular constituents [99]. The majority of commercial antifungal medications only have a few clinical uses, and they also have much more negative side effects and slow healing rates for microbial diseases. After using the drugs, some people experience side effects such as renal dysfunction, elevated body temperature, vomiting, liver disease, and indigestion. Other possible adverse effects include the consumption of commercially available drugs. The use of nanoparticles in the creation of a new and more effective drug against microbes was investigated. Fungi have a particularly high polymer content of fatty acids and proteins in their cell walls. The multipurpose silver nanoparticles have a potent activity against spore-producing fungal infection and severely damage to the fungal growth [78]. Treatment of fungi infection with nanoparticles resulted in significantly alterations to the membrane structure of the fungal species [56, 69].

10.3 Effect of Metallic Nanoparticles Exhibiting Anti-Plasmodial Action Effect

At this time, the most illnesses are being carried and transmitted all over the world by various vectors. In the event of a communicable disease, insect management is a necessity of the highest priority. The effectiveness of the enhanced anti-plasmodial species-specific control approach is diminished. This strategy has proven to be more cost-effective in terms of controlling the specific microorganisms in the healthcare sector, despite the fact that it has been more economical. Specifically, there is an immediate need for anti-malarial medications that are both effective and economical so that plasmodial activity can be brought under control. Plants have been exploited as conventional naturally derived products during the recent decades, and there are sufficient plant sources for the creation of NPs drugs to treat malaria and other tropical diseases [10]. It has been demonstrated that chemical components derived from plants, such as quinine, artemisinin, and aromatic compound, can be utilized effectively against malaria parasite strains that have developed resistance. Due to the great resistance of parasites, the alternative medicine is needed for regulating the resistant strains. Malaria can be efficiently controlled in the environment by using plant-made metallic nanoparticles including silver, platinum, and palladium nanoparticles. These nanoparticles are manufactured by plants. In addition, the bioactive manufacture of metallic silver nanostructures from bioactive compound has been employed to reduce the number of new cases of malaria [68].

10.4 Anti-Inflammatory Efficacy of Nanoparticles

A cascade mechanism that is an essential element of the immune system's reaction to an infection is the synthesis of inflammatory cytokines and auxins. This generation can occur in keratinocytes such as Lymphocytes, T—lymphocytes, and macrophages. The endocrine system secretes a number of inflammatory mediators, including enzymes and antibodies. The key immune organs also release cytokines, IL-1, and IL-2, which have the capacity to reduce inflammation. The healing process is initiated by these anti-inflammatory mediators. Additionally, inflammatory mediators play a role in biochemical processes and limit the spread of illnesses. Gold nanoparticles produced by biosynthesis were successful in stimulating the processes of cell therapy and tissue repair. The study provided conclusive evidence that biosynthesized nanomaterials of gold and platinum are effective alternatives to conventional anti-inflammatory treatments [67].

10.5 Research on the Anticancer Effects Mediated by Plant Based Green Synthesized Nanoparticles

Cancer is characterized by an abnormal cellular proliferation that is accompanied by dramatic shifts in the biochemical and enzymatic parameters of the cells. This is a trait that is shared by all tumor cells. Based on utilizing bio-based nanomaterials as innovative regulatory agents, the overexpression of cellular proliferation will be halted and managed with organized cell cycle mechanisms in malignant cells. Additionally, plant-mediated nanomaterials have a significant impact on a variety of tumor cell lines, including Hep 2, HCT 116, and He La cell lines [90]. Recent years have seen a proliferation of studies reporting that nanoparticles generated from plants have the ability to inhibit the growth of tumors. The enhanced cytotoxic effect can be attributed to the bioactive compounds as well as the other non-metal components that are present in the synthesis medium [11, 54]. Silver nanoparticles generated from plants have been shown to regulate the cell cycle as well as enzymes in the circulation. In addition, the nanoparticles that are produced by the plant have a relative control over the generation of free radicals by the cell. It is typical for free radicals to cause cell expansion as well as harm to the regular function of cells. Apoptosis is triggered in cancer cells by exposure to metal nanoparticles in concentrations found to be physiologically relevant [45]. The Ag nanoparticles treated MCF-7 cancer cell line retained intra-molecular concentration and regulating cellular metabolism [4]. Nanostructured materials have been shown to have a number of unique applications in the medical area, including the diagnosis and treatment of several different kinds of cancer and other retroviral disorders. Nanoparticles derived from biological sources are a novel and revolutionary approach to treating malignant deposits that do not interfere with the function of normal cells. According to research from the past, the environmentally friendly manufacturing of silver nanoparticles displayed a significantly greater cytotoxic effect in HeLa cell lines when compared to other synthetic medications that are based on chemicals [91].

10.6 Antiviral Effects of Metallic Nanoparticles

Alternative medications for treating and regulating the proliferation of viral infections are plants-mediated nanoparticles. Viral introduction into a host is criminally negligent, and it involves an accelerated translational process to increase the size of their colony. AgNPs nanoparticles can be biosynthesized to serve as potent, all-purpose antiviral agents that limit the operations of virus cells [57]. Previous research suggested that bio-AgNPs with convincing anti-HIV effect were tested at an early stage of the reverse transcription mechanism. Strong antiviral agents, the metallic NPs prevent the entry of viruses into the human system. To control the action of the virus, the biosynthesized metallic nanoparticles must bind to the viral membranes gp120, which has several binding sites [24]. In both cell-free and fibroblast viruses, the bio-based nanoparticles function as potent virucidal inhibitors. Consequently, the HIV-1 life cycle is continuously being inhibited by the silver and gold nanoparticles. These metal oxide nanoparticles will indeed function as an efficient antiviral drug against retroviruses [22].

10.7 Anti-Diabetic Management of Metallic Nanoparticles

The term "diabetes mellitus" (DM) refers to a set of metabolic disorders characterized by uncontrolled glucose levels in the patient. At certain dosages, particular foods, a balanced diet, or synthetic insulin medicines can prevent diabetes, however treating DM completely is a difficult task. The biosynthesized nanomaterials, however, might be a substitute for existing diabetic mellitus treatments [86]. Previous research shown that gold nanoparticles have positive therapeutic benefits on diabetic animals. Gold nanoparticles reduce liver enzymes in diabetic mice, including alanine transaminase, alkaline phosphatase, serum creatinine, and uric acid. The HbA (glycosylated hemoglobin) level decreased in the gold nanoparticles-treated diabetic model, maintaining it within the normal range [62]. Previous studies investigated the inhibition of a-amylase and acarbose sugar by *Sphaeranthus amaranthoides* biosynthesized silver nanoparticles in diabetes-induced animal studies. The majority of the ingredients in *S. amaranthoides* ethanolic extract are a-amylase inhibitory substances [75]. Similarly, earlier studies demonstrated that nanoparticles are an effective therapeutic agent with few side effects for the management of diabetes. In the group treated with silver nanoparticles in the clinical experiments on mice, the sugar level was controlled at 140 mg/dl [3].

10.8 Nanoparticles Generated from Plants Have Anti-Oxidative Mechanisms

Antioxidant compounds, which can be either enzymatic or non-enzymatic compounds, assist in regulating the generation of free radicals. Impairment to cells, including malignancy, atherosclerosis, and brain damage, can be attributed to the presence of free radicals. Reactive oxygen species (ROS) like superoxide dismutase, hydrogen peroxides, and hydrogen radicals produce free radicals, which are then released into the environment. Carbohydrate, polyphenolic compounds, quercetin, and bifunctional compounds including proteins and glycoproteins were found to exert a significant amount of control on the generation of free radicals. The ability of enzymatic and non-enzymatic antioxidants to scavenge free radicals is beneficial to the management and treatment of a wide range of chronic diseases, including diabetic mellitus, cancer, AIDS, nephropathy, autoimmune disease, and neurological conditions. The oxidizing impact of metal nanoparticles was found to be significantly greater than that of other synthetic recognitions, such as vitamin c and so on. While the tea leaf extract has a higher total phenolic and flavonoid content in the extracts, the nanoparticles demonstrated stronger antioxidant capacity based on the tea leaves extract [52].

10.9 The Role that Secondary Metabolites Play in the Bio-Reduction Process

The synthesis of nanostructured materials from the related ionic molecules has been relatively aided in the process by a number of secondary metabolites and catalysts. Carbohydrates (polysaccharides), proteins, chemical substances, carotenoids, and botanical resins were some of the plant biomolecules (bioactive compounds) that were predominantly engaged in the reduction reaction. Natural compounds derived from plants are utilized in the oxidation process that underpins the synthesis of green nanoparticles. The production of several chemical substances, including polyphenols, tannins, superoxide dismutase, terpenes, and alkaloid, is an important aspect of plants' protective factors. It is well known that these bioactive compounds are significant sources for regulating a variety of acute illnesses. Because natural products were shown to be the most important contributors to the manufacture of metallic nanoparticles by the reduction process that was proposed [63]. The phytoconstituents have a wide broad range of functional groups, including C–C (Alkenyl), C-N (amide), O–H (phenolic and alcohol), NAH (amine), CAH, and COOA (carboxylic group). The majority of the time, it is depicted as bioactive molecules of plants, although it might also be micro- or macro-biomolecules. The creation of nanoparticles would not be possible without the full participation of these chemical components. For instance, the plant extract of the *R. hymenosepalus* species enhances the synthesis of nanoparticles at ambient temperature, and the reaction kinetics accelerates. According to

Rodriguez-León et al. [72], the solvent extract of *R. hymenosepalus* is abundant in polyphenols such as catechins and stilbenes molecules. These polyphenols function as stabilizers and reducers during the formation of silver nanoparticles. Nanoparticles generated from secondary metabolites of plants like phenolic content, proteins, carbohydrates, polyphenols, and alkaloids were created using an environmentally friendly technique. Figure 3a–c illustrates how several plant compounds are utilized in the green synthetic process for the production of nanoparticles.

11 Utilization of Biologically Synthesized Nanostructures in the Commercial Sector

Nano-products have enormous potential uses in day-to-day living, including sewage treatment, which just emerged in recent years. In addition, there are many different eco-friendly nanoproducts that are currently accessible on the commercial market with a high level of efficiency. Some examples of these include a water filtration system, osteo and dental concrete, face lotion, and handcrafted products. For example, nanomaterials made of silver, silica, and platinum all have a variety of uses in the cosmetics and pharmaceutical industries. These nanostructures are also utilized as additives in a wide range of goods, including moisturizing ingredients, anti-aging creams, toothpastes, mouthwashes, hair-care products, and fragrances. In many different kinds of consumer goods, nanoparticles made of silica can be found functioning as an ingredient. In addition to this, modified silica nanomaterials are an excellent method for the control of pesticides, and it also has a wide range of applications outside of the agricultural sector [65].

12 Role of Nanotechnology in Cosmetics Production

The cosmetics sectors employ metal nanoparticles as a preservation and emulsion agent in their products. The new dimension of nanostructured materials is being utilized for a variety of commercial uses, most notably in the manufacture of cosmetics, materials for Industrial coatings. Nanomaterials of various metals, including gold, silver, and platinum, are increasingly being used in a broad range of commercial items, including cleanser, lotion, bleach, and polish based coating materials. The majority of the chemical components are man-made, and they have adverse impacts on humans [23, 26]. As a consequence of this, the green metallic nanoparticles can serve as a replacement for traditional preservation chemicals in the medical and food industries.

13 The Application of Nanoparticles in the Food Industry

Silver metal is a good conductor of heat, so nano-Ag is used in a wide range of mechanical devices. PCR lids and UV spectrophotometers are two examples of heat-sensitive instruments that employ it most frequently. The nanosilver, which is employed as coated materials, is used to make the instrumentation components [14, 15]. It is highly stable at elevated temperatures and does not interact with samples. The manufacturing and transportation of raw materials are all open-scale processes that contribute to the food industry's widespread usage of microbial contamination in the fresh produce. Therefore, a low-cost biosensor is required to assess the products quality [12]. The application of nanostructured materials as biosensors has enabled it to detect infections and track the many phases of contamination at a low cost production. This has been facilitated by the advent of nanotechnology.

14 Components that Play a Role in the Production of Metallic Nanoparticles

The creation of nanoparticles of varying sizes and shapes elicits a response from the system in the form of distinct concentrations of hydrogen ions. It has been reported that adjusting the pH of the media in which the Aloe vera extract was dissolved resulted in the production of Au–Agcore nanomaterials of varying dimensions and shapes [50]. In a similar fashion, the biosynthesis of nanoparticles by alfalfa plant extract of the pH is a retort for the size variation that occurs during the formation of nanoparticles. Moreover, temperature is one of the stimuli for the biogenesis of nanoparticles of varying sizes and forms [93]. In addition, the amount of time that the reduction process takes, measured in minutes or hours, is one of the elements that determines how the ions are transformed into different bulk metal forms. The optimal time duration yields a high absorbance peak value, which enables one to locate the NPs in the medium that are present in larger concentrations. Different sized NPs, including spherical, triangular, hexagonal, and rectangular, were formed under distinct growth conditions, which were established by this method [74].

15 Conclusion

In the past two decades, there has been a significant increase in research focused on the biosynthesis of metal nanoparticles utilizing plant derivatives. Metabolites derived from plants are responsible for the environmentally benign creation of metallic nanoparticles. The environmentally friendly synthesis of nanoparticles utilizing plant crude extracts and refined metabolites is an unique substrate for the creation of nanostructures on a large scale. This is an exciting potential. The plant-mediated nanoparticles may find applications in a variety of disciplines, including medications and treatments, sustainable and renewable energy, and other commercial products. The nanostructured materials generated from plants are anticipated to have an impact on the diagnosis and treatment of a variety of ailments while exhibiting regulated negative effects. In the not-too-distant future, there is a significant possibility that plants may play a significant role in the production of metallic nanoparticles for use in healthcare and industrial applications.

References

1. Acay H, Baran MF (2019) Biosynthesis and characterization of silver nanoparticles using king oyster (*Pleurotus eryngii*) extract: effect on some microorganisms. Appl Ecol Environ Res 17:9205–9214
2. Ahmad T, Iqbal J, Bustam MA, Irfan M, Asghar HMA (2021) A critical review on phytosynthesis of gold nanoparticles: issues, challenges and future perspectives. J Clean Prod 309:127460
3. Alkhalaf MI, Hussein RH, Hamza A (2020) Green synthesis of silver nanoparticles by Nigella sativa extract alleviates diabetic neuropathy through anti-inflammatory and antioxidant effects. Saudi J Biol Sci 27(9):2410–2419
4. Al-Nuairi AG, Mosa KA, Mohammad MG, El-Keblawy A, Soliman S, Alawadhi H (2020) Biosynthesis, characterization, and evaluation of the cytotoxic effects of biologically synthesized silver nanoparticles from *cyperus conglomeratus* root extracts on breast cancer cell line MCF-7. Biol Trace Elem Res 194(2):560–569
5. Alshameri AW, Owais M (2022) Antibacterial and cytotoxic potency of the plant-mediated synthesis of metallic nanoparticles Ag NPs and ZnO NPs: a review. OpenNano, 100077
6. Antony D, Yadav R, Kalimuthu R (2021) Accumulation of Phyto-mediated nano-CeO_2 and selenium doped CeO_2 on *Macrotyloma uniflorum* (horse gram) seed by nano-priming to enhance seedling vigor. Biocatal Agric Biotechnol 31:101923
7. Aromal SA, Philip D (2012) Green synthesis of gold nanoparticles using *Trigonella foenum-graecum* and its size-dependent catalytic activity. Spectrochim Acta Part A Mol Biomol Spectrosc 97:1–5
8. Aromal SA, Philip D (2012) Green synthesis of gold nanoparticles using Trigonella foenum-graecum and its size-dependent catalytic activity. Spectrochim. Acta A Mol Biomol Spectrosc 97:1–5
9. Azam A, Ahmed AS, Oves M, Khan MS, Habib SS, Memic A (2012) Antimicrobial activity of metal oxide nanoparticles against Gram-positive and Gram-negative bacteria: a comparative study. Int J Nanomed 7:6003
10. Barabadi H, Alizadeh Z, Rahimi MT, Barac A, Maraolo AE, Robertson LJ., Masjedi A, Shahrivar, Ahmadpour E F (2019) Nanobiotechnology as an emerging approach to combat malaria: a systematic review. Nanomed Nanotechnol Biol Med 18:221–233
11. Bello BA, Khan SA, Khan JA, Syed FQ, Anwar Y, Khan SB (2017) Antiproliferation and antibacterial effect of biosynthesized AgNps from leaves extract of *Guiera senegalensis* and its catalytic reduction on some persistent organic pollutants. J Photochem Photobiol B: Biol 175:99–108
12. Berekaa MM (2015) Nanotechnology in food industry; advances in food processing, packaging and food safety. Int J Curr Microbiol App Sci 4(5):345–357

13. Chelliah R, lHasan Madar I, Sultan G, Begum M, Pahi B, Tayubi IA, Selvakumar V, Oh DH (2023) Risk assessment and regulatory decision-making for nanomaterial use in agriculture. In: Engineered nanomaterials for sustainable agricultural production, soil improvement and stress management, pp 413–430
14. Chelliah R, Wei S, Daliri EBM, Rubab M, Elahi F, Yeon SJ, Yan P, Liu S, Oh DH (2021) Development of Nanosensors based intelligent packaging systems: food quality and medicine. Nanomaterials 11(6):1515
15. Chen H, Kou X, Yang Z, Ni W, Wang J (2008) Shape-and size-dependent refractive index sensitivity of gold nanoparticles. Langmuir 24(10):5233–5237
16. Cruz D, Falé PL, Mourato A, Vaz PD, Serralheiro ML, Lino ARL (2010) Preparation and physicochemical characterization of Ag nanoparticles biosynthesized by *Lippia citriodora* (Lemon Verbena). Colloids Surf B 81(1):67–73
17. Daisy P, Saipriya K (2012) Biochemical analysis of Cassia fistula aqueous extract and phytochemically synthesized gold nanoparticles as hypoglycemic treatment for diabetes mellitus. Int J Nanomed 7:1189
18. Das S, Das J, Samadder A, Bhattacharyya SS, Das D, Khuda-Bukhsh AR (2013) Biosynthesized silver nanoparticles by ethanolic extracts of Phytolacca decandra, Gelsemium sempervirens, *Hydrastis canadensis* and *Thuja occidentalis* induce differential cytotoxicity through G2/M arrest in A375 cells. Colloids Surf B 101:325–336
19. Deshmukh SP, Patil SM, Mullani SB, Delekar SD (2019) Silver nanoparticles as an effective disinfectant: a review. Mater Sci Eng C 97:954–965
20. Dipankar C, Murugan S (2012) The green synthesis, characterization and evaluation of the biological activities of silver nanoparticles synthesized from *Iresine herbstii* leaf aqueous extracts. Colloids Surf, B 98:112–119
21. El Shafey AM (2020) Green synthesis of metal and metal oxide nanoparticles from plant leaf extracts and their applications: a review. Green Process Synth 9(1):304–339
22. Fayaz AM, Ao Z, Girilal M, Chen L, Xiao X, Kalaichelvan PT, Yao X (2012) Inactivation of microbial infectiousness by silver nanoparticles-coated condom: a new approach to inhibit HIV-and HSV-transmitted infection. Int J Nanomedicine 7:5007
23. Fytianos G, Rahdar A, Kyzas GZ (2020) Nanomaterials in cosmetics: recent updates. Nanomaterials 10(5):979
24. Galdiero S, Falanga A, Vitiello M, Cantisani M, Marra V, Galdiero M (2011) Silver nanoparticles as potential antiviral agents. Molecules 16(10):8894–8918
25. Ghosh S, Patil S, Ahire M, Kitture R, Kale S, Pardesi K, Cameotra SS, Bellare J, Dhavale DD, Jabgunde A, Chopade BA (2012) Synthesis of silver nanoparticles using Dioscorea bulbifera tuber extract and evaluation of its synergistic potential in combination with antimicrobial agents. Int J Nanomed 7:483
26. Go MR, Bae SH, Kim HJ, Yu J, Choi SJ (2017) Interactions between food additive silica nanoparticles and food matrices. Front Microbiol 8:1013
27. Gurunathan S, Jeyaraj M, Kang MH, Kim JH (2019) The effects of apigenin-biosynthesized ultra-small platinum nanoparticles on the human monocytic THP-1 cell line. Cells 8(5):444
28. Huang F, Long Y, Liang Q, Purushotham B, Swamy MK, Duan Y (2019) Safed Musli (*Chlorophytum borivilianum* L.) callus-mediated biosynthesis of silver nanoparticles and evaluation of their antimicrobial activity and cytotoxicity against human colon cancer cells. J Nanomater
29. Husen A (2019) Natural product-based fabrication of zinc oxide nanoparticles and their application. In: Husen A, Iqbal M (eds) N*anomaterials* and plant potential. Springer International Publishing AG, Gewerbestrasse 11, 6330 Cham pp. 193–291. https://doi.org/10.1007/978-3-030-05569-1_7
30. Husen A (2020c) Introduction and techniques in nanomaterials formulation. In: Husen, A, Jawaid M (eds) Nanomaterials for agriculture and forestry applications. Elsevier Inc. 50 Hampshire St., 5th Floor, Cambridge, MA 02139, USA, pp 1–14
31. Husen A, Iqbal M (2019a) Nanomaterials and plant potential: an overview. In: Husen A, Iqbal M (eds) Nanomaterials and plant potential. Springer International Publishing AG, Gewerbestrasse 11, 6330 Cham, pp 3–29. https://doi.org/10.1007/978-3-030-05569-1_1

32. Husen A, Siddiqi KS (2014) Phytosynthesis of nanoparticles: concept, controversy and application. Nanoscale Res Lett 9(229):1–24
33. Husen A, Siddiqi KS (2014) Carbon and fullerene nanomaterials in plant system. J Nanobiotechnol 12(1):1–16
34. Husen A (2020a) Interactions of metal and metal-oxide nanomaterials with agricultural crops: an overview. In: Husen, A, Jawaid M (eds) Nanomaterials for agriculture and forestry applications. Elsevier Inc. 50 Hampshire St., 5th Floor, Cambridge, MA 02139, USA, pp 167–197. https://doi.org/10.1016/B978-0-12-817852-2.00007-X
35. Husen A (2020b) Carbon-based nanomaterials and their interactions with agricultural crops (Eds. Husen A, Jawaid M) Elsevier Inc. 50 Hampshire St., 5th Floor, Cambridge, MA 02139, USA, pp 199–218. https://doi.org/10.1016/B978-0-12-817852-2.00008-1
36. Husen A (2021) Harsh environment and plant resilience (Molecular and Functional Aspects). Springer Nature Switzerland AG, Gewerbestrasse 11, 6330 Cham, Switzerland. https://doi.org/10.1007/978-3-030-65912-7
37. Husen A (2022) Engineered nanomaterials for sustainable agricultural production, soil improvement and stress management. Elsevier Inc., 50 Hampshire St., 5th Floor, Cambridge, MA 02139, USA
38. Husen A, Iqbal M (2019b) Nanomaterials and plant potential. Springer International Publishing AG, Gewerbestrasse 11, 6330 Cham, Switzerland. https://doi.org/10.1007/978-3-030-05569-1
39. Husen A, Jawaid M (2020) Nanomaterials for agriculture and forestry applications. Elsevier Inc., 50 Hampshire St., 5th Floor, Cambridge, MA 02139, USA. https://doi.org/10.1016/C2018-0-02349-X
40. Jahan I, Erci F, Isildak I (2019) Microwave-assisted green synthesis of non-cytotoxic silver nanoparticles using the aqueous extract of *Rosa santana* (rose) petals and their antimicrobial activity. Anal Lett 52(12):1860–1873
41. Jain D, Daima HK, Kachhwaha S, Kothari SL (2009) Synthesis of plant-mediated silver nanoparticles using papaya fruit extract and evaluation of their anti microbial activities. Dig J Nanomater Biostructures 4(3):557–563
42. Jamdagni P, Khatri P, Rana JS (2018) Green synthesis of zinc oxide nanoparticles using flower extract of *Nyctanthes arbor-tristis* and their antifungal activity. J King Saud Univ. Sci 30(2):168–175
43. Jayaseelan C, Rahuman AA, Rajakumar G, Vishnu Kirthi A, Santhoshkumar T, Marimuthu S, Bagavan A, Kamaraj C, Zahir AA, Elango G (2011) Synthesis of pediculocidal and larvicidal silver nanoparticles by leaf extract from heartleaf moonseed plant, *Tinospora cordifolia Miers*. Parasitol Res 109(1):185–194
44. Kaviya S, Santhanalakshmi J, Viswanathan B, Muthumary J, Srinivasan K (2011) Biosynthesis of silver nanoparticles using Citrus sinensis peel extract and its antibacterial activity. Spectrochim. Acta A Mol Biomol Spectrosc 79(3):594–598
45. Ke Y, Al Aboody MS, Alturaiki W, Alsagaby SA, Alfaiz FA, Veeraraghavan VP, Mickymaray S (2019) Photosynthesized gold nanoparticles from Catharanthus roseus induces caspase-mediated apoptosis in cervical cancer cells (HeLa). Artif Cells Nanomed Biotechnol 47(1):1938–1946
46. Kora AJ, Sashidhar RB, Arunachalam J (2012) Aqueous extract of gum olibanum (Boswellia serrata): a reductant and stabilizer for the biosynthesis of antibacterial silver nanoparticles. Process Biochem 47(10):1516–1520
47. Krishnaraj C, Jagan EG, Rajasekar S, Selvakumar P, Kalaichelvan PT, Mohan NJ (2010) Synthesis of silver nanoparticles using Acalypha indica leaf extracts and its antibacterial activity against water borne pathogens. Colloids Surf B 76(1):50–56
48. Kumar A, Choudhary A, Kaur H, Guha S, Mehta S, Husen A (2022) Potential applications of engineered nanoparticles in plant disease management: a critical update. Chemosphere 295:133798
49. Kumar A, Choudhary A, Kaur H, Mehta S, Husen A (2021) Metal-based nanoparticles, sensors and their multifaceted application in food packaging. J Nanobiotechnol 19(256):1–25. https://doi.org/10.1186/s12951-021-00996-0

50. Kumar V, Yadav SK (2009) Plant-mediated synthesis of silver and gold nanoparticles and their applications. J Chem Technol Biotechnol Int Res Process Environ Clean Technol 84(2):151–157
51. Kuppusamy P, Yusoff MM, Maniam GP, Govindan N (2016) Biosynthesis of metallic nanoparticles using plant derivatives and their new avenues in pharmacological applications–an updated report. Saudi Pharm J 24(4):473–484
52. Machado S, Pinto SL, Grosso JP, Nouws HPA, Albergaria JT, Delerue-Matos C (2013) Green production of zero-valent iron nanoparticles using tree leaf extracts. Sci Total Environ 445:1–8
53. Maensiri S, Laokul P, Klinkaewnarong J, Phokha S, Promarak V, Seraphin S (2008) Indium oxide (In2O3) nanoparticles using *Aloe vera* plant extract: synthesis and optical properties. J Optoelectron Adv Mater 10(3):161–165
54. Mariadoss AVA, Saravanakumar K, Sathiyaseelan A, Venkatachalam K, Wang MH (2020) Folic acid functionalized starch encapsulated green synthesized copper oxide nanoparticles for targeted drug delivery in breast cancer therapy. Int J Biol Macromol 164:2073–2084
55. Matle KR, Mbatha E, Madoroba E (2020) A review of listeria monocytogenes from meat and meat products: epidemiology, virulence factors, antimicrobial resistance and diagnosis, Onderstepoort. J Vet Res 87:1–20
56. Mubeen B, Ansar AN, Rasool R, Ullah I, Imam SS, Alshehri S, Ghoneim MM, Alzarea SI, Nadeem MS, Kazmi I (2021) Nanotechnology as a novel approach in combating microbes providing an alternative to antibiotics. Antibiotics 10(12):1473
57. Mujtaba M, Wang D, Carvalho LB, Oliveira JL, Espirito Santo Pereira AD, Sharif R, Jogaiah S, Paidi MK, Wang L, Ali Q, Fraceto LF (2021) Nanocarrier-mediated delivery of miRNA, RNAi, and CRISPR-Cas for plant protection: current trends and future directions. ACS Agricultural Science & Technology 1(5):417–435
58. Munir H, Bilal M, Mulla SI, Abbas Khan H, Iqbal H (2021) Plant-mediated green synthesis of nanoparticles. Advances in Green Synthesis. Springer, Cham, pp 75–89
59. Nagati VB, Alwala J, Koyyati R, Donda MR, Banala R, Padigya PRM (2012) Green synthesis of plant-mediated silver nanoparticles using *Withania somnifera* leaf extract and evaluation of their antimicrobial activity. Asian Pac J Trop Biomed 2:1–5
60. Ng HS, Kee PE, Yim HS, Chen PT, Wei YH, Lan JCW (2020) Recent advances on the sustainable approaches for conversion and reutilization of food wastes to valuable bioproducts. Bioresour Technol 302:122889
61. Niraimathi KL, Sudha V, Lavanya R, Brindha P (2013) Biosynthesis of silver nanoparticles using Alternanthera sessilis (Linn.) extract and their antimicrobial, antioxidant activities. Colloids Surf B 102:288–291
62. Oladipo IC, Lateef A, Azeez MA, Asafa TB, Yekeen TA, Ogunsona SB, Irshad HM, Abbas SH (2020) Antidiabetic properties of phytosynthesized gold nanoparticles (AuNPs) from Datura stramonium seed. In: IOP conference series: materials science and engineering 805, No 1, p 012035). IOP Publishing
63. Ovais M, Khalil AT, Islam NU, Ahmad I, Ayaz M, Saravanan M, Shinwari ZK, Mukherjee S (2018) Role of plant phytochemicals and microbial enzymes in biosynthesis of metallic nanoparticles. Appl Microbiol Biotechnol 102(16):6799–6814
64. Padalia H, Moteriya P, Chanda S (2015) Green synthesis of silver nanoparticles from marigold flower and its synergistic antimicrobial potential. Arab J Chem 8(5):732–741
65. Peters RJ, Bouwmeester H, Gottardo S, Amenta V, Arena M, Brandhoff P, Marvin HJ, Mech A, Moniz FB, Pesudo LQ, Rauscher H (2016) Nanomaterials for products and application in agriculture, feed and food. Trends Food Sci Technol 54:155–164
66. Prema D, Prakash J, Vignesh S, Veluchamy P, Ramachandran C, Samal DB, Oh DH, Sahabudeen S, Devanand Venkatasubbu G (2020) Mechanism of inhibition of graphene oxide/zinc oxide nanocomposite against wound infection causing pathogens. Appl Nanosci 10(3):827–849
67. Qiu R, Xiong W, Hua W, He Y, Sun X, Xing M, Wang L (2021) A biosynthesized gold nanoparticle from *Staphylococcus aureus*–as a functional factor in muscle tissue engineering. Appl Mater Today 22:100905

68. Rahman K, Khan SU, Fahad S, Chang MX, Abbas A, Khan WU, Rahman L, Haq ZU, Nabi G, Khan D (2019) Nano-biotechnology: a new approach to treat and prevent malaria. Int J Nanomedicine 14:1401
69. Ranjani S, Shariq Ahmed M, MubarakAli D, Ramachandran C, Senthil Kumar N, Hemalatha S (2020) Toxicity assessment of silver nanoparticles synthesized using endophytic fungi against nosacomial infection. Inorg. Nano-Met Chem 51(8):1080–1085
70. Rasaee I, Ghannadnia M, Baghshahi S (2018) Biosynthesis of silver nanoparticles using leaf extract of *Satureja hortensis* treated with NaCl and its antibacterial properties. Microporous Mesoporous Mater 264:240–247
71. Razavi M, Salahinejad E, Fahmy M, Yazdimamaghani M, Vashaee D, Tayebi L (2015) Green chemical and biological synthesis of nanoparticles and their biomedical applications. Green Nanotechnol, 207–235
72. Rodríguez-León E, Iñiguez-Palomares R, Navarro RE, Herrera-Urbina R, Tánori J, Iñiguez-Palomares C, Maldonado A (2013) Synthesis of silver nanoparticles using reducing agents obtained from natural sources (*Rumex hymenosepalus* extracts). Nanoscale Res Lett 8(1):1–9
73. Roy S, Das TK (2015) Plant mediated green synthesis of silver nanoparticles-A. Int J Plant Biol Res 3:1044–1055
74. Saim AK, Kumah FN, Oppong MN (2021) Extracellular and intracellular synthesis of gold and silver nanoparticles by living plants: a review. Nanotechnol Environ Eng 6(1):1–11
75. Sana SS, Li H, Zhang Z, Sharma M, Usmani Z, Hou T, Netala VR, Wang X, Gupta VK (2021) Recent advances in essential oils-based metal nanoparticles: A review on recent developments and biopharmaceutical applications. J Mol Liq 333:115951
76. Saravanakumar K, Mariadoss AVA, Sathiyaseelan A, Wang MH (2020) Synthesis and characterization of nano-chitosan capped gold nanoparticles with multifunctional bioactive properties. Int J Biol Macromol 165:747–757
77. Sathishkumar M, Sneha K, Won SW, Cho CW, Kim S, Yun YS (2009) *Cinnamon zeylanicum* bark extract and powder mediated green synthesis of nano-crystalline silver particles and its bactericidal activity. Colloids Surf B 73(2):332–338
78. Sathiyaseelan A, Saravanakumar K, Mariadoss AVA, Wang MH (2020) Biocompatible fungal chitosan encapsulated phytogenic silver nanoparticles enhanced antidiabetic, antioxidant and antibacterial activity. Int J Biol Macromol 153:63–71
79. Satyavani K, Gurudeeban S, Ramanathan T, Balasubramanian T (2011) Biomedical potential of silver nanoparticles synthesized from calli cells of *Citrullus colocynthis* (L.) Schrad. J Nanobiotechnol 9:43
80. Selvakumar V, Muthusamy K, Mukherjee A, Chelliah R, Barathikannan K, Oh DH, Ayyamperumal R, Karuppannan S (2022) Antibacterial efficacy of Phytosynthesized multi-metal oxide nanoparticles against drug-resistant foodborne pathogens. J Nanomater
81. Shameli K, Bin Ahmad M, Jaffar Al-Mulla EA, Ibrahim NA, Shabanzadeh P, Rustaiyan A, Abdollahi Y, Bagheri S, Abdolmohammadi S, Usman MS, Zidan M (2012) Green biosynthesis of silver nanoparticles using Callicarpa maingayi stem bark extraction. Molecules 17(7):8506–8517
82. Sharma G, Nim S, Alle M, Husen A, Kim JC (2022) Nanoparticle-mediated delivery of flavonoids for cancer therapy: prevention and treatment. In: Kim JC, Alle M, Husen A (eds) Smart nanomaterials in biomedical applications. Nanotechnology in the Life Sciences. Springer, Cham. https://doi.org/10.1007/978-3-030-84262-8_3
83. Sharma P, Pandey V, Sharma MMM, Patra A, Singh B, Mehta S, Husen A (2021) A review on biosensors and Nanosensors application in agroecosystems. Nano Res Lett 16:136. https://doi.org/10.1186/s11671-021-03593-0
84. Siddiqi KS, Husen A (2020) Current status of plant metabolite-based fabrication of copper/copper oxide nanoparticles and their applications: a review. Biomater Res 24(11):1–15
85. Siddiqi KS, Husen A (2021) Plant response to silver nanoparticles: a critical review. Crit Rev Biotechnol 42(7):773–990. https://doi.org/10.1080/07388551.2021.1975091
86. Simos YV, Spyrou K, Patila M, Karouta N, Stamatis H, Gournis D, Dounousi E, Peschos D (2021) Trends of nanotechnology in type 2 diabetes mellitus treatment. Asian J Pharm Sci 16(1):62–76

87. Singh C, Sharma V, Naik PK, Khandelwal V, Singh H (2011) A green biogenic approach for synthesis of gold and silver nanoparticles using *Zingiber officinale*. Dig J Nanomater Biostructures 6(2):535–542
88. Singh S, Saikia JP, Buragohain AK (2013) A novel green synthesis of colloidal silver nanoparticles (SNP) using *Dillenia indica* fruit extract. Colloids Surf B 102:83–85
89. Song JY, Kim BS (2008) Biological synthesis of bimetallic Au/Ag nanoparticles using Persimmon (Diopyros kaki) leaf extract. Korean J Chem Eng 25(4):808–811
90. Sufyani A, Moslah N, Hussien NA, Hawsawi YM (2019) Characterization and anticancer potential of silver nanoparticles biosynthesized from *Olea chrysophylla* and *Lavandula dentata* leaf extracts on HCT116 colon cancer cells. J Nanomater
91. Suman TY, Rajasree SR, Kanchana A, Elizabeth SB (2013) Biosynthesis, characterization and cytotoxic effect of plant mediated silver nanoparticles using *Morinda citrifolia* root extract. Colloids Surf B 106:74–78
92. Suriyakalaa U, Antony JJ, Suganya S, Siva D, Sukirtha R, Kamalakkannan S, Pichiah PT, Achiraman S (2013) Hepatocurative activity of biosynthesized silver nanoparticles fabricated using *Andrographis paniculata*. Colloids Surf B 102:189–194
93. Tabande L, Sepehri M, Yasrebi J, Zarei M, Ghasemi-Fasaei R, Khatabi B (2022) A comparison between the function of *Serendipita indica* and *Sinorhizobium meliloti* in modulating the toxicity of zinc oxide nanoparticles in alfalfa (*Medicago sativa* L.). Environ Sci Pollut Res 29(6):8790–8803
94. Takooree H, Aumeeruddy MZ, Rengasamy KR, Venugopala KN, Jeewon R, Zengin G, Mahomoodally MF (2019) A systematic review on black pepper (*Piper nigrum* L.): from folk uses to pharmacological applications. Crit Rev Food Sci Nutr 59:S210–S243
95. Tharani M, Rajeshkumar S (2022) AntimicrobialaApplications of Nanodevices prepared from metallic nanoparticles and their role in controlling infectious diseases. In: Smart Nanodevices for point-of-care applications, 1–12. CRC Press
96. Tyagi P, Ranjan R (2021) Comparative study of the pharmacological, phytochemical and biotechnological aspects of *Tribulus terrestris* Linn. and *Pedalium murex* Linn: an overview. Acta Ecologica Sinica
97. Vankar PS, Bajpai D (2010) Preparation of gold nanoparticles from *Mirabilis jalapa* flowers
98. Vijayakumar M, Priya K, Nancy FT, Noorlidah A, Ahmed ABA (2013) Biosynthesis, characterisation and anti-bacterial effect of plant-mediated silver nanoparticles using *Artemisia nilagirica*. Ind Crops Prod 41:235–240
99. Vijayan S, Divya K, Varghese S, Jisha MS (2020) Antifungal efficacy of chitosan-stabilized biogenic silver nanoparticles against pathogenic *Candida* spp. isolated from human. Bionanoscience 10(4):974–982
100. Zahir AA, Rahuman AA (2012) Evaluation of different extracts and synthesised silver nanoparticles from leaves of *Euphorbia prostrata* against *Haemaphysalis bispinosa* and *Hippobosca maculata*. Vet Parasitol 187(3–4):511–520

Alkaloids: A Suitable Precursor for Nanomaterials Synthesis, and Their Various Applications

Noureddine Chaachouay, Abdelhamid Azeroual, Ouafae Benkhnigue, and Lahcen Zidane

Abstract Green chemistry refers to creating chemical products and methods that minimize or avoid producing harmful chemicals. In recent years, natural compounds, particularly secondary metabolites, have been investigated for their ability to synthesize various nanoparticles. Numerous bioactive compounds, such as alkaloids, terpenoids, polysaccharides, vitamins, proteins, and phenolic components, may be concerned in nanoparticle bio-reduction, stabilization, and development. Alkaloids are discovered in all species and belong to the most prominent family of bioactive substances. Their role in synthesizing alkaloids-based nanoparticles has attracted considerable attention in current years. The purpose of this chapter is to discuss the role of alkaloids in the creation of nanomaterials as capping and reducing agents for the formation of nanomaterials with consistent size and shape. Indeed, the various techniques for characterizing nanomaterials and their uses have also been presented and analyzed. Moreover, the future perspective for green nanomaterial manufacturing and its potential applications were discussed.

Keywords Alkaloids · Biological applications · Nanotechnology · Nanomaterials biosynthesis

N. Chaachouay (✉)
Agri-Food and Health Laboratory (AFHL), Hassan First University, 50 Rue Ibnou Lhaytham, B.P. 577, 26002 Settat, Morocco
e-mail: nour.chay@gmail.com

A. Azeroual
Agri-Food and Health Laboratory (AFHL), Faculty of Sciences and Techniques of Settat, Hassan First University, Po Box 382, 26000 Settat, Morocco

O. Benkhnigue
Department of Botany and Plant Ecology, Scientific Institute, University Mohammed V, B. P. 703, 10106 Rabat, Morocco

L. Zidane
Plant, Animal Productions and Agro-Industry Laboratory, Department of Biology, Faculty of Sciences, Ibn Tofail University, B.P. 133 14000, Kenitra, Morocco

A. Husen (ed.), *Secondary Metabolites Based Green Synthesis of Nanomaterials and Their Applications*, Smart Nanomaterials Technology,
https://doi.org/10.1007/978-981-99-0927-8_2

Abbreviations

AFM	Atomic Force Microscopy
Ag NPs	Silver nanoparticles
AgCuO bimetallic	Silver- Copper Oxide bimetallic
AuNPs	Gold nanoparticles
BET	Brunauer–Emmett–Teller
Cryo-TEM	Cryo-Transmission Electron Microscopy
DCS	Differential Centrifugal Sedimentation
DLS	Dynamic Light Scattering
EBSD	Electron Backscatter Diffraction
EELS	Electron Energy-Loss Spectroscopy
FAO	Food and Agriculture Organisation
FDA	Food and Drug Administration
FMR	Ferromagnetic Resonance
HRTEM	High Resolution Transmission Electron Microscopy
ICP-MS	Inductively Coupled Plasma–Mass Spectrometry
LEIS	Low-Energy Ion Scattering
Liquid TEM	Liquid-Cell Transmission Electron Microscopy
MFM	Magnetic Force Microscopy
NMR	Nuclear Magnetic Resonance
NTA	Nanoparticle Tracking Analysis
Pd NPs	Palladium nanoparticles
PL Spectroscopy	Photoluminescence Spectroscopy
Pt NPs	Platinum nanoparticles
SAXS	Small-Angle X-Ray Scattering
SQUID-nanoSQUID	Superconducting Quantum Interference Device
STEM	Scanning Transmission Electron Microscopy
TEM	Transmission Electron Microscopy
TGA	Thermogravimetric Analysis
VSM	Vibrating Sample Magnetometer
XPS	X-Ray Photoelectron Spectroscopy

1 Introduction

In the past few decades, numerous technology methods for environmental sustainability have been developed. In light of the increasing importance of the environment, technological resolutions are evaluated not only based on their cost-effectiveness but also on their potential to avoid releasing toxins into the environment [3, 142, 148]. Among the principal developed technological, nanomaterials have appeared as fascinating alternative options for environmental improvement in recent decades.

Recently, there has been a rise in the synthesis of nanomaterials by biologically mediated manufactured procedures [55, 60, 64, 72, 150]. It is simple to synthesize stable metal nanomaterials due to concentrating agents such as alkaloids, flavonoids, terpenoids, polyphenols, and stabilizing agents such as proteins and polysaccharides significant phytoconstituents of the plant extracts [5, 23]. These phytoconstituents engage in a continual redox process to produce environmentally favorable nanoparticles. Therefore, these chemicals are mostly accountable for decreasing ionic to bulk metallic nanoparticle production [8].

Numerous investigations have demonstrated that biosynthesized nanomaterials successfully suppress oxidative stress, apoptosis-related alterations, and genotoxicity. Moreover, nanomaterials have extensive applications in agriculture, food, cosmetics, electronics, renewable energy, and plant sciences [2, 3, 6].

Green synthesis approaches based on bioactive compounds such as alkaloids have been created as a choice to conventional chemical and physical techniques for synthesizing noble metal nanomaterials. Using alkaloids to produce nanomaterials is superior to using other biological materials since it removes the time-consuming procedure of maintaining cultured cells and can be expanded in a non-aseptic setting [154].

In contrast to other precursors, alkaloids are affordable, readily accessible, and industrially applicable because they do not need the utilization of cultures or purification [20]. These methods offer numerous benefits, including their eco-friendliness and lack of toxicity, low cost due to avoiding high-pressure and high-energy costs, and ability to make small-sized nanomaterials [1, 124, 179]. This chapter will emphasize and discuss the mechanistic role of alkaloids as a suitable precursor for nanomaterials synthesis and their various applications.

2 The Importance of Alkaloids

Alkaloids are a large class of nitrogen-containing chemical molecules found in nature. These nitrogen atoms make these molecules alkaline. Typically, these nitrogen atoms are located in a ring (cyclic) structure [98]. Alkaloids are mostly biosynthetically produced from amino acids, resulting in a vast diversity of chemical structures, and are obtained mainly from plants (Manske and [106]. Alkaloids are present in around 20% of herb species in minute amounts, and their production, processing, extraction and are key study fields [129, 163].

Alkaloids have been identified as one of the most significant phytochemical compounds derived from nature. Alkaloids significantly affect human medication and an organism's biological safety. Approximately 20% of the known secondary metabolites in plants are alkaloids [19]. Plant alkaloids defend them from predators and control their development [22]. Alkaloids are widely recognized as anesthetics, cardioprotective mechanisms, and anti-inflammatory drugs. Ephedrine, morphine, quinine, nicotine, and strychnine are well-known clinical alkaloids [98]. Recently,

Table 1 Chemical classification of alkaloids

Subgroups	Examples
Pyrrole and pyrrolidine derivatives	Stachydrine, Hygrine
Pyrrolizidine derivatives	Senecio alkaloids
Piperidine and pyridine derivatives	Anabasine, arecoline, coniine, lobeline, nicotine, pelletierine, piperine, ricinine, trigonelline
Tropane (N-methylpyrrolidine /Piperidine) derivatives	Hyoscyamine, hyoscine, atropine, meteloidine, cocaine and cinnamyl-cocaine
Quinoline derivatives	Cinchonodine, cinchonine, quinidine, quinine
Isoquinoline derivatives	Berberine, cephaëline, corydaline, emetine, hydrastine, papaverine, narcotine, tubocurarine
Aporphine derivatives	Boldine
Nor-lupinane derivatives	Cystine, laburnine, lupinine, sparteine
Indole derivatives	Ajmaline, aspidospermine, bruceine, ergometrine, ergotamine, physostigmine, reserpine, serpentine, strychnine, yohimbine, vincablastine
Imidazole derivatives	Pilocarpine
Purine derivatives	Caffeine, theobromine
Steroidal derivatives (some combined as glycosides)	Conessine, funtamine, Solanine, veratrum,
Terpenoids derivatives	Atisine, aconitine, lycaconitine
Tropolone derivatives	Colchicines

there has been a renaissance of interest in biomolecules, fueled by both a proactive action in popular therapy and their possibility for drug discovery [9].

Alkaloids are frequently classified according to their chemical composition; the two major classes are isoquinoline alkaloids and indole alkaloids (each more than 4000 compounds) [38]. Other significant alkaloid categories include steroidal alkaloids (B450 compounds), tropane alkaloids (B300 compounds), and pyrrolizidine and pyridine alkaloids (respectively, 570 and B250 compounds) [35, 50]. Alkaloids are classified depending on the carbon–nitrogen cycle contained in the molecule's arrangement. Furthermore, the Alkaloids can be categorized into 14 subgroups based on their ring structure [106] (Table 1).

3 Methods of Nanomaterials Synthesis

For the synthesis of nanoparticles, two principal methods are used: physical and chemically-mediated methods.

3.1 Chemical-Mediated Methods

Numerous chemical methods for synthesizing nanoparticles have been proposed, most commonly employed to create nanostructured materials (e.g., polyol synthesis, microemulsion, pyrolysis, chemical vapor deposition, electrochemical synthesis, sol–gel method, hydrothermal synthesis, chemical reduction) [26, 85]. Furthermore, using toxic materials and agents during the synthesis process and the generation of byproducts is detrimental to individuals and the environment [180]. Consequently, such nanoparticles are restricted for use in biological applications.

3.2 Physical Methods

Physical methods such as gamma radiation, sputtering, plasma-based technologies, microwave irradiation, pulsed laser method, deposition, ball milling, and sonochemical reduction are used to create nanomaterials [12]. In most of these methods, the rate of metal nanoparticle formation is exceptionally sluggish. For instance, the yield of nanomaterials from ball milling processes is 50% or less [29, 178]. Sputtering yields a broad particle size range, with only 6–8% of sputtered material estimated to be smaller than 100 nm in size [126]. Laser ablation and plasma methods need significant energy consumption. The extensive size range, sluggish manufacturing frequency, waste by-products, and significant energy usage render most physical techniques too costly for practical industrial use [147].

4 Synthesis of Nanomaterials Based on Alkaloids

4.1 Bio-Synthesis of Nanomaterials by Using Plant Derived Alkaloids

Preparing nanoparticles from plant metabolites is a crucial aspect of biosynthetic methods. The capacity of plants to decrease metal ions on their surface and in multiple organs and tissues distant from the source of ion penetration has been known for a very long time [111]. Using plant extract to synthesize various nanoparticles is advantageous not just due to its lower environmental impact but also because it may be utilized to generate vast numbers of nanoparticles [18]. The shape and size of the nanoparticles generated by the above techniques relied on the plant extract content, reaction time, temperature, pH, etc. Ingredient compounds in plant extracts, including alkaloids, polyphenols, terpenoids, phenolic acids, proteins, polysaccharides, vitamins, enzymes, proteins, and organic acids, may reduce metal ions [102]. In this respect, the last few years have successfully developed in vitro technologies that employ plant extracts to reduce metal ions and generate nanoparticles [20, 123].

The alkaloids extracted from many plant parts, including leaves, seeds, roots, barks, and flowers, have been utilized to manufacture different nanoparticles (Table 2).

In synthesizing Au, Ag, Pt, and Pd nanoparticles, plant extracts may function as capping factors and reducing. It has also been observed that these alkaloid-based plant extracts possess antioxidant, antibacterial, anti-inflammatory, antifungal, anti-diabetic, anti-HIV, snake venom neutralization, and larvicidal properties [4, 37, 160, 174].

Table 2 Synthesis of various nanoparticles utilizing plant extracts containing alkaloids as the primary capping and stabilizing agents

Plant (part)	Nanoparticle type	Application	References
Peganum harmala L. seed	palladium and Platinum nanoparticles (Pd NPs and Pt NPs)	Antioxydant and anticancer activities	[37]
Solanum tuberosum L	Silver nanoparticles (Ag NPs)	Fungicidal activity	[4]
Cordia sebestena L. leaves	AgCuO bimetallic	Utilization in the industry as a methylene blue disintegrating ingredient	[139]
Trachyspermum ammi (L.) Sprague	Silver nanoparticles (Ag NPs)	Pharmacological activities	[174]
Papaver somniferum L	Silver nanoparticles (Ag NPs)	Pharmacological activities	[174]
Bauhinia tomentosa L. leaves	Gold and silver nanoparticles (AuNPs and AgNPs)	Anticancer efficacy	[120]
Zingiber officinale Roscoe	Gold nanoparticles (AuNPs)	Anti-platelet agent	[95]
Andrographis paniculata (Burm.f.) Nees leaves	Silver nanoparticles (Ag NPs)	Hepatocurative activity	[167]
Clinacanthus nutans (Burm.f.) Lindau stem	Silver nanoparticles (Ag NPs)	Biomedical, biological probes, catalysis, dentistry, diagnostic and sensors, food industries,	[107]
Conocarpus lancifolius Engl	Silver nanoparticles (Ag NPs)	Antibacterial activity	[130]

4.2 Bio-Synthesis of Nanomaterials by Using Microorganisms

Nanomaterials may be generated by a range of physical and chemical techniques; biological organisms can increasingly be used [49, 162]. The creation of nanoparticles employing microorganisms is an alternative approach based chiefly on green chemistry principles [65, 76, 103]. In recent years, the mycogenesis of nanoparticles has emerged as a critical method through which algae, bacteria, and fungi may be employed to produce nanomaterials with desired size and shape, either extracellularly or intracellularly [59, 86, 169]. Ingredient compounds released by microbial biomass, such as alkaloids, enzymes and proteins, may act as capping and reducing agents throughout the synthesis. Green nanoparticle synthesis is a single-step or one-pot eco-friendly and cost-efficient bio-reduction method that uses less energy to start the reaction. Accordingly, these methods are termed green chemistry since they do not include harmful substances [125, 156].

4.2.1 Algae/Cyanobacteria

Algae are a varied category of unicellular and multicellular autotrophic organisms that perform photosynthesis when exposed to sunlight [56]. The utilization of algae in the production of nanoparticles is very prevalent now. Algae are used owing to their extraordinary ability to absorb metals and decrease metal ions, moderately inexpensive production costs, and, most significantly, their potential to manufacture nanoparticles on a massive scale [134]. The capacity of these microbes to survive extreme air circumstances more successfull than other microorganisms is an additional intriguing trait [84]. They are characterized as Bio-nanofactories because they may utilize both living and nonliving algae biomass for nanoparticle production [122]. The time needed to create silver nanoparticles is another additional benefit of employing algae [24].

Microalgal nanoparticles have several uses in medical therapy and chemical reaction catalysis. Due to alkaloids and hydrophilic surface classes such as thiol, amino, sulfate, amidic groups, carboxyl, and hydroxyl, algae-mediated nanoparticles have several uses [66, 89, 131]. The number of nanoparticles synthesized is dosage-dependent and is also affected by the species of algae employed [131]. Metal reduction is accomplished by various biomolecules, including polysaccharides, peptides, and pigments. Contrary to many other biosynthesizing methods, algal nanoparticle synthesis requires comparatively less time [25, 105, 133, 169]. Several cyanobacteria, macro, and microalgal taxa have been studied so far for the bio-manufacturing of metal nanoparticles such as gold, silver, iron, and palladium nanoparticles, as shown in Table 3. As a result, the investigations into algae-mediated biosynthesis of metallic nanoparticles may be directed toward a growing field known as "phyco nanotechnology".

Table 3 Synthesis of metallic nanoparticles from several biological entities

Biological entities/species	Nanoparticles	Applications	References
Algae			
Caulerpa racemosa	Silver	Catalytic degradation of methylene Blue	[34]
Chlorella vulgaris	Palladium	Catalytic activity	[36]
Gracilaria corticata	Silver	Antibacterial activity	[97]
Gelidiella acerosa	Silver	Antibacterial activity	[97]
Hypnea musciformis	Silver	Antibacterial activity	[172]
Jania rubens	Iron oxide	Antibiofilm activity	[145]
Sargassum vulgare	Iron oxide	Antibiofilm activity	[145]
Turbinaria conoides	Gold	Catalytic reduction of nitro compounds	[137]
Ulva fasciata	Iron oxide	Antibiofilm activity	[145]
Cyanobacteria			
Plectonema boryanum UTEX 485	Gold	Antibacterial activity	[99]
Plectonema boryanum	Palladium	Antibacterial activity	[100]
Desertifilum sp.	Silver	Antibacterial and cytotoxicity effects	[52]
Oscillatoria limnetica	Silver	Antibacterial activity	[53]
Phormidium tenue NTDM05	Cadmium Sulfide	Biolabel	[119]
Limnothrix sp. 37–2-1	Silver	Antibacterial activity	[125]
Leptolyngbya sp. WUC 59	Silver	Actions on seed germination and bacterial growth	[158]
Bacteria			
Bacillus cecembensis	Silver	Antibacterial activity against *A. gangotriensis*, *A. kerguelensis*, *B. indicus*, *E. coli*, *P. proteolytica*, and *P. antarctica*	[155]
Bacillus cereus	Silver	Antibacterial activity Against *Escherichia coli*, *Klebsiella pneumonia*, *Pseudomonas aeruginosa*, *Staphylococcus aureus*, and *Sal- monella typhi* bacteria	[166]
Bacillus indicus	Silver	Antimicrobial, catalysis	[117]
Bacillus megaterium	Gold	Catalysis, biosensing	[176]
Desulfovvibrio desulfuricans	Gold	Catalysis	[28]
E. coli DH 5α	Gold	Direct electrochemistry of hemoglobin	[32]

(continued)

Table 3 (continued)

Biological entities/species	Nanoparticles	Applications	References
E. coli	Cadmium sulfide	Wurtzite structures	[21]
Escherichia coli	Silver	Antibacterial activity	[30]
Enterobacter cloacae	Silver	Antimicrobial, receptors, electrical batteries, optical	[152]
Lactobacillus casei	Silver	Bio-labeling, cancer treatments, drug delivery	[90]
Klebsiella pneumonia	Silver	Antimicrobial, electrical batteries, optical receptors	[114]
Pseudomonas proteolytica	Silver	Antibacterial activity against *P. antarctica*, *P. proteolytica*, *E. coli*, *A. kerguelensis*, *B. indicus*, and *A. gangotriensis*	[155]
Rhodopseudomonas capsulate	Gold	Triangular	[54]
Fungus			
Aspergillus fumigates	Silver	Coating for intercalation material for electrical batteries and solar energy absorption	[14]
Aspergillusflavus	Silver	Isotropic	[173]
Aspergillus niger	Silver	Antibacterial agent	[45]
Aspergillus terreus	Zinc oxide	Biosensing, Catalysis, cell labeling, drug delivery, imaging, optoelectronics, molecular diagnostics, and solar cell,	[136]
Aspergillus flavus TFR7	Titanium dioxide	Plant nutrient fertilizer	[135]
Cariolus versicolor	Silver	Labels for living cells and tissues, water-soluble metallic catalysts	[146]
Fusarium oxysporum	Gold-silver alloy	Biomedical field	[149]
Fusarium semitectum	Silver	Biolabelling	[13]
Fusarium solani	Silver	Biolabeling, drug delivery, sensors	[74]
Phoma glomerata	Silver	Antimicrobial agent	[15]
Penicillium brecompactum	Silver	Antimicrobial agent	[153]
Penicillium fellutanum	Silver	Surface coating and thin film	[83]
Rhizopus nigricans	Silver	Bactericidal, catalytic	[140]
Trichoderma viride	Silver	Antimicrobial agent	[39]
Verticillium luteoalbum	Gold	Coatings, optics, sensor	[48]

(continued)

Table 3 (continued)

Biological entities/species	Nanoparticles	Applications	References
Yeast			
MKY3	Silver	Coatings for intercalation material for electrical batteries and solar energy absorption	[92]
Saccharimyces cerevisae broth	Gold, silver	Catalysis	[115]
Cryptococcus laurentii	Silver	Antifungal activity against phytopathogenic fungi	[42]
Rhodotorula glutinis	Silver	Antifungal activity against phytopathogenic fungi	[42]

4.2.2 Fungi

Fungi-mediated biosynthesis of metal and metal oxide nanoparticles is highly efficient for producing narrow-size distribution nanomaterials with well-defined morphological features [17, 33, 67, 78, 110]. They function better biologically in creating metal and metal oxide nanoparticles because they have a range of alkaloids and intracellular enzymes [21]. Comparatively speaking to bacteria, competent fungi can produce more nanoparticles [113]. Due to the prevalence of reducible compounds such as alkaloids, enzymes, and proteins on their cell surface membrane, the fungus has many advantages over other biological entities [121]. The most plausible mechanism for creating metallic nanoparticles is enzymatic reduction inside the cell wall of fungi. As shown in Table 3, several species of fungi are used to create metal and metal oxide nanoparticles, including silver, gold, titanium dioxide, and zinc oxide.

4.2.3 Bacteria

Various commercial biotechnological applications have extensively employed bacterial species, including genetic engineering, bioleaching, and bioremediation [48]. Bacteria can decrease metallic ions, making them ideal alternatives for nanoparticle synthesis [75]. Several bacterial strains are used to synthesize metallic and other new nanomaterials. Metal and metal oxide nanoparticles have been synthesized intensively using prokaryotic microorganisms [104]. The bacterial production of nanoparticles has been used because microorganisms are very simple to manipulate [169]. Considerable bacterial strains have been widely utilized to create bio-reduced metal nanoparticles with various shape and size characteristics (Table 3).

4.2.4 Yeast

Several research groups have documented effective yeast-based nanoparticle production [10, 17, 41, 159]. A silver-tolerant yeast strain and *Saccharomyces cerevisiae* broth have been reported to biosynthesize gold and silver nanoparticles [91, 181]. The biosynthesis of countless metallic nanoparticles employs a variety of yeast species, as shown in Table 3.

4.3 Nanomaterials Characterization Techniques

Characterizing novel nanomaterials with well-controlled crystalline phases, structures, sizes, shapes, and porosities is crucial for technological advancements in many fields [96]. Characterization of advanced nanomaterials is a crucial discipline that establishes their structure–property relationship and indicates their possible uses in modern nanotechnology and biotechnology [162]. In many circumstances, physical attributes may be assessed by more than one technique. Size and shape are two of the most investigated factors in the characterization of nanoparticles. In addition to measuring degree of aggregation, size distribution, surface area, and surface charge, we may also assess the surface chemistry to some extent [109]. Different benefits and drawbacks of each methodology complicate the selection of the best suitable method, while a combinatorial characterization strategy is sometimes necessary [81]. Among these methods, AFM, BET, Cryo-TEM, DCS, DLS, EBSD, Electron tomography, EELS, Electron diffraction, FMR, HRTEM, ICP-MS, Liquid TEM, LEIS, and other techniques have been utilized extensively to examine the characteristics of nanomaterials [49, 93, 94] (Table 4).

In addition, as the importance of nanoparticles in fundamental research and applications continues to grow, researchers from various disciplines must overcome the obstacles associated with the repeatable and dependable characterization of nanomaterials after their production and subsequent processing phases.

4.4 Applications of Nanomaterials

Nanomaterials are intriguing because distinct optical, magnetic, electrical, and other properties develop at this scale. These emergent features have numerous possible uses in various fields [3, 61–63, 68–71, 79]. In recent decades, nanomaterials have attracted much attention because of their distinctive features, such as oxidative conditions, antimicrobial activity, and resilience to extreme heat [68, 141, 151, 170]. Nanoparticles were utilized in a variety of sectors, including cosmetic industry, aeronautic industry, chemical industry, space and aviation, thermoelectric devices, automotive engineering, building industry, power generation, solar hydrogen, consumer electronics, fuel cell, optics, batteries, sensors, and pharmaceuticals (Fig. 1).

Table 4 An overview of the experimental methods used to characterize nanomaterials

Technique	Main derived data
Atomic Force Microscopy (AFM)	Accuracy in longitudinal nanoparticle sizes, nanoparticle distribution in cells and other matrices, diameter and form of nanoparticles in 3D form, evaluate the level of surface coverage with nanoparticle shape, rapid investigation structure [51]
Brunauer–Emmett–Teller (BET)	Area of the surface [127]
Cryo-TEM	Agglomeration routes, excellent for molecular biology and colloid chemistry to prevent the presence of artifacts or ruined samples, explore complicated growth mechanisms [77]
DCS	Nanoparticle shape and physical properties [116]
DLS	Agglomeration discovery, hydrodynamic diameter [77]
EBSD	crystal orientation, structure, and phase of materials in SEM. Examine grain morphology, microstructures, defects, deformation, reveal texture [77]
Electron tomography	Video, quantitative information down to the atomic scale, snapshots Realistic 3D particle visualization, [116]
EELS (EELS-STEM)	Bulk plasmon resonance, chemical state of atoms, collective interactions of atoms with neighbors, quantity and type of atoms present [77]
Electron diffraction	Lattice parameters, crystal structure, long-range order parameters, study order–disorder transformation [77]
FMR	Nanoparticle shape, size, size distribution, M values, surface composition, demagnetization field, crystallographic imperfection, magnetic anisotropic constant [77, 116]
HRTEM	Distinguish polycrystalline, monocrystalline and amorphous nanoparticle. All data via standard TEM and on the crystalline form of single particles [77]
ICP-MS	Nanoparticle concentration, size distribution, elemental composition, size [77]
Liquid TEM	Examine the growth process, superlattice development, and single particle movement, and visualize nanoparticle growth in the actual moment [77]
LEIS	Consistency and chemical design of self-assembled monolayers of nanoparticles [77]

(continued)

Table 4 (continued)

Technique	Main derived data
Mössbauer	Magnetic ordering of Fe atoms, distinguish between iron oxides, magnetic anisotropy energy, oxidation state, surface spins, symmetry, thermal unblocking [77]
MFM	Distinguish non-magnetic nanoparticles. Typical AFM imagery and data on the magnetic moments of individual nanoparticles. Investigate magnetic nanoparticles in the cellular interior [77]
NTA	Nanoparticle size and size distribution [77]
NMR (all types)	Atomic chemistry, electronic core structure, ligand density and organization, nanoparticle size, and ligand influence nanoparticle form [57]
PL spectroscopy	Optical qualities—relationship to structural characteristics, including flaws, composition, and size [116]
STEM	Study the atomic structure of hetero-interfaces. EDX for morphology study, combined with HAADF, elemental composition, crystal structure
Superparamagnetic relaxometry	Detect and localize superparamagnetic nanoparticles, Hydrodynamic size distribution, core properties [77]
SAXS	Size of the particles, size distribution, and rate of growth [116]
SQUID-nanoSQUID	Magnetization remanence, blocking temperature, magnetization saturation [77]
Thermogravimetric Analysis (TGA)	Composition and mass of stabilizers [27]
Transmission Electron Microscopy (TEM)	Nanoparticle size, aggregation state, shape, size monodispersity, study growth kinetics detect and quantify nanoparticles in matrices [182]
UV–Visible Spectroscopy	Agglomeration state, optical properties, size, hints on nanoparticle shape, concentration [175]
VSM	Similar to ZFC–FC curves and SQUID through M–H plots [116]
X-ray Photoelectron Spectroscopy (XPS)	Elemental composition, ligand binding, oxidation states, electronic structure [165]
XRD (group: X-ray based techniques)	Crystalline grain size, composition, crystal structure [144]
XAS (EXAFS, XANES)	X-ray absorption coefficient—interatomic lengths, Debye–Waller coefficients, as well as for non-crystalline nanoparticles, the chemical state of species [77]
XMCD	Magnetic values and site symmetry of transition metal ions in particular magnetic properties and ferrimagnetic substances [77]

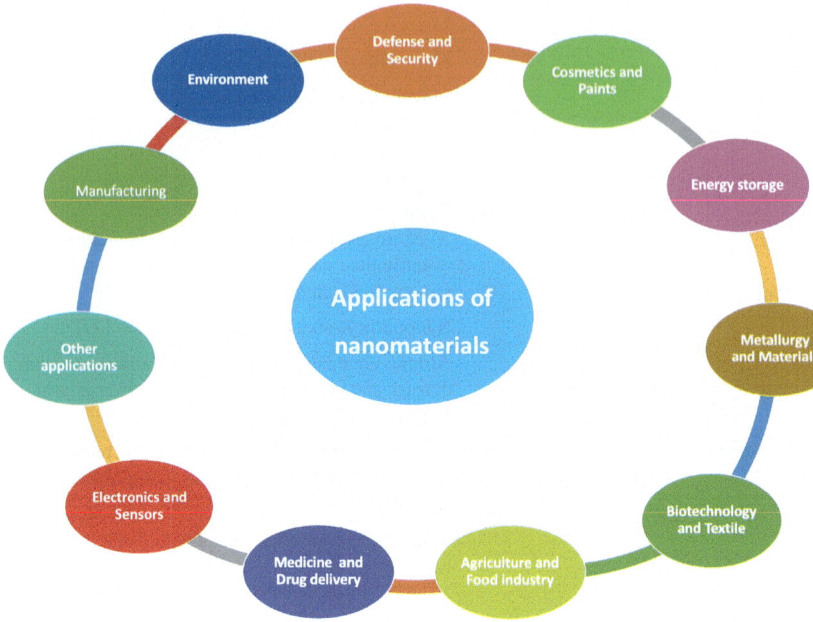

Fig. 1 The potential applications of nanomaterials

In the food, health care, environment, and biomedicine sectors, nanoparticles of gold and silver are widely utilized. Antiviral, anti-inflammatory, antifungal, anticancer, and antibacterial properties are among the applications of silver nanoparticles [82, 87, 88]. Besides, nanomaterials can be utilized as biosensors. Gold nanoparticles can be employed in treating disease and the discoloration of enamels and glasses [132]. Progressively, nanoparticles have been integrated into food packaging to maintain food's freshness and prevent microbial contamination by controlling the surrounding atmosphere [138]. These composites include clays and clay-like particles nanoflakes, which limit moisture entry and decrease gas transfer over the packing film. In addition, it is feasible to include nanoparticles with antimicrobial properties in such packaging [58].

As evidenced by their stress, strain properties, and prime young modulus, nanoparticles have many applications in mechanical industries, particularly in coating, adhesive, and lubrication applications. Additionally, this ability can generate mechanically more robust nanodevices for various applications. By inserting nanoparticles in metal and polymer matrices to boost their mechanical strength, tribological parameters can be manipulated at the nanoscale level. Because of the rolling mode of nanoparticles in the lubricated contact area, friction and wear can be kept to a minimum. In addition, nanoparticles have excellent sliding and delamination capabilities, which may also result in low friction and wear, enhancing the lubricating effect [164].

The nanotechnology-driven intelligent biosensor systems have been proven helpful in detecting infections, chemical pollutants, and pathogen inhibition. These nano-sensors apply to the agricultural, poultry, and food industries [112]. Chemical, physical, and biological components constitute the nano-sensor systems. In the agricultural sector, nano-sensors foresee incorporating precision and sustainable agriculture throughout pre-harvest and postharvest periods. Nano-sensors could be utilized for real-time soil health and quality monitoring during preharvest and postharvest [128]. Incorporating nanotechnology into cosmetic formulation is the most cutting-edge and burgeoning technology currently available. Cosmetic producers use nanoscale-sized chemicals to give enhanced UV protection, longer-lasting effects, enhanced color, deeper skin penetration, and superior finish quality, among other benefits [43]. Micellar nanoparticles are one of the most recent fields applied to cosmetics that are gaining popularity and being widely sold on regional, national, and international markets [46]. The capacity of nano-emulsion technology to generate nanoparticles with tiny size and a large surface area facilitates the efficient delivery of active ingredients to the skin [118].

Using nanoparticles and nano-vacuum tube arrays that improve efficiency and construct low-cost solutions, digital quantum batteries innovatively store energy and information [73]. Nanoparticles have many fantastic uses in this industry, including low electrode resistance and good electron transport, high capacitance with no electric breakdowns owing to quantization impacts, and high capacitance [47]. Metal and metal oxide nanoparticles have gained significant attention in biomedical applications such as biosensing, bioimaging, diagnostics, neurochemical monitoring, biomedical therapeutics, and implantable devices [108].

Innovative drug delivery is administering a drug so that the maximum concentration remains at the intended spot without impacting the rest of the body [7]. Superparamagnetic iron oxide nanoparticles are among the most often employed nanoparticles for selective medication delivery. The magnetic nature of iron oxide nanoparticles makes them suitable for medication delivery devices [171]. With the progress of technology and the need for precise knowledge in multiple fields, sensors play a crucial role in numerous fields, including environmental protection, aerospace, bioengineering, industrial manufacturing, medical diagnosis, and ocean exploration [11]. Georgios A. Sotiriou et al. [161] examined the use of silica-coated as biosensors for the detection of bovine serum albumin. Silica-coated demonstrated remarkable glucose sensing capability with a linear range from 1 to 8 mM and a low detection limit of 10 M [168].

The catalytic action of metal and metal oxide nanoparticles is an additional crucial and expansive field of study [7]. Thus, a plethora of data on the catalytic effect of green-produced nanoparticles employed for degrading hazardous dyes such as methyl orange, safranine, methyl red, crystal violet dye, methylene blue, etc. was available [80, 143]. Silver nanoparticles exhibit high catalytic activity for reducing a variety of toxic dyes.

4.5 Safety Considerations

Nanomaterials utilized in biological and physical applications may serve a variety of purposes. The individual properties of each specific nanomaterial that may result in the desired activity or property of the nanoparticle product may also constitute a danger to the user. Due to the range of impact data and nanoparticles, assessing the environmental risks of specific nanomaterials is difficult. There is very little data about the dangers of nanoparticles in soils and sediments, as most of the current material is found in aquatic environments [171]. The potential adverse effects of transformation products formed after introducing a nanomaterial into the environment garner increased consideration [31]. Due to the possibility of inhalation exposure, sprays or aerosols that may include nanomaterials should be subjected to a more thorough evaluation of their safety. Advancement in deploying computational technologies and approaches for evaluating and quantifying nano-characteristics in varied settings is necessary to gain insight into the existence and exposure to nanoparticles [101].

In light of this, a uniform examination of the safety of all nanomaterials, including nano-characteristics-related assays, is essential. FAO, FDA and other agencies should incorporate the physicochemical properties, size distribution of nanomaterials, stability, aggregation, density, shape, solubility, porosity, and impurities in their safety assessments following the latest guidance [40]. Although tiny nanomaterials have significantly larger molecular weights than known compounds that can permeate the skin, additional studies should be conducted on each nanomaterial used in different applications [44]. In addition, the potential pathways of exposure to nanomaterials should be determined, and in vivo and in vitro toxicological data, such as research on inhalation and dermal absorption, skin and eye irritation investigations, and genotoxicity tests, should be done [157, 177].

4.6 Challenges and Future Prospects

The biosynthesis of nanomaterials is an essential part of nanotechnology, and nanoparticles have applications in numerous areas that have never been seen before. With the introduction of new technology and more excellent scientific discoveries, the path has been created for the use of biological organisms in the creation of nanoparticles, among which the use of alkaloids can be advantageous over the use of other physical entities. Utilizing alkaloids to create nanoparticles is scalable, affordable, and ecologically friendly. It is ideally suited for producing nanoparticles devoid of harmful impurities, as needed for therapeutic applications. For example, despite the potential of nanomaterials synthesized by organism-extract-mediated techniques, the specific synthetic process parameters still need to be optimized. The insufficient data on the active ingredients involved in the production and stabilization of nanomaterials remains a formidable hurdle for scientists. In this approach, bio-template production can significantly impact the following decades by identifying and employing the bioactive compounds in the lifeforms responsible for nanoparticle creation in a single step. It can provide a new look to the product of nanoparticles based on the green approach.

4.7 Conclusion

Natural products are alternatives to potentially active ingredients to discover novel, less harmful, and environmentally friendly substances. Numerous bioactive components, including alkaloids, have been utilized as suitable precursors in synthesizing nanomaterials. Notably, as-synthesized nanomaterials have multiple applications in medicine, anticancer, catalysis, antibacterial, sensors, cosmetic sector, medical imaging, electronics industry, bioengineering, water treatment and many other sectors. The accumulation of these nanoparticles in the environment is a worry, and there is a need to focus on the in vivo toxicity of these nanomaterials and their long-term consequences on animals, humans, and the environment. We anticipate that more research into these concerns will enable us to utilize this synthesis process more effectively for the development and advancement of human society. Moreover, further investigation is required to provide additional insight into green synthesis, particularly for producing regulated shaped and size nanomaterials employing alkaloids. Using alkaloids can also bridge the gap in their synthesis process, facilitating the large-scale production of nanomaterials.

References

1. Ağçeli GK, Dulta K, Chauhan P, Chauhan PK (2021) Biological synthesis of nanomaterials and their advantages. In: nanomaterials in bionanotechnology, pp 69-88. CRC Press
2. Ahmadi MH, Ghazvini M, Alhuyi Nazari M, Ahmadi MA, Pourfayaz F, Lorenzini G, Ming T (2019) Renewable energy harvesting with the application of nanotechnology: a review. Int J Energy Res 43(4):1387–1410
3. Ali AS (2020) Application of nanomaterials in environmental improvement. Nanotechnol Environ
4. Almadiy AA, Nenaah GE (2018). Ecofriendly synthesis of silver nanoparticles using potato steroidal alkaloids and their activity against phytopathogenic fungi. Brazilian Arch Biol Technol, 61
5. Almadiy AA, Nenaah GE, Shawer DM (2018) Facile synthesis of silver nanoparticles using harmala alkaloids and their insecticidal and growth inhibitory activities against the khapra beetle. J Pest Sci 91(2):727–737
6. AlMasoud N, Alomar TS, Awad MA, El-Tohamy MF, Soliman DA (2020) Multifunctional green silver nanoparticles in pharmaceutical and biomedical applications. Green Chem Lett Rev 13(4):316–327
7. Annu AA, Ahmed S (2018) Green synthesis of metal, metal oxide nanoparticles, and their various applications. Handbook Ecomater 2018:1–45
8. Aromal SA, Philip D (2012) Green synthesis of gold nanoparticles using Trigonella foenum-graecum and its size-dependent catalytic activity. Spectrochim Acta Part A Mol Biomol Spectrosc 97:1–5

9. Atanasov AG, Zotchev SB, Dirsch VM, Supuran CT (2021) Natural products in drug discovery: advances and opportunities. Nat Rev Drug Discovery 20(3):200–216
10. Bachheti RK, Abate L, Deepti, Bachheti A, Madhusudhan A, Husen H (2021) Algae-, fungi-, and yeast-mediated biological synthesis of nanoparticles and their various biomedical applications. In: Kharisov B, Kharissova O (eds) Handbook of greener synthesis of nanomaterials and compounds, volume 1: fundamental principles and methods. Elsevier Inc. 50 Hampshire St., 5th Floor, Cambridge, MA 02139, USA, pp 701–734. https://doi.org/10.1016/B978-0-12-821938-6.00022-0
11. Bai Y, Mora-Sero I, De Angelis F, Bisquert J, Wang P (2014) Titanium dioxide nanomaterials for photovoltaic applications. Chem Rev 114(19):10095–10130
12. Baig N, Kammakakam I, Falath W (2021) Nanomaterials: a review of synthesis methods, properties, recent progress, and challenges. Mater Adv 2(6):1821–1871
13. Basavaraja S, Balaji SD, Lagashetty A, Rajasab AH, Venkataraman A (2008) Extracellular biosynthesis of silver nanoparticles using the fungus Fusarium semitectum. Mater Res Bull 43(5):1164–1170
14. Bhainsa KC, D'souza SF (2006) Extracellular biosynthesis of silver nanoparticles using the fungus Aspergillus fumigatus. Colloids Surf B 47(2):160–164
15. Birla SS, Tiwari VV, Gade AK, Ingle AP, Yadav AP, Rai MK (2009) Fabrication of silver nanoparticles by Phoma glomerata and its combined effect against Escherichia coli, Pseudomonas aeruginosa and Staphylococcus aureus. Lett Appl Microbiol 48(2):173–179
16. Biswas P, Wu C-Y (2005) Nanoparticles and the environment. J Air Waste Manag Assoc 55(6):708–746
17. Boroumand Moghaddam A, Namvar F, Moniri M, Md Tahir P, Azizi S, Mohamad R (2015) Nanoparticles biosynthesized by fungi and yeast : a review of their preparation, properties, and medical applications. Molecules 20(9):16540–16565
18. Chaachouay N, Douira A, Hassikou R, Brhadda N, Dahmani J, Belahbib N, Ziri R, Zidane L (2020) Etude floristique et ethnomédicinale des plantes aromatiques et médicinales dans le Rif (Nord du Maroc) [Phdthesis, Département de Biologie - Université Ibn Tofail - Kénitra]. https://tel.archives-ouvertes.fr/tel-03376377
19. Chaachouay N, Douira A, Zidane L (2021) Herbal medicine used in the treatment of human diseases in the Rif, Northern Morocco. Arab J Sci Eng 47:131–153. https://doi.org/10.1007/s13369-021-05501-1
20. Chaachouay N, Douira A, Zidane L (2021) COVID-19, prevention and treatment with herbal medicine in the herbal markets of Salé Prefecture, North-Western Morocco. Eur J Integr Med 42(2), Urban & Fischer, p 101285
21. Chen Y-L, Tuan H-Y, Tien C-W, Lo W-H, Liang H-C, Hu Y-C (2009) Augmented biosynthesis of cadmium sulfide nanoparticles by genetically engineered Escherichia coli. Biotechnol Prog 25(5):1260–1266
22. Chik SC, Or TC, Luo D, Yang CL, Lau AS (2013) Pharmacological effects of active compounds on neurodegenerative disease with gastrodia and uncaria decoction, a commonly used poststroke decoction. Scientific World J
23. Chowdhury A, Ara J, Islam MS (2021) Green synthesis of phytochemical nanoparticles and their antimicrobial activity, a review study. Biomed J Scientific Techn Res 34(4):26929–26935
24. Chugh D, Viswamalya VS, Das B (2021) Green synthesis of silver nanoparticles with algae and the importance of capping agents in the process. J Genetic Eng Biotechnol 19(1):1–21
25. Dahoumane SA, Mechouet M, Wijesekera K, Filipe CD, Sicard C, Bazylinski DA, Jeffryes C (2017) Algae-mediated biosynthesis of inorganic nanomaterials as a promising route in nanobiotechnology–a review. Green Chem 19(3):552–587
26. Darroudi M, Ahmad MB, Zamiri R, Zak AK, Abdullah AH, Ibrahim NA (2011) Time-dependent effect in green synthesis of silver nanoparticles. Int J Nanomed 6:677
27. De Blasio C (2019) Fundamentals of biofuels engineering and technology. Springer
28. Deplanche K, Macaskie LE (2008) Biorecovery of gold by Escherichia coli and Desulfovibrio desulfuricans. Biotechnol Bioeng 99(5):1055–1064

29. Dhand C, Dwivedi N, Loh XJ, Ying ANJ, Verma NK, Beuerman RW, Lakshminarayanan R, Ramakrishna S (2015) Methods and strategies for the synthesis of diverse nanoparticles and their applications: a comprehensive overview. RSC Adv 5(127):105003–105037
30. Divya K, Kurian LC, Vijayan S, Manakulam Shaikmoideen J (2016) Green synthesis of silver nanoparticles by Escherichia coli: analysis of antibacterial activity. J Water Environ Nanotechnol 1(1):63–74
31. Dreher KL (2004) Health and environmental impact of nanotechnology: toxicological assessment of manufactured nanoparticles. Toxicol Sci 77(1):3–5
32. Du L, Jiang H, Liu X, Wang E (2007) Biosynthesis of gold nanoparticles assisted by Escherichia coli DH5α and its application on direct electrochemistry of hemoglobin. Electrochem Commun 9(5):1165–1170
33. DurÁn N, Marcato PD, Ingle A, Gade A, Rai M (2010) Fungi-mediated synthesis of silver nanoparticles: characterization processes and applications. In: Progress in mycology, pp 425-449. Springer
34. Edison TNJI, Atchudan R, Kamal C, Lee YR (2016) Caulerpa racemosa: a marine green alga for eco-friendly synthesis of silver nanoparticles and its catalytic degradation of methylene blue. Bioprocess Biosyst Eng 39(9):1401–1408
35. Eguchi R, Ono N, Hirai Morita A, Katsuragi T, Nakamura S, Huang M, Altaf-Ul-Amin M, Kanaya S (2019) Classification of alkaloids according to the starting substances of their biosynthetic pathways using graph convolutional neural networks. BMC Bioinformatics 20(1):1–13
36. Eroglu E, Chen X, Bradshaw M, Agarwal V, Zou J, Stewart SG, Duan X, Lamb RN, Smith SM, Raston CL (2013) Biogenic production of palladium nanocrystals using microalgae and their immobilization on chitosan nanofibers for catalytic applications. RSC Adv 3(4):1009–1012
37. Fahmy SA, Fawzy IM, Saleh BM, Issa MY, Bakowsky U, Azzazy HME-S (2021) Green synthesis of platinum and palladium nanoparticles using Peganum harmala L. Seed alkaloids : Biological and computational studies. Nanomaterials 11(4):965
38. Fattorusso E, Taglialatela-Scafati O (2007) Modern alkaloids: structure, isolation, synthesis, and biology. Wiley
39. Fayaz AM, Balaji K, Girilal M, Yadav RKalaichelvan PT, Venketesan R (2010) Biogenic synthesis of silver nanoparticles and their synergistic effect with antibiotics: a study against gram-positive and gram-negative bacteria. Nanomed Nanotechnol Biol Med 6(1):103–109
40. FDA U (2014) Final Guidance for industry : safety of nanomaterials in cosmetic products. Docket FDA-2011-D-0489. Washington, DC: US Food & Drug Administration
41. Feng H, Liu S, Huang X, Ren R, Zhou Y, Song C, Qian D (2017) Green biosynthesis of CdS nanoparticles using yeast cells for fluorescence detection of nucleic acids and electrochemical detection of hydrogen peroxide. Int J Electrochem Sci 12(1):618–628
42. Fernández JG, Fernández-Baldo MA, Berni E, Camí G, Durán N, Raba J, Sanz MI (2016) Production of silver nanoparticles using yeasts and evaluation of their antifungal activity against phytopathogenic fungi. Process Biochem 51(9):1306–1313. https://doi.org/10.1016/j.procbio.2016.05.021
43. Fukui H (2018) Development of new cosmetics based on nanoparticles. In: Nanoparticle technology handbook, pp 399–405. Elsevier
44. Fytianos G, Rahdar A, Kyzas GZ (2020) Nanomaterials in cosmetics: recent updates. Nanomaterials 10(5):979
45. Gade AK, Bonde PP, Ingle AP, Marcato PD, Duran N, Rai MK (2008) Exploitation of Aspergillus niger for synthesis of silver nanoparticles. J Biobased Mater Bioenergy 2(3):243–247
46. Gajbhiye S, Sakharwade S (2016) Silver nanoparticles in cosmetics. J Cosmetics Dermatol Sci Appl 6(1):48–53
47. Gensler R, Gröppel P, Muhrer V, Müller N (2002) Application of nanoparticles in polymers for electronics and electrical engineering. particle & particle systems characterization: measurement and description of particle properties and behavior in powders and other disperse systems 19(5):293–299

48. Gericke M, Pinches A (2006) Microbial production of gold nanoparticles. Gold bulletin 39(1):22–28
49. Goyal RK (2017) Nanomaterials and nanocomposites: Synthesis, properties, characterization techniques, and applications. CRC Press
50. Greger H (2019) Structural classification and biological activities of Stemona alkaloids. Phytochem Rev 18(2):463–493
51. Hafiz SS, Labadini D, Riddell R, Wolff EP, Xavierselvan M, Huttunen PK, Mallidi S, Foster M (2020) Surfaces and interfaces of liquid metal core–shell nanoparticles under the microscope. Part Part Syst Charact 37(5):1900469
52. Hamida RS, Abdelmeguid NE, Ali MA, Bin-Meferij MM, Khalil MI (2020) Synthesis of silver nanoparticles using a novel cyanobacteria desertifilum sp. extract: their antibacterial and cytotoxicity effects. Int J Nanomed 15:49
53. Hamouda RA, Hussein MH, Abo-elmagd RA, Bawazir SS (2019) Synthesis and biological characterization of silver nanoparticles derived from the cyanobacterium Oscillatoria limnetica. Sci Rep 9(1):13071. https://doi.org/10.1038/s41598-019-49444-y
54. He S, Guo Z, Zhang Y, Zhang S, Wang J, Gu N (2007) Biosynthesis of gold nanoparticles using the bacteria Rhodopseudomonas capsulata. Mater Lett 61(18):3984–3987
55. Herlekar M, Barve S, Kumar R (2014) Plant-mediated green synthesis of iron nanoparticles. J Nanoparticles
56. Hoek C, Mann D, Jahns HM, Jahns M (1995) Algae : an introduction to phycology. Cambridge University Press
57. Holzgrabe U (2017) Nuclear magnetic resonance spectroscopyl quantitative NMR, methods and applications
58. Hoseinnejad M, Jafari SM, Katouzian I (2018) Inorganic and metal nanoparticles and their antimicrobial activity in food packaging applications. Crit Rev Microbiol 44(2):161–181
59. Hulkoti NI, Taranath TC (2014) Biosynthesis of nanoparticles using microbes—a review. Colloids Surf B 121:474–483
60. Husen A (2019) Natural product-based fabrication of zinc oxide nanoparticles and their application. In: Husen A, Iqbal M (eds) N*anomaterials* and plant potential. Springer International Publishing AG, Gewerbestrasse 11, 6330 Cham, pp 193–291. https://doi.org/10.1007/978-3-030-05569-1_7
61. Husen A (2020c) Introduction and techniques in nanomaterials formulation. In: Husen, A, Jawaid M (eds) Nanomaterials for agriculture and forestry applications. Elsevier Inc. 50 Hampshire St., 5th Floor, Cambridge, MA 02139, USA, pp 1–14
62. Husen A (2022c) Engineered nanomaterials for sustainable agricultural production, soil improvement and stress management. Elsevier Inc., 50 Hampshire St., 5th Floor, Cambridge, MA 02139, USA
63. Husen A, Jawaid, M (2020) Nanomaterials for agriculture and forestry applications. Elsevier Inc., 50 Hampshire St., 5th Floor, Cambridge, MA 02139, USA. https://doi.org/10.1016/C2018-0-02349-X
64. Husen A, Siddiqi KS (2014) Phytosynthesis of nanoparticles: Concept, controversy and application. Nanoscale Res Lett 9(229):1–24
65. Husen A, Siddiqi KS (2014) Plants and microbes assisted selenium nanoparticles: characterization and application. J Nanobiotechnology 12(28):1–10
66. Siddiqi KS, Husen A (2016) Fabrication of metal and metal oxide nanoparticles by algae and their toxic effects. Nanoscale Res Lett 11(363):1–11
67. Siddiqi KS, Husen A (2016) Fabrication of metal nanoparticles from fungi and metal salts: scope and application. Nanoscale Res Lett 11(98):1–15
68. Husen A, Siddiqi KS (2022) Advances in smart nanomaterials and their applications. Elsevier Inc., 50 Hampshire St., 5th Floor, Cambridge, MA 02139, USA
69. Husen A (2020a) Interactions of metal and metal-oxide nanomaterials with agricultural crops: an overview. In: Husen, A, Jawaid M (eds) Nanomaterials for agriculture and forestry applications. Elsevier Inc. 50 Hampshire St., 5th Floor, Cambridge, MA 02139, USA, pp 167–197. https://doi.org/10.1016/B978-0-12-817852-2.00007-X

70. Husen A (2020b) Carbon-based nanomaterials and their interactions with agricultural crops (Eds. Husen A, Jawaid M) Elsevier Inc. 50 Hampshire St., 5th Floor, Cambridge, MA 02139, USA, pp 199–218. https://doi.org/10.1016/B978-0-12-817852-2.00008-1
71. Husen A (2021) Harsh environment and plant resilience (Molecular and Functional Aspects). Springer Nature Switzerland AG, Gewerbestrasse 11, 6330 Cham, Switzerland. https://doi.org/10.1007/978-3-030-65912-7
72. Husen A, Iqbal M (2019) Nanomaterials and plant potential. Springer International Publishing AG, Gewerbestrasse 11, 6330 Cham, Switzerland. https://doi.org/10.1007/978-3-030-05569-1
73. Ibrahim ID, Jamiru T, Sadiku ER, Hamam Y, Alayli Y, Eze AA (2019) Application of nanoparticles and composite materials for energy generation and storage. IET Nanodielectrics 2(4):115–122
74. Ingle A, Rai M, Gade A, Bawaskar M (2009) Fusarium solani: a novel biological agent for the extracellular synthesis of silver nanoparticles. J Nanopart Res 11(8):2079–2085
75. Iravani S (2014). Bacteria in nanoparticle synthesis : Current status and future prospects. Int Scholarly Res Notices
76. Jadoun S, Chauhan NPS, Zarrintaj P, Barani M, Varma RS, Chinnam S, Rahdar A (2022) Synthesis of nanoparticles using microorganisms and their applications: a review. Environ Chem Lett, 1-45
77. Jayawardena HSN, Liyanage SH, Rathnayake K, Patel U, Yan M (2021) Analytical methods for characterization of nanomaterial surfaces. Anal Chem 93(4):1889–1911
78. Jeevanandam J, Chan YS, Danquah MK (2016) Biosynthesis of metal and metal oxide nanoparticles. ChemBioEng Rev 3(2):55–67
79. Jin-Chul K, Madhusudhan A, Husen A (2021) Smart nanomaterials in biomedical applications. Springer Nature Switzerland AG, Gewerbestrasse 11, 6330 Cham, Switzerland
80. Jyoti K, Singh A (2016) Green synthesis of nanostructured silver particles and their catalytic application in dye degradation. J Genetic Eng Biotechnol 14(2):311–317
81. Kalantar-zadeh K, Fry B (2008) Characterization techniques for nanomaterials. Nanotechnol Enabled Sensors, 211–281
82. Kathiravan V, Ravi S, Ashokkumar S (2014) Synthesis of silver nanoparticles from Melia dubia leaf extract and their in vitro anticancer activity. Spectrochim Acta Part A Mol Biomol Spectrosc 130:116–121
83. Kathiresan K, Manivannan S, Nabeel MA, Dhivya B (2009) Studies on silver nanoparticles synthesized by a marine fungus, Penicillium fellutanum isolated from coastal mangrove sediment. Colloids Surf B 71(1):133–137
84. Khan AU, Khan M, Malik N, Cho MH, Khan MM (2019) Recent progress of algae and blue–green algae-assisted synthesis of gold nanoparticles for various applications. Bioprocess Biosyst Eng 42(1):1–15
85. Khan FA (2020) Synthesis of nanomaterials: methods & technology. In: Applications of nanomaterials in human health, pp 15-21. Springer
86. Khandel P, Shahi SK (2018) Mycogenic nanoparticles and their bio-prospective applications: current status and future challenges. J Nanostr Chem 8(4):369–391
87. Khurana C, Vala AK, Andhariya N, Pandey OP, Chudasama B (2014) Antibacterial activity of silver: the role of hydrodynamic particle size at nanoscale. J Biomed Mater Res Part A 102(10):3361–3368
88. Kim K-J, Sung W-S, Moon S-K, Choi J-S, Kim J-G, Lee D-G (2008) Antifungal effect of silver nanoparticles on dermatophytes. J Microbiol Biotechnol 18(8):1482–1484
89. Korbekandi H, Iravani S (2013) Biological synthesis of nanoparticles using algae. Green biosynthesis of nanoparticles: mechanisms and applications, 53-60
90. Korbekandi H, Iravani S, Abbasi S (2012) Optimization of biological synthesis of silver nanoparticles using Lactobacillus casei subsp. Casei J Chem Technol Biotechnol 87(7):932–937
91. Kowshik M, Ashtaputre S, Kharrazi S, Vogel W, Urban J, Kulkarni SK, Paknikar KM (2002) Extracellular synthesis of silver nanoparticles by a silver-tolerant yeast strain MKY3. Nanotechnology 14(1):95

92. Kowshik M, Vogel W, Urban J, Kulkarni SK, Paknikar KM (2002) Microbial synthesis of semiconductor PbS nanocrystallites. Adv Mater 14(11):815–818
93. Kumar CS (2017) Magnetic characterization techniques for nanomaterials. Springer
94. Kumar CS (2018) In-situ Characterization Techniques for Nanomaterials. Springer
95. Kumar KP, Paul W, Sharma CP (2011) Green synthesis of gold nanoparticles with Zingiber officinale extract: Characterization and blood compatibility. Process Biochem 46(10):2007–2013
96. Kumar PS, Pavithra KG, Naushad M (2019) Characterization techniques for nanomaterials. In: Nanomaterials for solar cell applications, pp 97–124. Elsevier
97. Kumar P, Senthamil Selvi S, Govindaraju M (2013) Seaweed-mediated biosynthesis of silver nanoparticles using Gracilaria corticata for its antifungal activity against Candida spp. Appl Nanosci 3(6):495–500
98. Kurek J (2019) Alkaloids: their importance in nature and human life. BoD–Books on Demand
99. Lengke MF, Fleet ME, Southam G (2006) Morphology of gold nanoparticles synthesized by filamentous cyanobacteria from gold (I)-thiosulfate and gold (III)- chloride complexes. Langmuir 22(6):2780–2787
100. Lengke MF, Fleet ME, Southam G (2007) Synthesis of palladium nanoparticles by reaction of filamentous cyanobacterial biomass with a palladium (II) chloride complex. Langmuir 23(17):8982–8987
101. Linkov I, Steevens J, Adlakha-Hutcheon G, Bennett E, Chappell M, Colvin V, Davis JM, Davis T, Elder A, Foss Hansen S (2009) Emerging methods and tools for environmental risk assessment, decision-making, and policy for nanomaterials: summary of NATO advanced research workshop. J Nanopart Res 11(3):513–527
102. Makarov VV, Love AJ, Sinitsyna OV, Makarova SS, Yaminsky IV, Taliansky ME, Kalinina NO (2014) "Green" nanotechnologies : Synthesis of metal nanoparticles using plants. Acta Naturae (англоязычная версия), 6(1 (20)):35–44
103. Mandal D, Bolander ME, Mukhopadhyay D, Sarkar G, Mukherjee P (2006) The use of microorganisms for the formation of metal nanoparticles and their application. Appl Microbiol Biotechnol 69(5):485–492
104. Manimaran M, Kannabiran K (2017) Actinomycetes-mediated biogenic synthesis of metal and metal oxide nanoparticles: progress and challenges. Lett Appl Microbiol 64(6):401–408
105. Manivasagan P, Kim S-K (2015) Biosynthesis of nanoparticles using marine algae : a review. Marine algae extracts: processes, products, and applications, 295–304
106. Manske RHF, Holmes HL (2014) The alkaloids: chemistry and physiology. Elsevier
107. Mat Yusuf SNA, Che Mood CNA, Ahmad NH, Sandai D, Lee CK, Lim V (2020) Optimization of biogenic synthesis of silver nanoparticles from flavonoid-rich Clinacanthus nutans leaf and stem aqueous extracts. Royal Soc Open Sci 7(7):200065
108. Matijević E (2012) Fine particles in medicine and pharmacy. Springer
109. Minelli C (2016) Talk on 'Measuring nanoparticle properties : are we high and dry or all at sea?'at 'Nanoparticle Characterisation–Challenges for the Community'event–IOP. Institute of Physics), book of abstracts
110. Mittal, S., and Roy, A. (2021). Fungus and plant-mediated synthesis of metallic nanoparticles and their application in degradation of dyes. In: Photocatalytic degradation of dyes, pp 287–308. Elsevier
111. Mohammadlou M, Maghsoudi H, Jafarizadeh-Malmiri H (2016) A review on green silver nanoparticles based on plants: synthesis, potential applications and eco-friendly approach. Int Food Res J 23(2):446
112. Mohana NC, Mithun PR, Rao HY, Mahendra C, Satish S (2020) Nanoparticle applications in sustainable agriculture, poultry, and food : trends and perspective. In: Nanotoxicity, pp 341–353. Elsevier
113. Mohanpuria P, Rana NK, Yadav SK (2008) Biosynthesis of nanoparticles: technological concepts and future applications. J Nanopart Res 10(3):507–517
114. Mokhtari N, Daneshpajouh S, Seyedbagheri S, Atashdehghan R, Abdi K, Sarkar S, Minaian S, Shahverdi HR, Shahverdi AR (2009) Biological synthesis of very small silver nanoparticles

by culture supernatant of Klebsiella pneumonia: the effects of visible-light irradiation and the liquid mixing process. Mater Res Bull 44(6):1415–1421
115. Mourato A, Gadanho M, Lino AR, Tenreiro R (2011) Biosynthesis of crystalline silver and gold nanoparticles by extremophilic yeasts. Bioinorg Chem Appl
116. Mourdikoudis S, Pallares RM, Thanh NT (2018) Characterization techniques for nanoparticles: comparison and complementarity upon studying nanoparticle properties. Nanoscale 10(27):12871–12934
117. Mouxing FU, Qingbiao LI, Daohua SUN, Yinghua LU, Ning HE, Xu D, Huixuan W, Huang J (2006) Rapid preparation process of silver nanoparticles by bioreduction and their characterizations. Chin J Chem Eng 14(1):114–117
118. Mu L, Sprando RL (2010) Application of nanotechnology in cosmetics. Pharm Res 27(8):1746–1749
119. MubarakAli D, Gopinath V, Rameshbabu N, Thajuddin N (2012) Synthesis and characterization of CdS nanoparticles using C-phycoerythrin from the marine cyanobacteria. Mater Lett 74:8–11. https://doi.org/10.1016/j.matlet.2012.01.026
120. Mukundan D, Mohankumar R, Vasanthakumari R (2017) Comparative study of synthesized silver and gold nanoparticles using leaves extract of Bauhinia tomentosa Linn and their anticancer efficacy. Bull Mater Sci 40(2):335–344
121. Narayanan KB, Sakthivel N (2011) Synthesis and characterization of nano-gold composite using Cylindrocladium floridanum and its heterogeneous catalysis in the degradation of 4-nitrophenol. J Hazard Mater 189(1–2):519–525
122. Omar HH, Bahabri FS, El-Gendy AM (2017) Biopotential application of synthesis nanoparticles as antimicrobial agents by using Laurencia papillosa. Int J Pharmacol 13(3):303–312
123. Parsons JG, Peralta-Videa JR, Gardea-Torresdey JL (2007) Use of plants in biotechnology: synthesis of metal nanoparticles by inactivated plant tissues, plant extracts, and living plants. Develop Environ Sci 5:463–485
124. Parveen K, Banse V, Ledwani L (2016) Green synthesis of nanoparticles: Their advantages and disadvantages. AIP Conf Proc 1724(1):020048
125. Patel V, Berthold D, Puranik P, Gantar M (2015) Screening of cyanobacteria and microalgae for their ability to synthesize silver nanoparticles with antibacterial activity. Biotechnol Reports 5:112–119
126. Patil S, Chandrasekaran R (2020) Biogenic nanoparticles: a comprehensive perspective in synthesis, characterization, application and its challenges. J Genetic Eng Biotechnol 18(1):1–23
127. Peralta ME, Jadhav SA, Magnacca G, Scalarone D, Mártire DO, Parolo ME, Carlos L (2019) Synthesis and in vitro testing of thermoresponsive polymer-grafted core-shell magnetic mesoporous silica nanoparticles for efficient controlled and targeted drug delivery. J Colloid Interface Sci 544:198–205
128. Prasad R, Kumar V, Prasad KS (2014) Nanotechnology in sustainable agriculture: present concerns and future aspects. Afr J Biotech 13(6):705–713
129. Raffauf RF (2018) Plant alkaloids: a guide to their discovery and distribution. Routledge
130. Raheema RH, Shoker RM (2020) Phytochemicals screening and antibacterial activity of silver nanoparticles, phenols and alkaloids extracts of conocarpus lancifolius. EurAsian J BioSciences 14(2)
131. Rahman A, Kumar S, Nawaz T (2020) Biosynthesis of nanomaterials using algae. In: Microalgae cultivation for biofuels production, pp 265–279. Elsevier
132. Rai M, Yadav A, Gade A (2009) Silver nanoparticles as a new generation of antimicrobials. Biotechnol Adv 27(1):76–83
133. Rajathi FAA, Parthiban C, Kumar VG, Anantharaman P (2012) Biosynthesis of antibacterial gold nanoparticles using brown alga, Stoechospermum marginatum (kützing). Spectrochim Acta Part A Mol Biomol Spectrosc 99:166–173
134. Rajkumar R, Ezhumalai G, Gnanadesigan M (2021) A green approach for the synthesis of silver nanoparticles by Chlorella vulgaris and its application in photocatalytic dye degradation activity. Environ Technol Innov 21:101282

135. Raliya R, Biswas P, Tarafdar JC (2015) TiO2 nanoparticle biosynthesis and its physiological effect on mung bean (Vigna radiata L.). Biotechnol Reports 5:22–26
136. Raliya R, Tarafdar JC (2014) Biosynthesis and characterization of zinc, magnesium and titanium nanoparticles: an eco-friendly approach. Int Nano Lett 4(1):1–10
137. Ramakrishna M, Rajesh Babu D, Gengan RM, Chandra S, Nageswara Rao G (2016) Green synthesis of gold nanoparticles using marine algae and evaluation of their catalytic activity. J Nanostr Chem 6(1):1–13
138. Rastogi S, Kumari V, Sharma V, Ahmad FJ (2021) Gold Nanoparticle-based sensors in food safety applications. Food Analyt Methods, 1–17
139. Ravi L, Selvaraj V, Shankar S, Sundar RDV, Segaran G, Kumaresan S, Sadhana V (2020) Methylene blue degradation by AgCuO bimetallic nanomaterial, green synthesized using Cordia sebestena leaves. Res J Pharm Technol 13(7):3122–3128
140. Ravindra BK, Rajasab AH (2014) A comparative study on biosynthesis of silver nanoparticles using four different fungal species
141. Rizwan M, Singh M, Mitra CK, Morve RK (2014) Ecofriendly application of nanomaterials: nanobioremediation. J Nanoparticles
142. Roco MC (2005) Environmentally responsible development of nanotechnology. Environ Sci Technol 39(5):106A-112A
143. Saha J, Begum A, Mukherjee A, Kumar S (2017) A novel green synthesis of silver nanoparticles and their catalytic action in reduction of Methylene Blue dye. Sustain Environ Res 27(5):245–250
144. Sakthivel NA, Shabaninezhad M, Sementa L, Yoon B, Stener M, Whetten RL, Ramakrishna G, Fortunelli A, Landman U, Dass A (2020) The missing link: Au191 (SPh-t Bu) 66 Janus nanoparticle with molecular and bulk-metal-like properties. J Am Chem Soc 142(37):15799–15814
145. Salem DM, Ismail MM, Tadros HR (2020) Evaluation of the antibiofilm activity of three seaweed species and their biosynthesized iron oxide nanoparticles (Fe3O4-NPs). Egypt J Aquatic Res 46(4):333–339
146. Sanghi R, Verma P (2009) Biomimetic synthesis and characterisation of protein capped silver nanoparticles. Biores Technol 100(1):501–504
147. Seetharaman PK, Chandrasekaran R, Gnanasekar S, Chandrakasan G, Gupta M, Manikandan DB, Sivaperumal S (2018) Antimicrobial and larvicidal activity of eco-friendly silver nanoparticles synthesized from endophytic fungi Phomopsis liquidambaris. Biocatal Agric Biotechnol 16:22–30
148. Sen M (2020) Nanotechnology and the environment. BoD–Books on Demand
149. Senapati S, Ahmad A, Khan MI, Sastry M, Kumar R (2005) Extracellular biosynthesis of bimetallic Au–Ag alloy nanoparticles. Small 1(5):517–520
150. Siddiqi KS, Husen A (2020) Current status of plant metabolite-based fabrication of copper/copper oxide nanoparticles and their applications: a review. Biomater Res 24(11):1–15
151. Siddiqi KS, Husen A, Rao RAK (2018) A review on biosynthesis of silver nanoparticles and their biocidal properties. J Nanobiotechnol 16(14):1–28
152. Shahverdi AR, Minaeian S, Shahverdi HR, Jamalifar H, Nohi A-A (2007) Rapid synthesis of silver nanoparticles using culture supernatants of Enterobacteria: a novel biological approach. Process Biochem 42(5):919–923
153. Shaligram NS, Bule M, Bhambure R, Singhal RS, Singh SK, Szakacs G, Pandey A (2009) Biosynthesis of silver nanoparticles using aqueous extract from the compactin producing fungal strain. Process Biochem 44(8):939–943
154. Shankar SS, Rai A, Ahmad A, Sastry M (2004) Rapid synthesis of Au, Ag, and bimetallic Au core–Ag shell nanoparticles using Neem (Azadirachta indica) leaf broth. J Colloid Interface Sci 275(2):496–502
155. Shivaji S, Madhu S, Singh S (2011) Extracellular synthesis of antibacterial silver nanoparticles using psychrophilic bacteria. Process Biochem 46(9):1800–1807
156. Siddiqi KS, Husen A (2016) Fabrication of metal nanoparticles from fungi and metal salts: scope and application. Nanoscale Res Lett 11(1):1–15

157. Singh N, Jenkins GJ, Asadi R, Doak SH (2010) Potential toxicity of superparamagnetic iron oxide nanoparticles (SPION). Nano Rev 1(1):5358
158. Singh Y, Kaushal S, Singh Sodhi R (2020) Biogenic synthesis of silver nanoparticles using cyanobacterium Leptolyngbya sp. WUC 59 cell-free extract and their effects on bacterial growth and seed germination. Nanoscale Adv 2(9):3972–3982. https://doi.org/10.1039/D0NA00357C
159. Skalickova S, Baron M, Sochor J (2017) Nanoparticles biosynthesized by yeast: a review of their application. Kvasny Prumysl 63(6):290–292
160. Song CE (2009) Cinchona alkaloids in synthesis and catalysis: Ligands, immobilization and organocatalysis
161. Sotiriou GA, Sannomiya T, Teleki A, Krumeich F, Vörös J, Pratsinis SE (2010) Non-toxic dry-coated nanosilver for plasmonic biosensors. Adv Func Mater 20(24):4250–4257
162. Srivastava R (2012) Synthesis and characterization techniques of nanomaterials. Int J Green Nanotechnol 4(1):17–27
163. Srivastava S, Srivastava AK (2013) Biotechnology and genetic engineering for alkaloid production. Natural products: phytochemistry, botany and metabolism of alkaloids, Phenolics and terpenes; Ramawat, GK, Mérillon, J.-M., Eds, 213–250
164. Stark WJ, Stoessel PR, Wohlleben W, Hafner A (2015) Industrial applications of nanoparticles. Chem Soc Rev 44(16):5793–5805
165. Stevie FA, Donley CL (2020) Introduction to x-ray photoelectron spectroscopy. J Vac Sci Technol A: Vac Surf Films 38(6):063204
166. Sunkar S, Nachiyar CV (2012) Biogenesis of antibacterial silver nanoparticles using the endophytic bacterium Bacillus cereus isolated from Garcinia xanthochymus. Asian Pac J Trop Biomed 2(12):953–959
167. Suriyakalaa U, Antony JJ, Suganya S, Siva D, Sukirtha R, Kamalakkannan S, Pichiah PT, Achiraman S (2013) Hepatocurative activity of biosynthesized silver nanoparticles fabricated using Andrographis paniculata. Colloids Surf B 102:189–194
168. Tabrizi MA, Varkani JN (2014) Green synthesis of reduced graphene oxide decorated with gold nanoparticles and its glucose sensing application. Sens Actuators B Chem 202:475–482
169. Thakkar KN, Mhatre SS, Parikh RY (2010) Biological synthesis of metallic nanoparticles. Nanomed Nanotechnol Biol Med 6(2):257–262
170. Tian J, Xu J, Zhu F, Lu T, Su C, Ouyang G (2013) Application of nanomaterials in sample preparation. J Chromatogr A 1300:2–16
171. Ulbrich K, Hola K, Subr V, Bakandritsos A, Tucek J, Zboril R (2016) Targeted drug delivery with polymers and magnetic nanoparticles: covalent and noncovalent approaches, release control, and clinical studies. Chem Rev 116(9):5338–5431
172. Vadlapudi V, Amanchy R (2017) Synthesis, characterization and antibacterial activity of Silver Nanoparticles from Red Algae. Hypnea musciformis. Adv Biol Res 11(5):242–249
173. Vigneshwaran N, Kathe AA, Varadarajan PV, Nachane RP, Balasubramanya RH (2006) Biomimetics of silver nanoparticles by white rot fungus, Phaenerochaete chrysosporium. Colloids Surf B 53(1):55–59
174. Vijayaraghavan K, Nalini SK, Prakash NU, Madhankumar D (2012) One step green synthesis of silver nano/microparticles using extracts of Trachyspermum ammi and Papaver somniferum. Colloids Surf B 94:114–117
175. Wang X, Ramstrom O, Yan M (2010) Quantitative analysis of multivalent ligand presentation on gold glyconanoparticles and the impact on lectin binding. Anal Chem 82(21):9082–9089
176. Wen L, Lin Z, Gu P, Zhou J, Yao B, Chen G, Fu J (2009) Extracellular biosynthesis of monodispersed gold nanoparticles by a SAM capping route. J Nanopart Res 11(2):279–288
177. Wolfram J, Zhu M, Yang Y, Shen J, Gentile E, Paolino D, Fresta M, Nie G, Chen C, Shen H (2015) Safety of nanoparticles in medicine. Curr Drug Targets 16(14):1671–1681
178. Yadav TP, Yadav RM, Singh DP (2012) Mechanical milling: a top down approach for the synthesis of nanomaterials and nanocomposites. Nanosci Nanotechnol 2(3):22–48
179. Yousefzadi M, Rahimi Z, Ghafori V (2014) The green synthesis, characterization and antimicrobial activities of silver nanoparticles synthesized from green alga Enteromorpha flexuosa (wulfen). J Agardh Mater Lett 137:1–4

180. Zhang M, Yang J, Cai Z, Feng Y, Wang Y, Zhang D, Pan X (2019) Detection of engineered nanoparticles in aquatic environments: Current status and challenges in enrichment, separation, and analysis. Environ Sci Nano 6(3):709–735
181. Zhang X, Qu Y, Shen W, Wang J, Li H, Zhang Z, Li S, Zhou J (2016) Biogenic synthesis of gold nanoparticles by yeast Magnusiomyces ingens LH-F1 for catalytic reduction of nitrophenols. Colloids Surf A 497:280–285
182. Zhou W, Greer HF (2016) What can electron microscopy tell us beyond crystal structures? Eur J Inorg Chem 2016(7):941–950

Flavonoids Mediated Nanomaterials Synthesis, Characterization, and Their Applications

Muhittin Kulak and Canan Gulmez Samsa

Abstract Phytochemicals are widely assayed for their potent biological activities and successful findings have been reported. However, due to their poor water solubility and non-targeted site accumulation in addition to the diseased site, the potent of the phytochemicals is not at the desired level. In addition, quickly metabolized properties of the compounds are also critical obstacles in reaching the potent of the compound. In order to overcome the mentioned problem, flavonoids have been incorporated/loaded/encapsulated/conjugated to the nanomaterials as nano-carriers. Herewith in the chapter, we used VOSviewer software to reduce the dimension of the topics according to the networks constructed. The core and common flavonoid compounds were retrieved and used for literature reports and discussion section of the current chapter. In this regard, compounds such as hesperidin, rutin, quercetin, tannin/ tannic acid, naringin, naringenin, and kaemferol were reported. According to the in vivo and in vitro reports, flavonoid-mediated nanoparticles exhibited enhanced/higher activities with no appreciable toxic symptoms in comparison to the bare/pure compound or bare nanoparticle.

Keywords Secondary metabolites · Nano-engineered materials · Nanoparticles · Nanostructures · Flavonoid conjugation

M. Kulak (✉)
Department of Herbal and Animal Production, Vocational School of Technical Sciences, Igdir University, 76000 Igdir, Türkiye
e-mail: muhittin.kulak@igdir.edu.tr

C. Gulmez Samsa
Department of Pharmacy and Pharmaceutical Services, Tuzluca Vocational School, Igdir University, 76000 Igdir, Türkiye

A. Husen (ed.), *Secondary Metabolites Based Green Synthesis of Nanomaterials and Their Applications*, Smart Nanomaterials Technology,
https://doi.org/10.1007/978-981-99-0927-8_3

1 Introduction

Nanomaterials (NPs) are particles with at least one dimension ranging between 1 and 100 nm. Due to their large surface linked to ratio of area-volume, higher/enhanced catalytic reactivity, thermal conductivity, chemical stability, and biological effectivity are of the attained properties of nanomaterials. Such some potent properties of those materials have led to a broad spectra of uses, viz., biomedical, pharmaceutical, chemical, cosmetic, agricultural, and forestry industries [25–32, 41, 42, 71, 70, 80–10]. Corresponding to their formulations, the composition and thereby potential application of nanomaterials remarkably differ. According to their physical and chemical properties, nanomaterials are categorized into two distinct groups [organic based (fullerenes and carbon nanotubes) and inorganic based (metal oxides, metals, quantum dots)] [54]. In addition, the fabrication of nanoparticles is based on three methods such as chemical methods (redox process, sol–gel process, hydrothermal), physical methods (laser ablation, radiation, sonication, electrodeposition), and biological methods (plants, bacteria and fungi) [20]. For characterization of the synthesized nanomaterials, several techniques including scanning electron microscope (SEM), field emission scanning electron microscopic (FESEM), transmission electron microscopy (TEM), high-resolution transmission electron microscopy (HR-TEM), X-ray diffraction (XRD), X-ray photoelectron spectroscopy (XPS), Fourier transform infrared spectroscopy (FTIR), thermogravimetric analyzer (TGA), selected area electron diffraction (SAED), zeta potential and particle size analyzer (Zeta-PALS), photon correlation spectroscopy (PCS), energy dispersive X-ray spectroscopy (EDX), differential scanning calorimetry (DSC), ^{1}H nuclear magnetic resonance (^{1}H NMR), atomic force microscope (AFM), and dynamic light scattering (DLS) are widely employed.

Of the fabrication methods and sources, plant-based nanomaterial synthesis has attracted researchers due to the metabolites such as flavonoids, alkaloids, tannins, saponins available. In addition to their significant contribution to the multiple industrial branches, the plant-based nanomaterial synthesis is safer and easier [51]. Importantly, enhanced/attained properties were provided to the common nanomaterials and vice-versa. The current chapter is based on flavonoid-based nanomaterials.

2 Flavonoids: A Large Group of Phenolic Compounds

Being confined to the plant kingdom, a great number and diversity of low molecular weight compounds are produced. Up to now the approximately 50.000 compounds have already been isolated, characterized, and identified. However, the relevant number of compounds are considered to be over hundreds of thousands. Those compounds are categorized into two major groups [57]. The first major group is "primary metabolites, which constitute the small parts of the elucidated compounds. Primary metabolites are common to all organisms, being capable of division and

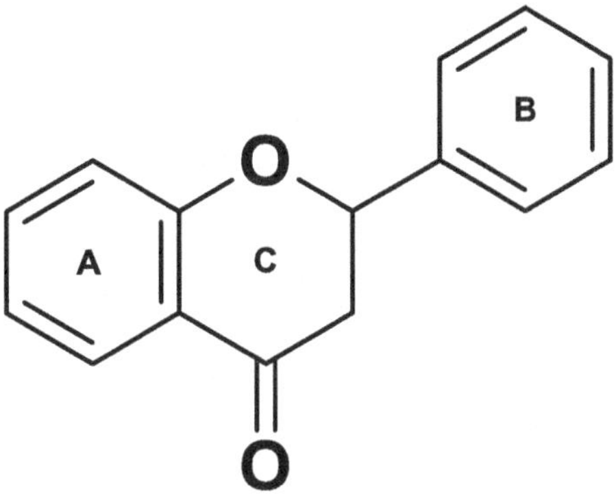

Fig. 1 The basic structure of flavonoids ([13]; with approved permission)

required for primary functions of living organisms, while secondary metabolites are confined to the plant kingdom, in general. Secondary metabolites are not essential for plant life, critically exhibiting variations such as species-specific, organ-specific, and developmental stage-specific [23, 63]. The specificity in biosynthesis has been emerged through the evolutionary processes of plant lineages in order to meet and address the specific needs [57]. It is worthy to note that plant secondary metabolites are derived from primary metabolites and thereby secondary metabolites act as crucial primary metabolites with respect to growth and survival of the plants under sub-optimal environmental conditions. Corresponding to the biosynthetic pathway, secondary metabolites are sorted into three major groups. Those categories are alkaloids (nitrogen-containing compounds), terpenoids, and phenylpropanoids/allied phenolic compounds [19].

Of the secondary metabolites' cosmos, phenolic compounds are biosynthesized from pentose phosphate, shikimate, and phenylpropanoids pathway [61]. A plethora of critical functions have been attributed to those specialized compounds in stress-submitted plants [3, 8]. As a large class of phenolics, flavonoids are reputed metabolites with an estimated number of more than 5000 compounds described in six major sub-classes, viz., flavonols, flavanones, flavanols, anthocyanidins, and isoflavones. Two benzene rings joined by a 3-carbon bridge (C_6–C_3–C_6) constitute the carbon skeleton of flavonoids (Fig. 1) [78]. As of the most of the phenolics, natural and synthetic flavonoids situate at great interest because of their astonishing and prevalent properties in fields of physiology and medicine [44].

3 VOS Viewer Aided Analysis of Flavonoid-Mediated Nanomaterials

After doing a basic search on SCOPUS using TITLE-ABS-KEY (flavonoid OR flavonoids), 157,136 documents were recorded. Then, limitation criteria as (TITLE-ABS-KEY (flavonoid OR flavonoids)) AND (nanomaterial OR nanomaterials

OR nanoparticle OR nanoparticles OR nanostructures OR nanostructures) were employed and accordingly, 13.506 documents were noted from a wide array of subject areas such as Chemistry (N = 4.503), Biochemistry, Genetics and Molecular Biology (N = 3.972), Pharmacology, Toxicology and Pharmaceutics (N = 3.363), Agricultural and Biological Sciences (N = 2.903), Materials Science (N = 2.518), Chemical Engineering (N = 2.513) and Medicine (N = 2.019), etc. (Accessed on September 7, 2022). Such a high number of documents discloses the remarkable uses and potential of flavonoids. Furthermore, in addition to reveal the core topics in disseminated research reports in 2022 with 2.357 document results, a visualized network was constructed using VOSviewer software for the keywords and terms of those documents (Figs. 2 and 3, respectively). Of the phenolics cosmos identified after analysis; hesperidin/hesperitin, rutin, quercetin, tannin/tannic acid, naringin, kaemferol, baicalin, baicalein, catechin, fisetin, and morin are of the most pronounced keywords in the nanoparticle related reports (Fig. 3). The following sections are based on those phyto-molecules and their conjugation with nanostructures. In addition to the keyword analysis, terms analysis was also performed and the network was constructed as given in Fig. 3. According to the network analysis, five major clusters were observed. The cluster-1 is linked to the uses of nanoparticles in plants, being composed of plant growth, antioxidant enzymes, and stress indexes. Cluster-2 is associated with the reactive oxygen species, apoptosis, and cancer cells, in general. Cluster-3 is comprised of bioactive molecules, anti-inflammatory properties, and medicinal plants. Cluster-4 is related to the phenolics and antioxidant as well as antimicrobial activity indexes. Cluster-5 is linked to the characterization instruments applied for nanoparticles (Fig. 3).

4 Synthesis and Characterization of Flavonoid-Mediated Nanomaterials

Nanoparticle synthesis is sharply increasing with a broad of applications in fields of science, engineering, and technology [55], as SCOPUS data supports the idea with a disseminated high number of documents (approximately 850.000 on September 9, 2022). In particular, plant-based nanomaterials provide cost-effective, environment friendly, safer, less toxic properties as well as ease of synthesis in one step. Nanomaterials synthesized with various metal ions (i.e., gold, silver, zinc, titanium, and nickel) using plants as a source are quite new in the literature. In recent years, green synthesis of nanoparticles is quite remarkable, especially since it provides an opportunity for new and superior features that will meet the needs of fields such as agriculture, biomedicine, nanobiotechnology, biosensors, electrochemistry, and cosmetics. Green synthesis provides the desired characteristics and significant advantages to meet the needs of these industrial processes [6, 51, 49].

Fig. 2 Keywords retrieved from flavonoid and nanoparticles documents

Fig. 3 Terms retrieved from flavonoid and nanoparticles documents

For synthesis of nanoparticles, an array of physical and chemical methods has been employed. However, those methods are not eco-friendly approach and subsequently impose harmful/adverse impacts because of involving chemicals. In order to overcome or reduce the potential risk, the researches have been addressed on green synthesis of nanoparticles. The synthesis of nanoparticles generally follows the steps of reduction, stabilization, nucleation, aggregation, capping, and characterization. There are "top-down" and "bottom-up" approaches for nanomaterial synthesis. Top-down methods include mechanical milling, laser ablation, etching, sputtering, and electro-explosion. Bottom-up methods are chemical vapor deposition, solvothermal and hydrothermal methods, the sol–gel method, soft and hard templating methods, as well as reverse micelle methods [9, 28, 38].

In this regard, the plant-derived compounds/extracts are often as reductant and stabilizers during synthesis of nanoparticles [35, 55, 76]. Of those mediators, flavonoids are of the widely preferred reducing agents [69] (Table 1). According to data retrieved documents from SCOPUS, more pronounced flavonoids and their conjugation with nanoparticles were listed in Table 1 in detail with respect to the synthesis, characterization, and applications.

Table 1 Synthesis and characterization of flavonoid-mediated nanomaterials

Flavonoid	Nanoparticle	Synthesis	Characterization	Application	Authors
Hesperidin	Gold nanoparticles	Chemical reduction method	UV spectroscopy, FTIR, TEM, SAED	Antioxidant activity	Pradhan et al. [58]
Hesperidin	Silver nanoparticles	Microwave-assisted process	UV–vis spectra, SEM, XRD, XPS, FTIR, TGA	Antimicrobial activity	Zhao et al. [87]
Hesperidin	Hyaluronic acid with metal organic frameworks	Chemical synthesis method	TEM, SEM, Zeta-PALS, FTIR	Antimicrobial activity	Yang et al. [82]
Hesperidin	Hesperidin-poly lactic co-Glycolic acid (hesperidin -PLGA-Poloxamer 407)	Nanoparticipation technique	XRD, FTIR, PCS, SEM	Anticancer (antioxidant, blood compatibility, MCF-7 cells, cytotoxicity)	Ali et al. [4]
Hesperidin	Gold nanoparticles	Chemical synthesis method	UV–VIS pectroscopy, FTIR, XRD, FESEM, TEM and EDX, Zeta-PALS	Anticancer, anti-inflammatory, and phagocytosis inducer model	Sulaiman et al. [72]
Rutin	Chitosan nanoparticles	Ionic gelation method	SEM, TEM, Zeta, UPLC–MS/MS, DSC	Brain-drug uptake and cerebral Ischemia	Ahmad et al. [2]
Rutin	Chitosan nanoparticles	Ionic crosslinking with tripolyphosphate	UV–VIS spectroscopy, FTIR, DSC, FESEM, Zeta-PALS	Antimicrobial activity (dental disorders)	Patil and Jobanputra [53]
Rutin	Silver nanoparticles	Chemical reduction method	FTIR, TGA, Zeta-PALS, TEM	Antithrombotic Function	Wu et al. [80]
Quercetin	Eudragit and polyvinyl alcohol as carriers	Nanoprecipitation technique	DSC, XRD, FTIR, ^{1}H NMR, Zeta	Antioxidant activity	Wu et al. [81]
Quercetin	Polylactic-co-glycolic acid	Double emulsion-solvent evaporation method	UV–vis, SEM, TEM, particle size, AFM	Alzheimer's disease	Sun et al. [73]

(continued)

Table 1 (continued)

Flavonoid	Nanoparticle	Synthesis	Characterization	Application	Authors
Tannic acid	Silver nanoparticles	Chemical reduction method	FTIR, EDS, Zeta, HR-TEM	Antidiabetic, antioxidant potential and antimicrobial activity	Saratale et al. [68]
Tannic acid	Silver nanoparticles	Chemical synthesis method	FTIR, AFM, Zeta, HR-TEM	Antibacterial activity	Kim et al. [39]
Tannic acid	Silver and gold nanoparticles	Chemical reduction method	UV–Vis, TEM, Zeta	Anti-leishmanial and cytotoxic activities	Lopes et al. [46]
Hesperidin, naringin and diosmin	Silver nanoparticles	Chemical reduction method	XRD, FTIR, TEM	Antibacterial effects and cytotoxicity	Sahu et al. [67]
Naringenin	Silk fibroin nanoparticles	Chemical reduction method	DLS, FESEM, TGA, ATR-FTIR	Anticancer therapy	Fuster et al. [20]
Kaempferol	Poly (ethyleneoxide)-poly(propylene oxide)-poly(ethylene oxide) nanoparticles	Chemical synthesis method	Zeta sizer	Inhibition of ovarian cancer cell viability	Lu et al. (2012)
Kaempferol	Chitosan nanoparticles	Electrostatic self-assembly method	Dynamic light scattering, FTIR, XRD	Anti-quorum sensing activity	Ilk et al. [34]

SEM: Scanning Electron Microscope; **TEM**: Transmission electron microscopy; **XRD**: X-ray diffraction; **XPS**: X-ray Photoelectron Spectroscopy; **FTIR**: Fourier transform infrared spectroscopy; **TGA**: Thermogravimetric analyser; **SAED**: Selected area electron diffraction; **Zeta-PALS**: Zeta potential and particle size analyser; **PCS**: Photon correlation spectroscopy; **FESEM**: Field Emission Scanning Electron Microscopic; **EDX**: Energy dispersive X-ray spectroscopy; **DSC**: Differential scanning calorimetry; **^{1}H NMR**: ^{1}H nuclear magnetic resonance; **AFM**: Atomic Force Microscope; **HR-TEM**: High resolution transmission electron microscopy; **DLS**: Dynamic Light Scattering

5 Together is Better: Flavonoids to Meet the Demand for Multifunctional Nanomaterials

5.1 Hesperidin

Hesperidin ($C_{28}H_{34}O_{15}$) is a flavanone glycoside with a plethora of biological effects, being found in peels of sweet orange, lemon, and grapefruits [26, 45, 58]. As noted above, highly effective nanoparticles might be obtained through conjugation with flavonoids. For instance, hesperidin was isolated from the orange peel and the isolated and characterized hesperidin was used for functionalization of gold nanoparticles via chemical reduction method. The novel hesperidin gold nanoparticles were assayed for their antioxidant activities by scavenging of free radicals of DPPH, ABTS, and hydroxyl radicals of hydrogen peroxide, exhibiting higher activities in comparison to ascorbic acid, a common antioxidant molecule [58]. Hesperidin and pectin-based biosynthesis of silver nanoparticle composite (HP-AgNPs) was performed by microwave-assisted process [87]. The novel nanostructures were assayed for their antibacterial activities. The findings suggested that the conjugation and synergistic effect of hesperidin and pectin contributed to a critical enhancement of antibacterial properties of silver nanoparticles, suggesting that the new nanomaterial can be considered and developed as an antibacterial nanomaterial. Yang et al. [82] reported a multifunctional therapeutic nanomaterial. In this context, hesperidin was used as a load material. Being dependent on the dose, in vitro and in vivo antibacterial tests suggested that the novel material enhanced the antibacterial activity by increasing the membrane permeability and subsequently destroying the integrity of bacteria. In order to increase the stability and bioactive potentials of hesperidin as well as to minimize its absorption problem, Ali et al. [4] prepared modified-nanohesperidin structures, revealing that the nanohesperidin materials were found to be highly effective as novel chemotherapeutic agents due to their activities with respect to exerting cell growth arrest, activation of DNA fragmentation and induction of apoptotic cell through via caspase-3 and p53-dependent pathways. The authors clearly uttered that modified hesperidin was found to be cyto-compatible and more potent in comparison to the native hesperidin. Even the proven and documented quite number of health-promoting effects of hesperidin, the absorption problems of hesperidin due to the poor solubility and bioavailability interrupt its potential. Similar to Ali et al. [4], an efficient delivery system was suggested in order to reach hesperidin therapeutic target [72], being more potent in antioxidant activities and protecting against DNA damage in a cyto-compatible and no appreciable toxic manner.

5.2 Rutin

Rutin (also known as quercetin-3-O-rutinoside and vitamin P) is polyphenolic flavonoid with an identified chemical structure as (2-(3,4-dihydroxyphenyl)-4,5-dihydroxy-3- [3,4,5trihy-droxy-6- [(3,4,5-trihydroxy-6 methyl-oxan-2-yl) oxymethyl] oxan-2-yl] oxy-chromen-7-one. As of other flavonoid-based potential drug candidates, rutin exerts extensive biological effects [52]. Owing to the beneficial biological activities, the compound and its nanoformulation/conjugations were widely employed for multiple purposes. For instance, being a potent antioxidant agent, rutin was reported to be effective in reducing the risk of ischemic disease [2]. However, factors such as low water solubility and low bioavailability and exposing to the chemical degradation might critically affect the therapeutic potentials of rutin. In order to reach the potent of rutin, a novel nano-engineered drug delivery system was suggested to eliminate the factors that cause low amount uptake/access of rutin to the brain [2]. Along with high ratio access of rutin to brain, it was aimed to reduce/avoid the rutin distribution to the non-targeted sites and thereby reducing the potential peripheral side effects. In this regard, rutin-loaded Mucoadhesive polymeric nanostructures were synthesized using ionic gelation method. The findings prove that rutin-mediated nanoformulation was more efficiently accessed and more targeted to the brain [2]. Furthermore, for applications in dental disorders, rutin-chitosan nanoparticles were fabricated and assayed for their antibacterial activities against dental pathogens *Bacillus pumilus, Bacillus thuringiensis, Pseudomonas aeruginosa, Enterococcus faecalis, Acinetobacter junii.* The loading/entrapments of rutin onto chitosan provided the rutin from an enzymatic degradation. The protection was manifested/observed in superiority of new nanostructure over the native rutin and chitosan according to the values of minimum inhibitory concentration (MIC) and minimum bactericidal concentration (MBC) [53]. For potential applications regarding antithrombotic therapy of rutin-conjugated nanomaterials, as noted above, low stability and aqueous solubility of rutin hamper its potential uses in clinical applications [80, 24]. In this context, in order to make rutin an effective anti-coagulant and more bioavailable in blood, Wu et al. [80] designed and fabricated rutin-loaded silver nanoparticles to reveal the potential anti-coagulant. In comparison to the native rutin, the authors reported promising findings for rutin-mediated silver nanoparticles corresponding to appreciable biocompatibility and in long-term antithrombotic therapy.

5.3 Quercetin

As a crucial flavonol, quercetin is of the potent antioxidant agent among polyphenols [50], which are translated into the wide activities, viz., antiviral [56, 40], antibacterial [36, 83], anti-carcinogenic [18, 48], and anti-inflammatory [37, 43]. Those properties have pioneered the researcher to focus on the modified structure of quercetin to enhance its potential uses, as the case we have mentioned above. For example, as of other mentioned compounds, the compound is characterized with poor water solubility. During application of the compounds, the poor solubility was overcome by using dimethylsulfoxide (DMSO) [81]. As reported by Wu et al. [81], higher DMSO might bring about the potential risk linked to vasoconstrictor impact and neurological toxicity [79]. In this context, a quercetin-loaded nanoparticle system was developed and employed by Wu et al. [81]. Promisingly, the antioxidant activity of novel nanostructure was higher than the pure/native quercetin according to an array of assays such as DPPH scavenging, anti-superoxide formation, superoxide anion scavenging, and anti-lipid peroxidation. Furthermore, quercetin conjugation/functionalization of polylactic-co-glycolic acid (PLGA) was designed and examined for the potential uses in Alzheimer's disease [73]. In order to increase solubility and eliminate the extensive first pass metabolism of quercetin, which are the consistent with targets of many researchers, a nano-delivery system consisting of quercetin with a single carrier was prepared. Accordingly, histological data encouragingly revealed no toxicity of nanoparticles, which were attributed to nature of quercetin. In particular, therapeutic index was enhanced and side effects were attenuated with quercetin-mediated delivery system. Being parallel with targets of the previous reports, Zang et al. [85] remarkably reviewed and reported the core of quercetin nanoformulations regarding tumor therapy. The same issues have been addressed in this review study, as well. The chemical structure of quercetin, such as low solubility, less bioavailability, and plausible biotransformation, is of the critical known obstacles to potential uses of quercetin in vivo applications. In addition, the distribution to the non-targeted sites instead of diseased sites because of its quickly metabolized properties and excretion through urinary system is of the questionable topics, in the case of applications of native quercetin [17, 85].

5.4 *Tannin and Tannic Acid*

Tannins are polyphenolic compounds, which are composed of two distinct types as hydrolyzable tannins (gallic acid and some individual sugars) and condensed types (polymers of flavonoids). Tannins bind proteins by hydrogen bonding [11]. Tannic acid, as a specific form of tannin, is of naturally occurring polyphenols found in all aerial plan tissues [64]. As of other natural compounds, tannic acid has been characterized with antioxidant [7, 25], antimutagenic [15, 66], and anticancer properties [84]. In order to enhance the reported biological activities of tannin, the compound has been conjugated with many nanoparticles. For instance, Saratale et al. [68] prepared a novel particle including tannic acid and silver nanoparticles. Then, the multi-biogenic activities (i.e., antidiabetic, antioxidant, and antimicrobial activities) were assayed. Tannin-mediated fabricated silver nanoparticles significantly inhibited α-amylase and α-glucosidase and were also potent antioxidant and antimicrobial. Regarding the searches linked to the novel antibacterial agents, the functionalization of nanoparticles or flavonoids is in vogue, as well. For example, Kim et al. [39] designed a tannic acid-mediated green synthesis of silver nanoparticles, reporting that conjugation of silver nanoparticles and tannic acid enhanced the antibacterial activity, which were attributed to the synergism of the both materials. In a similar manner, tannic acid and silver/gold conjugation resulted in higher antileishmanial activities in comparison to the solo uses of tannic acid and nanoparticles [46].

5.5 Naringin and Naringenin

Naringin is of the flavanone glycosides found in citrus fruits, being effective in human health due to their antioxidant [12], anti-inflammatory [74], anti-apoptotic [77], anti-osteoporotic [21], and anti-carcinogenic [62] properties. Corresponding to the naringin conjugation with nanoparticles in vogue, silver nanoparticles-naringin conjugation exhibited higher stability and subsequently, higher antibacterial and cytotoxic activities were observed [67]. Naringenin, an aglycone flavonoid and metabolite of naringin, was loaded on silk fibroin nanoparticles for improving anti-cancer therapy [20]. For their potential applications, cytotoxic effects of native/free naringenin, silk fibroin nanoparticles, and the interaction of both materials were assayed in comparison. As the case observed for other flavonoid compounds and their synergistic activities, conjugated form of materials demonstrated higher anticancer potential on HeLa cancer cells, in relation to the native/free naringenin [20]. Similarly, naringenin-conjugated polyvinyl alcohol (Na/PVA) nanoparticles exhibited broad antibacterial activities against both Gram-positive and Gram-negative food-borne. Remarkably, a-100% reduction in bacterial load was observed with synthesized Na/PVA nanostructures and consequently, a long-term shelf life was achieved with the relevant treatment [1].

5.6 Kaemferol

Kaempferol is of the widely studied polyphenol ubiquitously in fruits and vegetables, being reputed anticancer agent as a potent promoter of apoptosis [14, 60]. The compound exhibits less toxic to the normal cells, in relation to the common chemotherapy drugs [14, 86]. In comparison to the aforementioned flavonoids, a quite number of reports linked to the kaemferol-mediated nanoparticles are available for various applications [16, 27, 33, 34, 47, 59, 75]. For example, Kaempferol nanoparticles achieved an effective, strong, and selective inhibition of ovarian cancer cell viability [47]. Because of quantity and poor dissolution characteristics of kaempferol, a kaempferol nanoparticle formulation was proposed by Tzeng et al. [75]. The authors indicated that the nanoformulation demonstrated higher antioxidant activity in comparison to bare kaempferol, suggesting potential uses of nanoformulation in health and food searches [75]. Moreover, encapsulated kaempferol into chitosan nanoparticles was more effective in comparison to native/bare kaempferol. Those findings suggest that kaempferol functions might be enhanced with chitosan nanoparticles [34]. Also, enhanced antifungal activities [33], augmented ophthalmological properties [16] and enhanced cytotoxic activities [59] of kaemferol were achieved through conjugation with nanoparticles.

6 Conclusion

Plants have been used for multiple purpose since pre-historic times. Corresponding to the documented beneficial biological activities of the crude plant extracts, high throughput screening of the crude extracts were done and individual compounds were isolated, purified, characterized, and then assayed for their potential roles in combating the diseases. The chemical, clinical, and pharmaceutical reports have indicated that the compounds exhibit poor solubility and non-targeted site accumulation even though they have remarkable properties. Of the cosmos of metabolites, flavonoids are of the largest groups of the secondary metabolites. Among flavonoids, hesperidin/hesperitin, rutin, quercetin, tannin/tannic acid, naringin, kaemferol, baicalin, baicalein, catechin, fisetin, and morin were of the most investigated compounds. In the relevant reports of nanomaterial and flavonoids, it was aimed to increase the solubility and targeted site accumulation of flavonoids through loading/incorporation/encapsulation/conjugation. Significantly, plant-based nanomaterials are cost-effective, environment friendly, safe, and less toxic. On the other hand, the toxic symptoms of the nanomaterials have been noted during treatment of diseases. However, the combination of both materials provided desired activities with no appreciable toxicity according to the clinical findings.

References

1. Ab Rashid S, Rosli NSM, Hway TS, Yenn TW, Ring LC, Yusof FAM, Wen-Nee T (2022) Naringenin-grafted polyvinyl alcohol (Na/PVA) nanoparticles: synthesis, characterisation and *in vitro* evaluation of its antimicrobial efficiency on fresh beef. Trop Life Sci Res 33(1):143
2. Ahmad N, Ahmad R, Naqvi AA, Alam MA, Ashafaq M, Samim M, Iqbal Z, Ahmad FJ (2016) Rutin-encapsulated chitosan nanoparticles targeted to the brain in the treatment of Cerebral Ischemia. Int J Biol Macromol 91:640–655
3. Akula R, Ravishankar GA (2011) Influence of abiotic stress signals on secondary metabolites in plants. Plant Signal Behav 6(11):1720–1731
4. Ali SH, Sulaiman GM, Al-Halbosiy MM, Jabir MS, Hameed AH (2019) Fabrication of hesperidin nanoparticles loaded by poly lactic co-Glycolic acid for improved therapeutic efficiency and cytotoxicity. Artif Cells Nanomed. Biotechnol 47(1):378–394
5. Altinkaynak C, Gulmez C, Atakisi O, Özdemir N (2020) Evaluation of organic-inorganic hybrid nanoflower's enzymatic activity in the presence of different metal ions and organic solvents. Int J Biol Macromol 164:162–171
6. Anand U, Carpena M, Kowalska-Góralska M, Garcia-Perez P, Sunita K, Bontempi E, Dey A, Prieto MA, Proćków J, Simal-Gandara J (2022) Safer plant-based nanoparticles for combating

antibiotic resistance in bacteria: a comprehensive review on its potential applications, recent advances, and future perspective. Sci Total Environ 821:153472
7. Andrade RG Jr, Dalvi LT, Silva JMC Jr, Lopes GK, Alonso A, Hermes-Lima M (2005) The antioxidant effect of tannic acid on the in vitro copper-mediated formation of free radicals. Arch Biochem Biophys 437(1):1–9
8. Ashraf MA, Iqbal M, Rasheed R, Hussain I, Riaz M, Arif MS (2018) Environmental stress and secondary metabolites in plants: an overview. In: Plant metabolites and regulation under environmental stress. Academic Press, pp 153–167
9. Aslam M, Abdullah AZ, Rafatullah M (2021) Recent development in the green synthesis of titanium dioxide nanoparticles using plant-based biomolecules for environmental and antimicrobial applications. J Ind Eng Chem 98:1–16
10. Barreto JA, O'Malley W, Kubeil M, Graham B, Stephan H, Spiccia L (2011) Nanomaterials: applications in cancer imaging and therapy. Adv Mater 23(12):18–40
11. Broderick GA, Wallace RJ, Ørskov ER (1991) Control of rate and extent of protein degradation. In: Tsuda T, Sasaki Y, Kawashima R (eds) Physiological aspects of digestion and metabolism in ruminants: proceedings of the seventh international symposium on ruminant physiology. Academic Press, Waretown, NJ, USA, pp 541–592
12. Cavia-Saiz M, Busto MD, Pilar-Izquierdo MC, Ortega N, Perez-Mateos M, Muñiz P (2010) Antioxidant properties, radical scavenging activity and biomolecule protection capacity of flavonoid naringenin and its glycoside naringin: a comparative study. J Sci Food Agric 90(7):1238–1244
13. Cetinkaya H, Kulak M, Karaman M, Karaman HS, Kocer F (2017) Flavonoid accumulation behavior in response to the abiotic stress: can a uniform mechanism be illustrated for all plants. Flavonoids-From Biosynthesis Hum Health 151–65
14. Chen AY, Chen YC (2013) A review of the dietary flavonoid, kaempferol on human health and cancer chemoprevention. Food Chem 138(4):2099–2107
15. Chen SC, Chung KT (2000) Mutagenicity and antimutagenicity studies of tannic acid and its related compounds. Food Chem Toxicol 38(1):1–5
16. Chuang YL, Fang HW, Ajitsaria A, Chen KH, Su CY, Liu GS, Tseng CL (2019) Development of kaempferol-loaded gelatin nanoparticles for the treatment of corneal neovascularization in mice. Pharmaceutics 11(12):635
17. Dabeek WM, Marra MV (2019) Dietary quercetin and kaempferol: Bioavailability and potential cardiovascular-related bioactivity in humans. Nutrients 11(10):2288
18. Dajas F (2012) Life or death: neuroprotective and anticancer effects of quercetin. J Ethnopharmacol 143(2):383–396
19. Fang X, Yang C, Wei Y, Ma Q, Yang L, Chen X (2011) Genomics grand for diversified plant secondary metabolites. Plant Diversity Resour 33(1):53–64
20. Fuster MG, Carissimi G, Montalbán MG, Víllora G (2020) Improving anticancer therapy with naringenin-loaded silk fibroin nanoparticles. Nanomaterials 10(4):718
21. Gera S, Sampathi S, Maddukuri S, Dodoala S, Junnuthula V, Dyawanapelly S (2022) Therapeutic potential of naringenin nanosuspension: *in vitro* and *in vivo* anti-osteoporotic studies. Pharmaceutics 14(7):1449
22. Gullon B, Lú-Chau TA. Moreira MT, Lema JM, Eibes G (2017) Rutin: a review on extraction, identification and purification methods, biological activities and approaches to enhance its bioavailability. Trends Food Sci Technol 67:220–235
23. Gulmez C, Kulak M (2021) New insights to enhance the desired anti-diabetic compounds in medicinal and aromatic plants exposed to abiotic stress factors. In: Gantait S, Verma SK, Sharangi AB (eds) Biotechnology of anti-diabetic medicinal plants. Springer, Singapore
24. Gulmez C, Altinkaynak C, Özdemir N, Atakisi O (2018) Proteinase K hybrid nanoflowers (P-hNFs) as a novel nanobiocatalytic detergent additive. I J Biol Macromol 119:803–810
25. Gülçin İ, Huyut Z, Elmastaş M, Aboul-Enein HY (2010) Radical scavenging and antioxidant activity of tannic acid. Arab J Chem 3(1):43–53
26. Hajialyani M, Hosein Farzaei M, Echeverría J, Nabavi SM, Uriarte E, Sobarzo-Sánchez E (2019) Hesperidin as a neuroprotective agent: a review of animal and clinical evidence. Molecules 24(3):648

27. Halder A, Das S, Bera T, Mukherjee A (2017) Rapid synthesis for monodispersed gold nanoparticles in kaempferol and anti-leishmanial efficacy against wild and drug resistant strains. Rsc Adv 7(23):14159–14167
28. Hano C, Abbasi BH (2021) Plant-based green synthesis of nanoparticles: Production, characterization and applications. Biomolecules 12(1):31
29. Husen, A, Iqbal, M (2019). Nanomaterials and plant potential. Springer International Publishing AG, Gewerbestrasse 11, 6330 Cham, Switzerland. https://doi.org/10.1007/978-3-030-05569-1
30. Husen A (2019) Natural product-based fabrication of zinc oxide nanoparticles and their application. In: Husen A, Iqbal M (eds) N*anomaterials* and plant potential. Springer International Publishing AG, Gewerbestrasse 11, 6330 Cham, pp 193–291. https://doi.org/10.1007/978-3-030-05569-1_7
31. Husen A (2022) Engineered nanomaterials for sustainable agricultural production, soil improvement and stress management. Elsevier Inc., 50 Hampshire St., 5th Floor, Cambridge, MA 02139, USA
32. Husen A, Jawaid, M (2020) Nanomaterials for agriculture and forestry applications. Elsevier Inc., 50 Hampshire St., 5th Floor, Cambridge, MA 02139, USA. https://doi.org/10.1016/C2018-0-02349-X
33. Ilk S, Saglam N, Özgen M (2017) Kaempferol loaded lecithin/chitosan nanoparticles: preparation, characterization, and their potential applications as a sustainable antifungal agent. Artif Cells Nanomed Biotechnol 45(5):907–916
34. Ilk S, Sağlam N, Özgen M, Korkusuz F (2017) Chitosan nanoparticles enhances the anti-quorum sensing activity of kaempferol. Int J Biol Macromol 94:653–662
35. Jain S, Mehata MS (2017) Medicinal plant leaf extract and pure flavonoid mediated green synthesis of silver nanoparticles and their enhanced antibacterial property. Sci Rep 7(1):1–13
36. Jaisinghani RN (2017) Antibacterial properties of quercetin. Microbiol Res 8(1):6877
37. Karuppagounder V, Arumugam S, Thandavarayan RA, Sreedhar R, Giridharan VV, Watanabe K (2016) Molecular targets of quercetin with anti-inflammatory properties in atopic dermatitis. Drug Discov 21(4):632–639
38. Khan T, Ullah N, Khan MA, Nadhman A (2019) Plant-based gold nanoparticles; a comprehensive review of the decade-long research on synthesis, mechanistic aspects and diverse applications. Adv Colloid Interface Sci 272:102017
39. Kim TY, Cha SH, Cho S, Park Y (2016) Tannic acid-mediated green synthesis of antibacterial silver nanoparticles. Arch Pharm Res 39(4):465–473
40. Kim CH, Kim JE, Song YJ (2020) Antiviral activities of quercetin and isoquercitrin against human herpesviruses. Molecules 25(10):2379
41. Kumar A, Choudhary A, Kaur H, Guha S, Mehta S, Husen A (2022) Potential applications of engineered nanoparticles in plant disease management: a critical update. Chemosphere 295:133798
42. Kumar A, Choudhary A, Kaur H, Mehta S, Husen A (2021) Metal-based nanoparticles, sensors and their multifaceted application in food packaging. J Nanobiotechnol 19(256):1–25. https://doi.org/10.1186/s12951-021-00996-0
43. Lesjak M, Beara I, Simin N, Pintać D, Majkić T, Bekvalac K, Orčić D, Mimica-Dukić N (2018) Antioxidant and anti-inflammatory activities of quercetin and its derivatives. J Funct Foods 40:68–75
44. Liu ZQ (2022) What about the progress in the synthesis of flavonoid from 2020? Eur J Med Chem 243:114671
45. Londoño-Londoño J, de Lima VR, Lara O, Gil A, Pasa TBC, Arango GJ, Pineda JRR (2010) Clean recovery of antioxidant flavonoids from citrus peel: optimizing an aqueous ultrasound-assisted extraction method. Food Chem 119(1):81–87
46. Lopes L, Brito LM, Bezerra TT, Gomes KN, Carvalho FA, Chaves MH (2018) Silver and gold nanoparticles from tannic acid: synthesis, characterization and evaluation of antileishmanial and cytotoxic activities. An Acad Bras Cienc 90:2679–2689
47. Luo H, Jiang B, Li B, Li Z, Jiang BH, Chen YC (2012) Kaempferol nanoparticles achieve strong and selective inhibition of ovarian cancer cell viability. Int J Nanomedicine 7:3951

48. Maalik A, Khan FA, Mumtaz A, Mehmood A, Azhar S, Atif M, Karim S, Altaf Y, Tariq I (2014) Pharmacological applications of quercetin and its derivatives: a short review. Tropical J Pharm Res 13(9):1561–1566
49. Marslin G, Siram K, Maqbool Q, Selvakesavan RK, Kruszka D, Kachlicki P, Franklin G (2018) Secondary metabolites in the green synthesis of metallic nanoparticles. Materials 11(6):940
50. Materska M (2008) Quercetin and its derivatives: chemical structure and bioactivity-a review. Polish J Food Nutr Sci 58(4):407–413
51. Mohammadzadeh V, Barani M, Amiri MS, Yazdi MET, Hassanisaadi M, Rahdar A, Varma RS (2022) Applications of plant-based nanoparticles in nanomedicine: a review. Sustain Chem Pharm 25:100606
52. Negahdari R, Bohlouli S, Sharifi S, Maleki Dizaj S, Rahbar Saadat Y, Khezri K, Jafari A, Ahmadian E, Gorbani Jahandizi N, Raeesi S (2021) Therapeutic benefits of rutin and its nanoformulations. Phytother Res 35(4):1719–1738
53. Patil AG, Jobanputra AH (2015) Rutin-chitosan nanoparticles: fabrication, characterization and application in dental disorders. Polym Plast Technol Eng 54(2):202–208
54. Peralta-Videa JR, Zhao L, Lopez-Moreno ML, de la Rosa G, Hong J, Gardea-Torresdey JL (2011) Nanomaterials and the environment: a review for the biennium 2008–2010. J Hazard Mater 186(1):1–15
55. Periasamy S, Jegadeesan U, Sundaramoorthi K, Rajeswari T, Tokala VNB, Bhattacharya S, Muthusamy S, Sankoh M, Nellore MK (2022) Comparative analysis of synthesis and characterization of silver nanoparticles extracted using leaf, flower, and bark of hibiscus rosasinensis and examine its antimicrobicidal activity. J Nanomater 8123854
56. Di Petrillo A, Orr G, Fais A, Fantini MC (2022) Quercetin and its derivates as antiviral potentials: a comprehensive review. Phytother Res 36(1):266–278
57. Pichersky E, Gang DR (2000) Genetics and biochemistry of secondary metabolites in plants: an evolutionary perspective. Trends Plant Sci 5(10):439–445
58. Pradhan SP, Swain S, Sa N, Pilla SN, Behera A, Sahu PK, Si SC (2022) Photocatalysis of environmental organic pollutants and antioxidant activity of flavonoid conjugated gold nanoparticles. Spectrochim Acta A Mol Biomol Spectrosc 282:121699
59. Raghavan BS, Kondath S, Anantanarayanan R, Rajaram R (2015) Kaempferol mediated synthesis of gold nanoparticles and their cytotoxic effects on MCF-7 cancer cell line. Process Biochemi 50(11):1966–1976
60. Ramos S (2007) Effects of dietary flavonoids on apoptotic pathways related to cancer chemoprevention. J Nutr Biochem 18(7):427–442
61. Randhir R, Lin YT, Shetty K, Lin YT (2004) Phenolics, their antioxidant and antimicrobial activity in dark germinated fenugreek sprouts in response to peptide and phytochemical elicitors. Asia Pac J Clin Nutr 13(3):295–307
62. Rauf A, Shariati MA, Imran M, Bashir K, Khan SA, Mitra S, Zengin G (2022) Comprehensive review on naringenin and naringin polyphenols as a potent anticancer agent. Environ Sci Pollut Res 29:31025–31041
63. Rhodes MJC (1994) Physiological roles for secondary metabolites in plants: some progress, many outstanding problems. Plant Mol Biol 24(1):1–20
64. Robles H (2014) Tannic Acid. Encyclopedia of toxicology, 3rd edn. Elsevier, Amsterdam, The Netherlands, pp 474–475
65. Roduner E (2006) Size matters: why nanomaterials are different. Chem Soc Rev 35(7):583–592
66. Sahiner N, Sagbas S, Sahiner M, Silan C, Aktas N, Turk M (2016) Biocompatible and biodegradable poly (tannic acid) hydrogel with antimicrobial and antioxidant properties. Int J Biol Macromol 82:150–159
67. Sahu N, Soni D, Chandrashekhar B, Satpute DB, Saravanadevi S, Sarangi BK, Pande RA (2016) Synthesis of silver nanoparticles using flavonoids: hesperidin, naringin and diosmin, and their antibacterial effects and cytotoxicity. Int Nano Lett 6(3):173–181
68. Saratale RG, Saratale GD, Ahn S, Shin HS (2021) Grape pomace extracted tannin for green synthesis of silver nanoparticles: assessment of their antidiabetic, antioxidant potential and antimicrobial activity. Polymers 13(24):4355

69. Sathishkumar P, Gu FL, Zhan Q, Palvannan T, Yusoff ARM (2018) Flavonoids mediated 'Green' nanomaterials: a novel nanomedicine system to treat various diseases–current trends and future perspective. Mater Lett 210:26–30
70. Sharma P, Pandey V, Sharma MMM, Patra A, Singh B, Mehta S, Husen A (2021) A review on biosensors and nanosensors application in agroecosystems. Nano Res Lett 16:136. https://doi.org/10.1186/s11671-021-03593-0
71. Sharma G, Nim S, Alle M, Husen A, Kim JC (2022) Nanoparticle-mediated delivery of flavonoids for cancer therapy: prevention and treatment. In: Kim JC, Alle M, Husen A (eds) Smart nanomaterials in biomedical applications. Nanotechnology in the life sciences. Springer, Cham. https://doi.org/10.1007/978-3-030-84262-8_3
72. Sulaiman GM, Waheeb HM, Jabir MS, Khazaal SH, Dewir YH, Naidoo Y (2020) Hesperidin loaded on gold nanoparticles as a drug delivery system for a successful biocompatible, anti-cancer, anti-inflammatory and phagocytosis inducer model. Sci Rep 10(1):1–16
73. Sun D, Li N, Zhang W, Zhao Z, Mou Z, Huang D, Liu J, Wang W (2016) Design of PLGA-functionalized quercetin nanoparticles for potential use in Alzheimer's disease. Colloids Surf B: Biointerfaces 148:116–129
74. Tutunchi H, Naeini F, Ostadrahimi A, Hosseinzadeh-Attar MJ (2020) Naringenin, a flavanone with antiviral and anti-inflammatory effects: a promising treatment strategy against COVID-19. Phytother Res 34(12):3137–3147
75. Tzeng CW, Yen FL, Wu TH, Ko HH, Lee CW, Tzeng WS, Lin CC (2011) Enhancement of dissolution and antioxidant activity of kaempferol using a nanoparticle engineering process. J Agric Food Chem 59(9):5073–5080
76. Umavathi S, Subash M, Gopinath K, Alarifi S, Nicoletti M, Govindarajan M (2021) Facile synthesis and characterization of ZnO nanoparticles using Abutilon indicum leaf extract: an eco-friendly nano-drug on human microbial pathogens. J Drug Deliv Sci Technol 66:102917
77. Wang K, Chen Z, Huang J, Huang L, Luo N, Liang X, Liang M, Xie W (2017) Naringenin prevents ischaemic stroke damage via anti-apoptotic and anti-oxidant effects. Clin Exp Pharmacol Physiol 44(8):862–871
78. Wang S, Zhang J, Chen M, Wang Y (2013) Delivering flavonoids into solid tumors using nanotechnologies. Expert Opin Drug Deliv 10(10):1411–1428
79. Windrum P, Morris TCM, Drake MB, Niederwieser D, Ruutu T (2005) Variation in dimethyl sulfoxide use in stem cell transplantation: a survey of EBMT centres. Bone Marrow Transplant 36(7):601–603
80. Wu H, Su M, Jin H, Li X, Wang P, Chen J, Chen J (2020) Rutin-loaded silver nanoparticles with antithrombotic function. Front Bioeng Biotechnol 8:598977
81. Wu TH, Yen FL, Lin LT, Tsai TR, Lin CC, Cham TM (2008) Preparation, physicochemical characterization, and antioxidant effects of quercetin nanoparticles. Int J Pharm 346(1–2):160–168
82. Yang J, Wang Y, Gao Y, Wang Z, Yin C, Ding X, Yang E, Sun D, Wang W, Guo F (2022) Efficient sterilization system combining flavonoids and hyaluronic acid with metal organic frameworks as carrier. J Biomed Mater Res Part B Appl Biomater 110:1887–1898
83. Yi S, Yi L, Li L, Jin F, Beiyan L, Xuedong Z, Hongkun W (2011) Antibacterial activity of quercetin on oral infectious pathogens. Afr J Microbiol Res 5(30):5358–5361
84. Youness AR, Kamel RA, Elkasabgy N, Shao P, Farag MA (2021) Recent advances in tannic acid (gallotannin) anticancer activities and drug delivery systems for efficacy improvement; a comprehensive review. Molecules 26(5):1486
85. Zang X, Cheng M, Zhang X, Chen X (2021) Quercetin nanoformulations: a promising strategy for tumor therapy. Food Funct 12(15):6664–6681
86. Zhang Y, Chen AY, Li M, Chen C, Yao Q (2008) Ginkgo biloba extract kaempferol inhibits cell proliferation and induces apoptosis in pancreatic cancer cells. J Surg Res 148(1):17–23
87. Zhao ZY, Li PJ, Xie RS, Cao XY, Su DL, Shan Y (2022) Biosynthesis of silver nanoparticle composites based on hesperidin and pectin and their synergistic antibacterial mechanism. Int J Biol Macromol 214:220–229

Synthesis, Characterization, and Applications of Nanomaterials from Carotenoids

Manisha Lakhanpal, Amisha Kamboj, Antul Kumar, Radhika Sharma, Anuj Choudhary, Anand Sonkar, Satyakam Guha, and Sahil Mehta

Abstract There has been a manifold increase in the demand for carotenoids globally over the past few years due to the plethora of uses associated with them. These natural bioactive substances have a myriad of health-promoting properties. They can be synthesized from chemical or biological sources. The biological synthesis of nanomaterials is considered superficial over chemical production because it is more feasible, rapid, gives good returns, and has lower production costs than the latter. Carotenoids have low bioavailability, solubility, and excessive chemical instability which make them unsuitable for industrial use. To enhance their distinctive features and effectiveness, nano-encapsulation is a crucial mechanism through which their size can be reduced to the nanoscale range. These nanoparticles (NPs) showed greater bioavailability, chemical stability, better absorption, and longer shelf life after being entrapped. From the stability perspective of NPs, carotenoids act as vital reducing and capping agents for them. There have been multiple metals, metal oxides, and carbon-based NPs that have been generated from carotenoid-rich extracts. These capping agents maintain the stability of the junction where NPs get interchanged with their preparation media. These capped carotenoid-based NPs uphold their physical, chemical, and biological attributes. In this chapter, we aim to enlighten the multifarious benefits of carotenoid-based nanomaterials, which show immense potential in various sectors of health and the economy.

M. Lakhanpal · A. Kamboj
Department of Forest Products, Dr. Yashwant Singh Parmar University of Horticulture and Forestry, Nauni, Solan 173230, Himachal Pradesh, India

A. Kumar · A. Choudhary
Department of Botany, Punjab Agricultural University, Ludhiana 141004, Punjab, India

R. Sharma
Department of Soil Science, Punjab Agricultural University, Ludhiana 141004, Punjab, India

A. Sonkar · S. Guha · S. Mehta (✉)
Department of Botany, Hansraj College, University of Delhi, New Delhi 110007, India
e-mail: sahilmehtasm21@gmail.com

A. Husen (ed.), *Secondary Metabolites Based Green Synthesis of Nanomaterials and Their Applications*, Smart Nanomaterials Technology,
https://doi.org/10.1007/978-981-99-0927-8_4

Keywords Carotenoids · Chemical stability · Green synthesis · Nanoparticles · Nano-encapsulation

Abbreviations

AgNPs	Silver NPs
AuNPs	Gold NPs
CuNPs	Copper NPs
FeNPs	Iron NPs
NAFLD	Nonalcoholic fatty liver disease
NPs	Nanoparticles
PdNPs	Palladium NPs
ROS	Reactive oxygen species
SeNPs	Selenium NPs
TiO_2NPs	Titanium dioxide NPs
ZnONPs	Zinc oxide NPs
α	Alpha
β	Beta
γ	Gamma

1 Introduction

Instantaneous advancements and technical advances in science and technology fields have greatly stoked the curiosity of the scientific community throughout the world in investigating unconventional features of nanoparticles (NPs)-based materials [49–53, 58, 143]. It was contemplated that NPs form the foundation of nanotechnology [17], and are described as particles having a minimum of 1 dimension in the nanoscale range, i.e., 1–100 nm [65, 141]. Over the past few decades, carotenoids have garnered the worldwide attention of the food and feed sector, textiles chemical and pharmaceutical firms, and cosmetic industries [98]. The word "carotenoid" originates from the botanical name—*Daucus carota* (Carrot), recognized by Wackenroder [163] as the first carotene source [61].

Carotenoids are defined as a diverse class of isoprenoid, lipophilic naturally occurring coloring pigments that are fat soluble and contain 40 carbon skeleton (C40) with eight isoprene molecules [21, 67, 98]. These compounds are known to have a broad scope in the field of human health as they possess strong antioxidant activity [101], anticancerous [144], anti-diabetic [27], anti-tumor [135, 139], anti-inflammatory [64], and anti-aging properties [107]. They are also used as coloring agents and are crucial for producing vitamin A, which is necessary for the development, functioning of the immune system, reproduction, and vision [95, 99]. There are over 600 known carotenoids that have been extracted and are well recognized in natural sources to date [165].

These coloring pigments can be produced chemically or biologically [41]. The biological or green synthesis of carotenoids is preferred over chemical synthesis because it is seen to be more sustainable, cost-effective, reliable, consumes less energy, eco-friendly, easy, and effective sources for high levels of productivity and purity [81, 117]. On contrary to this, the chemical synthesis of carotenoids is a complex and energy-intensive process that uses toxic chemicals and is expensive and inefficient [115, 123, 142].

2 Sources of Carotenoids

They are naturally synthesized by a variety of plants and micro-organisms [86]. Among different plant sources colored fruits, flowers, and dark green colored vegetables are the primary sources of natural carotenoids [166]. The abundant sources of α and β carotenes are cantaloupes, apricots, carrots, lettuce, pumpkin, tomatoes, spinach, broccoli, and sweet potatoes [171]. While persimmons, grapes, orange juice, yellow bell peppers, broccoli, kiwi fruit, spinach, squash, tangerines, maize, mandarins, and zucchini have the highest levels of lutein and zeaxanthin [82, 153]. Tomatoes, watermelon, carrot, guava, papaya, pink grapefruit, apricot, pitanga ripe fruit, autumn olive, red cabbage, and bitter melon are rich sources of lycopene [25, 30, 97, 121, 140].

The microbial production of carotenoids is an efficient way, through which consumer demand for safety can be satisfied, and procuring these natural pigments at the level of industries, expeditiously and efficiently. They are synthesized by a variety of fungi, yeast, lichen, bacteria, and algae [22, 32, 72]. Carotenoids are mostly obtained from fungal and yeast strains such as Mucorales, *Rhodotorula* genera, *Phaffia rhodozyma, Phycomycus blakesleeanus, Blakeslea trispora*, and *Choanephora cucurbitarum* [131]. The industrial production of β-carotene and lycopene is governed by the mold *Blakeslea trispora.* Several species of *Rhodotorula* yeasts, including *R. minuta, R. glutinis, R. mucilaginosa, R. graminis*, and *R. achenniorum*, are known to produce the carotenoids β-carotene, torulene, and torularhodin. Other yeasts that synthesize carotenoids include *Sporobolomyces patagonicus, S. roseus*, and *S. salmonicolor* [4, 160].

Algae are uni- or multi-cellular minute organisms, found in both seawater and freshwater. They have been deemed as ecologically sound and reliable resources for the production of carotenoids [11]. The production of lutein, β-carotene, canthaxanthin, fucoxanthin, and astaxanthin is mostly carried out by the algal strains *Botryococcus braunii, Chlorella* sp., *Haematococcus pluvialis, Dunaliella salina, Scenedesmus* sp. and *Spirulina platensis* [90]. *Gracilaria damaecornis* and *Porphyridium cruentum* (red algae) and *Macrocystis pyrifera* (brown algae) are the main sources of Zeaxanthin. While the major producers of lutein are green (*Chlorophyta*) and red (*Eucheuma isiforme*) algae species [31, 77].

The bacterial synthesis of carotenoids has gained considerable attention due to the notion that it is said to be a sustainable, cost-effective, faster, and economic technique for obtaining carotenoids [32]. Carotenoids namely zeaxanthin, canthaxanthin, and astaxanthin are synthesized by *Paracoccus zeaanthinifaciens, Flavobacterium* sp., *Agrobacterium aurantiacum, Paracoccus carotinifaciens*, and *Bradyrhizobium* sp. Enterobacter and Halobacteria (*Halobacterium salinarium* and *Halobacterium sarcina*) species P_{41} both produce a notable amount of α, β bacterioruberin, and β-carotene, respectively [9, 47, 156]. Cyanobacteria (blue-green algae) sp. such as *Thermosynechococcus elongates* and *Prochlorococcus marinus* are the leading producers of β-carotene and zeaxanthin [24, 66].

Carotenoids are abundant in lichens. Lichens are slow-growing symbiotic organisms that are different from vascular plants and comprised of a fungal and an algal association. β-carotene, zeaxanthin, and violaxanthin are isolated from the thallus of the lichen *Lobaria retigera*. The carotenoid capochrome, a derivative of capsanthin, was produced from the lichens of *Peltigem* sp. *including, P. polydactyla, P. practextata, P. rufescms*, and *P. dolichorhza* [28, 60] (Table 1).

3 Classification of Carotenoids

Carotenoids are generally categorized into carotenes and xanthophylls depending upon characteristic molecular structures. Carotenes are defined as a class of carotenoids which is free of oxygen, purely hydrocarbons with molecular formula $C_{40}H_{56}$. Carotenes comprises of α, β, γ-carotenes, and lycopene [108, 149]. There are about 50 different kinds of carotenes present in nature. β-carotene is a thermolabile, tetra-terpenoid, orange-colored pigment, with a C_{40} structure that is light, oxygen-sensitive, and regarded as a precursor of vitamin A. It shows curative and preventative effects on heart disease, cancer, fibrosis, apoptosis, hepatic steatosis, oxidative stress, and inflammation [18, 110, 168]. Lycopene another vital carotene imparts tomatoes and other red fruits and vegetables their color. It is a nonoxygenated carotenoid which is a polyene consisting of an eight-membered isoprene unit (C5), chemically bonded head to tail, except for a central unit, which has a reverse connection [25, 128]. They have therapeutic effects on cardiovascular diseases, chronic degenerative diseases, prostate, lung, and ovarian malignancies. In addition to this, they also act as biological suppressors of free radicals and modulate gene expression [43, 129, 152].

Xanthophylls are oxygenated and hydroxylated carotenoid pigments, with a molecular formula $C_{40}H_{54}OH_2$, and are separated (based on polarity) from carotenes. There are over 800 different forms of xanthophylls occurring in nature [94, 159]. They comprise mainly astaxanthin, zeaxanthin, lutein, neoxanthin, and fucoxanthin. Astaxanthin is a red, lipophilic xanthophyll that shows antioxidant activity 10 times higher than the β-carotene and 500 times more than vitamin E [14, 45, 63]. They have diverse curative properties for treating conditions like cancer, liver ailments, cardiovascular diseases, neurotoxin-induced neurotoxicity, free radicals damage, and nervous system diseases [46, 169].

Table 1 Major classes and sources of carotenoids

Major carotenoid sources		Carotenoid class	Key references
Plant sources Fruit, vegetables, and plant parts: flowers, leaves, seeds, roots, and shoots	Apricots, carrots, cantaloupe, lettuce, pumpkin, tomatoes, spinach, sweet potatoes, and broccoli	α and β carotenes	Fordham et al. [30], Mayer-Miebach et al. [97], Porcu et al. [121], Shegokar and Mitri [145], Sajilata et al. [140], Ciriminna et al. [25], Zakynthinos and Varzakas [171], Korani et al. [75], Becerra et al. [15]
	Tomatoes, watermelon, guava, papaya, pink grapefruit, apricot, autumn olive, pitanga ripe fruit, red cabbage, carrot, and bitter melon	Lycopene	
	Broccoli, peas, watercress, spinach, and Aztec marigold flowers	Lutein	
	Alfalfa sprouts, corn, mandarin oranges, red bell pepper, and marigold flowers	Zeaxanthin	
	Spinach and yellow bell pepper	Violaxanthin	
	Rocket plant, corn salad, and leek	Neoxanthin	
	Stigmas of saffron flowers	Crocin and Crocetin	
Microbial sources Fungi, yeast, bacteria, and algae	**Algae**–*Botryococcus braunii, Cymodocea nodosa, Dunaliella salina, Gracilaria birdiae,* and *Posidonia oceanic*	β-carotene	Cerda-Olmedo [23], Tinoi et al. [160], Raja et al. [124], Almeida and Cerda-Olmedo [5], Papaioannou and Liakopoulou-Kyriakides [114], Indrayani et al. [55]
	Yeast–*Blakeslea trispora, Phycomyces blakesleeanus, Rhodotorula* sp., and *Sporidiobolus pararoseus*		
	Fungi–*Aspergillus giganteus, Choanephora cucurbitaru,* and *Phycomycus blakesleeanu*		
	Bacteria–Strains of*Paracoccus* sp. DSM 11574and *Enterobacter* sp. P_{41}		
	Algae–*Botryococcus braunii, Chlorophyta* sp., *Coccomyxa acidophila, Chlorella zofingiensis, Eucheuma isiforme, Muriellopsis* sp. and *Scenedesmus almeriensis*	Lutein	Krubasik and Sandmann [77], Frigard et al. [31]

(continued)

Table 1 (continued)

Major carotenoid sources		Carotenoid class	Key references
	Algae–*Botryococcus braunii, Chlorella vulgaris, Chlorella zofingiensis, Chlorococcum* sp., and *Haematococcus pulvialis*	Astaxanthin	Rodriguez-Saiz et al. [134], Liu et al. [87], Kim et al. [70], Gervasi et al. [34], Hayashi et al. [42]
	Yeast–*Xanthophyllomyces dendrorhous, Phaffia rhodozyma*		
	Fungi–*Peniophora* sp., Strains of *Aurantiochytrium sp.* KH105, *Thraustochytriidae sp.* AS4A1 and *Thraustochytrium* ONC-T18 and CHN-1		
	Bacteria–*Agrobacterium aurantiacum, Paracoccus sp.* StrainDSM 1157 and *Paracoccus carotinifaciens*		
	Algae–*Botryococcus braunii, Chaetoceros gracilis, Dunaliella salina, Porphyridium cruentum, Gracilaria damaecornis,* and *Macrocystis pyrifera*	Zeaxanthin	Masetto et al. [96], Berry et al. [16], Pollmann et al. [120], Nascimento et al. [105]
	Bacteria–*Flavobacterium* sp., *Paracoccus zeaxanthinifaciens, Synechococcus* sp. and *Thermosynechococcus elongate*		
	Bacteria–*Haloarchaea*	Lycopene	Mantzouridou and Tsimidou [93], Rodrigo-Banos et al. [133]
	Fungi–*Blakeslea trispora*		
	Algae–Chlorophyta sp., *Alaria crassifolia, Heterokontophyta* sp., *Sargassm* sp.*, Laminaria japonica, Odontellaaurita, Phaeodactylum tricornutum,* and *Undari pinnatifidia*	Neoxanthin and Fucoxanthin	Takaichi [154], Kholany et al. [69]

Zeaxanthin is a non-provitamin-A carotenoid with eleven conjugated double-bonds that are scattered between the polyene chain and the ionone rings which have an OH group that can bind to the fatty acids during esterification [140]. These play a principal role in reducing oxidative stress induced by reactive oxygen species (ROS) [147], preventing metabolic disorders nonalcoholic fatty liver disease (NAFLD) which leads to obesity and diabetes [73], atherosclerosis [161], lessening skin inflammation [112] and inhibiting sunburn [38].

Lutein is a stereoisomer of zeaxanthin, a fat-soluble pigment with 40 carbons and a prominent series of conjugated double-bonds [19, 102]. It has been observed for many useful therapeutic benefits such as hepatoprotective [84], antioxidant [148], anti-inflammatory [71], cardioprotection [111], anti-diabetic vision loss [54], anti-cancer [174], optic nerve injury [62], and traumatic brain injury [155]. Zeaxanthin and lutein are potent xanthophylls that are often referred to as macular pigments accumulated primarily in human retinas, protecting them from short-wavelength visible light and cataract [119, 150].

Neoxanthin is one of the major carotenoids found in plant leaves and was first discovered in barley leaves. It belongs to the β-branch of xanthophylls and has one 5, 6-epoxy, and three hydroxyl groups in its structure [35, 36]. It is the constituent of light-harvesting complexes that play a significant role in carotenoid biosynthesis in green plants and is considered an abscisic acid precursor [20, 57].

Fucoxanthin is the marine carotenoid that contains an allenic bond, 5, 6-monoepoxide, and 9 conjugated double bounds, in its structure. They account for over 10% of the overall production of carotenoids that occurs in nature. According to Willstatter and Page, it was initially derived from the seaweed sp. *Laminaria, Fucus,* and *Dictyota* (in 1914) [68, 164, 172]. They have several health-promoting properties such as anti-diabetic, antioxidant, anti-cancer, anti-proliferative, anti-angiogenic, anti-obesity, anti-malarial, and anti-inflammatory [74, 76, 104, 158].

4 Carotenoid-Mediated Green Synthesis of Nanomaterials

The inimitable properties of NPs lead focus on more sustainable and efficient methods for their synthesis. The conventional process of nanomaterial generation includes physical and chemical means. Consumption of energy and space along with the use of toxic solvents makes these procedures unviable. The biological approach for NPs synthesis involves the use of bioactive constituents of micro-organisms and plants as a reducing and stabilizing medium for a more sustainable and safe biosynthesis of NPs.

Bioactive constituents such as saponins, polyphenols, carotenoids, terpenes, and alkaloids are responsible for the green assembly of NPs synthesis. Carotenoids are pervasive bioactive constituents having a broad range of applications in bio-systems. The use of carotenoids for the production of NPs is an efficient approach with the propitious stability of NPs and sustainability.

4.1 Metal-Based NPs

4.1.1 Silver NPs (AgNPs)

AgNPs possess several industrial applications. Due to their physical and chemical properties, silver NPs are utilized in agriculture, and wastewater treatment [85], microbiology [136], and biomedical field (Zhao et al. 2020). Efficiency-focused green synthesis of silver NPs includes the use of active ingredients of biological origin as reducing and capping agents. Carotenoids containing biological extracts are being used for the stabilized synthesis of AgNPs. Active ingredients enrich extract of *Coriander sativum* containing polyphenolic compounds, lignins, flavonoids, and carotenoids used for photosynthesis of novel AgNPs. Synthesized NPs showed significant antimicrobial, antioxidant, and cytotoxicity against cancer cells [6].

Canthaxanthin is a red-colored carotenoid that occurs in some fungi and bacteria [29] and acts as a good reductant for AgNPs synthesis [162]. Green synthesized AgNPs from Canthaxanthin extracted from bacteria *Ditezia maris* showed potential cytotoxicity against keratinocyte cells [162]. *Chlorella vulgaris* (green algae) is abundant in active constituents including carotenoids, carbohydrates, lipids, and other related secondary compounds [79]. An eco-friendly approach to AgNPs synthesis using microalgae (*Chlorella vulgaris*) was established. Instead of chemical agents extract of *Chlorella vulgaris* is used as stabilizing, reducing, and capping agent for Ag NPs synthesis [91].

Carotenoids-rich seed extracts of *Calendula officinalis* and *Persea americana* are exploited for the synthesis of relatively stable AgNPs [13]. Another sustainable methodology for AgNPs synthesis was given by Kumar et al. [80] by using fruit extract of *Eugenia stipitata;* it is, therefore, justified that carotenoids, malic, and citric acid were mainly responsible for the synthesis of AgNPs [80] (Fig. 1).

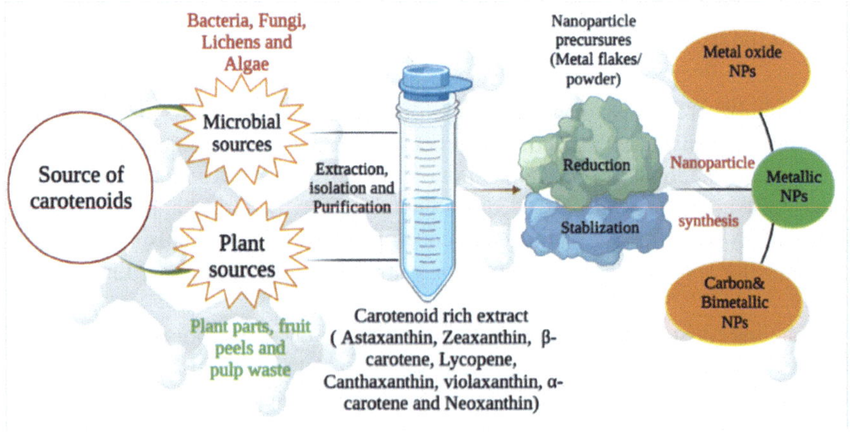

Fig. 1 Carotenoid-mediated green synthesis of nanoparticles

4.1.2 Gold NPs (AuNPs)

The use of AuNPs is gaining keen interest in biomedical [132, 173], bioengineering [92], and agricultural applications [39, 122]. Pigment-mediated synthesis of novel Au NPs is cost-effective, sustainable, and more efficient than chemical synthesis. Fucoxanthin, a major carotenoid from diatoms (*Nanofrustulum shiloi*) and seaweeds have tremendous potential for reducing gold ions into gold nanospheres [138]. Carotenoid-rich extract from *Cucurbita pepo* leaves resulted in the formation of Au NPs in the size range of 10–15 nm [37]. Bioactive compounds such as carotenoids, polyphenols, vitamins, proteins, flavonoids, and minerals are chiefly distributed in fruit peels acting as safe bioreagents for NP synthesis. Biogenesis of AuNPs from fruit peel extract of *Passiflora edulis* and *Ananas comosus* didn't possess any cytotoxicity and was relatively safe for use [118]. Roychoudhury compared the efficiency of pigments (carotenoids, polysaccharides, and chlorophyll) against the reduction of auric chloride solution to AuNPs. Results showed that among other pigments extracted from *Anabaena sphaerica* and *Chlorococcum infusionum*, carotenoids were more efficient in NP synthesis [137]. Remarkable antibacterial activity was shown by spherical AuNPs derived from green algal biomass of *Rhizoclonium fontinale* [116].

4.1.3 Palladium NPs (PdNPs)

Due to the unique optical and catalytic activity of PdNPs, they have diverse applications in the field of electrochemistry [8, 83, 167] and bioengineering [89]. The action of β-carotenes and ascorbic acid in the stem of *Cissus quadrangularis* as a reductant for the synthesis of AuNPs, AgNPs, and PdNPs is reported [8].

4.1.4 Copper NPs (CuNPs)

CuNPs possess various novel chemical, optical, and surface characteristics and a broader range of applications in agriculture [56, 88], bioengineering [10] and biomedical [113, 127] field. A more efficient sustainable process of CuNPs synthesis using bud extract of *Syzygium aromaticum* was developed. The presence of carotenoids, phytosterols, and proteins is the main reasons for the stabilization and reduction of CuNPs and the resultant nanomaterial possesses good antimicrobial activities against *Pseudomonas* and *Bacillus* sp. [125].

4.1.5 Iron NPs (FeNPs)

Iron NPs are consistently being used in drug delivery and environment clean-up systems. Green approaches for the synthesis of iron-structured nanomaterials are a good alternative for the safe and cost-effective biogenesis of NPs. Lycopene is an antioxidant carotenoid generally found in fruits and vegetables [130]. Lycopene can be naturally extracted from tomatoes, along with all other nutritional and health benefits. Lycopene is a good reducing and stabilizing carotenoid for NP synthesis. Green synthesis of FeNPs, AuNPs, and AgNPs was exhibited by lycopene extracted from tomato fruits and resultant NPs possessed good anticancerous activity [146].

4.2 Metal Oxide NPs

4.2.1 Titanium Dioxide NPs (TiO_2NPs)

TiO_2NPs is a metal oxide NPs with specialized physiochemical properties which make their use efficient in industrial applications, i.e., environment [40, 59]. The biogenesis of TiO_2NPs is gaining more focus due to its low cost, safety, and efficiency with more stable nanomaterials. Active content enriches (carotenoids) extract of *Cucurbita pepo* seeds successfully reduced titanium trichloride solution into TiO_2NPs [2].

4.2.2 Zinc Oxide NPs (ZnONPs)

Carotenoid present in lemon peel was able to synthesize thin film zinc oxide nanostructures by spin-coating method [26]. ZnONPs procured from peel extract of *Citrus sinesis* had good antibacterial and antifungal activity as well as can be used for strawberry preservation [33]. Combined peel extract from four different fruits, i.e., *Citrus aurantifolia, Lycopersicon esculentum, Citrus sinensis*, and *Citrus paradise* collectively able to reduce zinc nitrate into ZnONPs [106].

4.3 Carbon-Based NPs

Readily available β-carotene was used as a bio-reductant for the synthesis of colloidal carotene-carbon NPs with anticancerous potential [100].

4.3.1 Graphene Oxide NPs

Biogenesis of graphene nanosheets from graphene powder by β-carotene as reductant and their use in supercapacitor electrodes is reported [170]. The reduction of graphene flakes into graphene NPs by β-carotene from wild carrot root was confirmed by spectroscopic evidences [78].

4.4 Selenium NPs (SeNPs)

Biogenic synthesis of SeNPs by phytoactive constituents, i.e., carotenoids is gaining more focus due to improved applications. SeNP assembly with good antibacterial and antioxidant potential was successfully synthesized by carotenoid-rich extract of *Calendula officinalis* [44] (Table 2).

5 Applications of Carotenoid-Based NPs

5.1 Biomedical

The exponential growth of the world's population at an alarming rate leads to disease incidences on a larger scale. The majority of premature deaths worldwide are the result of these abrupt disease epidemics. Therefore, to tackle the emergence and re-emergence of diseases in a population, we need to explore the new horizons of biomedical research to develop an effective approach that can be used to reduce disease prevalence. Bioactive compounds such as carotenoids can be efficiently used for this purpose. Carotenoids have numerous health implications in various sectors of healthcare as they have the potential to obviate many appalling disorders.

Carotenoids are considered one of the highest-value bio-products that can be extensively used to regress many of the deadliest diseases. They have attracted enormous attention in the field of healthcare as they possess antibacterial [1, 8, 12, 33, 44, 103, 116, 125, 157], antimicrobial [6], antioxidant [6, 44, 80, 151], antifungal [33], anticancerous [7, 100, 146], and anticoagulation activity [1].

5.2 Environment

Degradation of the environment is a major issue for developing countries. Expeditious expansion in population and high-speed exploitation of resources need sustainable solutions for environmental safety. Nanotechnology is an emerging solution for the remediation of environment-related issues. Nanomaterials can degrade harmful and toxic pollutants and can be used as a new-generation solution for environment-related problems. Carotenoid-derived ZnONPs had the potential to degrade methylene blue which is produced mainly by textile industries as a toxic pollutant [106, 109] Carotenoid-based TiO_2NPs can be used in wastewater treatment [126] (Fig. 2).

Table 2 Size, type, sources, and application of nano-particle derived from carotenoids

Source of carotenoids	NPs type and size (nm)	Phyto-chemicals acting as reducing or stabilizing agent	Applications	References
Wild carrot root	Graphene NPs 2.5	β-carotene	–	Kuila et al. [78]
Natural β-carotene	Graphene nanosheets 2.4	β-carotene	Supercapacitor electrode	Zaid et al. [170]
Rhizoclonium fontinale	Au NPs 16	Carotenoids, protein, and chlorophyll	Antibacterial	Parial and Pal [116]
Calendula officinalis	Ag NPs 5–10	Carotenoids, triterpenoids, flavonoids, coumarins, quinones, volatile oil, and amino acids	–	Baghizadeh et al. [13]
Moringa olifera	Au NPs 3–5	Carotenoids, sterols, tannins, and phenolic compounds	Anticancerous and catalytic activity	Anand et al. [7]
Anabaena sphaerica and *Chlorococcum infusionum*	Au NPs 1–50	Carotenoids, polysaccharides, and chlorophyll	–	Roychoudhury et al. [137]
Eugenia stipitata	Ag NPs 15–45	Carotenoids, malic and citric acid	Antioxidant activity	Kumar et al. [80]
Gordonia amicalis	Au, Ag NPs 25–50	Carotenoid B and Carotenoid K	Antioxidant	Sowani et al. [151]
β-carotene	C NPs 5–20	β-carotene	Anticancerous activity	Misra et al. [100]
Citrus aurantifolia	ZnO 50	Carotenoids, limonoids, and coumarins	Electrochemical	Colak and Karakose [26]
Citrus aurantifolia, Lycopersicon esculentum, Citrus sinensis, and *Citrus paradise*	ZnO 9.06–19.66	Flavonoids, Limonoids, and Carotenoids	Degradation of methylene blue (pollutant)	Nava et al. [106]

(continued)

Table 2 (continued)

Source of carotenoids	NPs type and size (nm)	Phyto-chemicals acting as reducing or stabilizing agent	Applications	References
Juglans regia	Ag NPs 3–50	Carotenoids, tannins, reducing sugars, and phenolic compounds	Antibacterial and anticoagulation activity	Abbasi et al. [1]
Chlorella vulgaris	Ag NPs 40–90	Carotenoids, lipids, carbohydrates	Quinolones synthesis	Mahajan et al. [91]
Cucurbita pepo	Au NPs 10–15	Carotenoids	–	Gonnelli et al. [37]
Syzygium aromaticum	Cu NPs 15	Tannins, flavonoids, alkaloids, and carotenoids	Antibacterial	Rajesh et al. [125]
Cissus quadrangularis	Au, Ag, Pd NPs 4.99–26, 4–16.7, 12–26	Ascorbic acid, β-carotene, and phenolic compounds	Antibacterial and electrochemical activity	Anjana et al. [8]
Cucurbita pepo	TiO_2	Carotenoids, proteins, phytosterols, etc.	–	Abisharani et al. [2]
Persea americana	Ag NPs 2–50	Carotenoids (α-carotene, β-carotene, tocopherols, zeaxanthin, violaxanthin, neoxanthin, and tocopherols)	Antimicrobial and Surface-Enhanced Raman Scattering (SERS)	Vazqueza et al. (2019)
Sonchus asper	TiO_2 9–22	Carotenoids	Antibacterial	Babu et al. [12]
Citrus sinesis	ZnO 11.2–10.8	Vitamin C and Carotenoids	Antibacterial, antifungal, and strawberry preservation	Gao et al. [33]
Citrus sinesis	ZnO 200–230	Flavonoids, Limonoids, and Carotenoids	Antibacterial activity	Thi et al. [157]
Dietzia maris	Ag NPs 40–50	Canthaxanthin	Biomedical application	Venil et al. [162]
Lycopene capsule	Ag NPs 40	Lycopene	Antibacterial	Murthykumar et al. [103]

(continued)

Table 2 (continued)

Source of carotenoids	NPs type and size (nm)	Phyto-chemicals acting as reducing or stabilizing agent	Applications	References
Morchella esculenta	Au NPs 16.51	β-carotene and antioxidant compounds	Biomedical applications	Acay [3]
Calendula officinalis	Se NPs 40–60	Astaxanthin, zeaxanthin, β-carotene, lycopene, and canthaxanthin	Antibacterial and antioxidant	Hernandez-Diaz et al. [44]
Coriandrum sativum	Ag NPs 1	Polyphenolic compounds, lignins, flavonoids, and carotenoids	Antimicrobial, antioxidant and cytotoxicity against cancer cells	Alsubki et al. [6]
Date pulp waste	ZnO 30	Carotenoids and anthocyanins, etc.	Wastewater treatment	Rambabu et al. [126]
Passiflora edulis and *Ananas comosus*	Au NPs 20.71 ± 7.44	Carotenoids, polyphenols, vitamins, proteins, flavonoids, and minerals	–	Pechyen et al. [118]
Nanofrustulum shiloi	Au NPs 2–35 nm	Primary carotenoid (Fucoxanthin)	–	Roychoudhury et al. [138]
Solanum lycopersicum	Ag, Au Fe NPs 50–100	Lycopene	Anticancerous activity	Shejawal et al. [146]
Citrus sinesis and *Musa acuminate*	TiO_2 7.3–27.3	Carotenoids, steroids, fatty acids, vitamin B and pectin, etc.	Degradation of methylene blue	Olana et al. [109]

5.3 Electrochemical

The diverse catalytic, physical, and chemical properties of NPs make them suitable for electrochemical use [8, 26]. NPs can be used to fulfill the future energy demand more sustainably. Carotenoid-mediated NPs are used for the construction of batteries, fuel cells, diodes, and capacitors [170], etc.

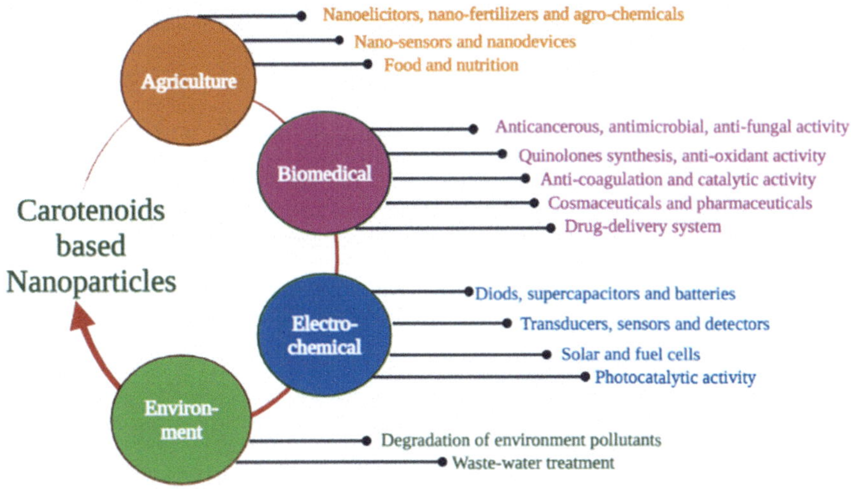

Fig. 2 Application of carotenoid-based nanoparticles

6 Conclusion

NPs are revolutionary materials in the environment, energy, biomedical, and agriculture sectors. Bioactive ingredient-based synthesis of NPs has great scope in research and development. Novel reducing and stabilizing potential of carotenoids can be exploited for sustainable and low-cost synthesis of NPs. Carotenoids are mainly procured from plants and micro-organisms and their extract is being used for the reduction of metal salts into NPs. In particular, the isolation of carotenoids and their further use for NP synthesis instead of combined plant extract is still limited. Their potential for NPs synthesis like aluminum, iron oxide, magnesium, and other metalloids NPs can be explored. Carotenoid-derived NPs have diverse applications in biomedical, electrochemical, and environment but their use in agriculture, i.e., elicitation and nano-fertilizer is limited. More areas for NP application can be explored for future research.

References

1. Abbasi Z, Feizi S, Taghipour E, Ghadam P (2017) Green synthesis of silver nanoparticles using aqueous extract of dried *Juglans regia* green husk and examination of its biological properties. Green Process Synth 6:477–485
2. Abisharani JM, Devikala S, Kumar R, Arthanareeswari M, Kamaraj P (2019) Green synthesis of TiO_2 nanoparticles using *Cucurbita pepo* seeds extract. Mater Today Proc 14:302–307
3. Acay H (2020) Utilization of *Morchella esculenta*-mediated green synthesis golden nanoparticles in biomedicine applications. Prep Biochem Biotechnol 51:1–10

4. Aksu Z, Eren AT (2007) Production of carotenoids by isolated yeast of *Rhodotorula glutinis*. Biochem Eng J 35:107–113
5. Almeida ERA, Cerda-Olmedo E (2008) Gene expression in the regulation of carotene biosynthesis in Phycomyces. Curr Genet 53:129–137
6. Alsubki R, Tabassum H, Abudawood M, Rabaan A, Alsobaie SF, Ansar S (2021) Green synthesis, characterization, enhanced functionality and biological evaluation of silver nanoparticles based on *Coriander sativum*. Saudi J Biol Sci 28:2102–2108
7. Anand K, Gengan RM, Phulukdaree A, Chuturgoon A (2015) Agroforestry waste *Moringa oleifera* petals mediated green synthesis of gold nanoparticles and their anti-cancer and catalytic activity. J Ind Eng Chem 21:1105–1111
8. Anjana PM, Bindhu MR, Umadevi M, Rakhi RB (2019) Antibacterial and electrochemical activities of silver, gold, and palladium nanoparticles dispersed amorphous carbon composites. Appl Surf Sci 479:96–104
9. Asker D, Beppu T, Ueda K (2007) *Sphingomonas jaspsi sp.* A novel carotenoid-producing bacterium isolated from Misasa, Tottori, Japan. Int J Syst Evol Microbiol 57:1435–1441
10. Athanassiou EK, Grass RN, Stark WJ (2006) Large-scale production of carbon-coated copper nanoparticles for sensor applications. Nanotechnology 17:1668–1673
11. Babadi FE, Boonnoun P, Nootong K, Powtongsook S, Goto M, Shotipruk A (2020) Identification of carotenoids and chlorophylls from green algae *Chlorococcum humicola* and extraction by liquefied dimethyl ether. Food Bioprod Process 123:296–303
12. Babu N, Pathak VM, Singh A, Navneet A (2019) Sonchus asper leaves aqueous extract mediated synthesis of titanium dioxide nanoparticles. J Pharm Innov 8(4):817–822
13. Baghizadeh A, Ranjbar S, Gupta VK, Asif M, Pourseyedi S, Karimi JM, Mohammadinejad R (2015) Green synthesis of silver nanoparticles using seed extract of *Calendula officinalis* in liquid phase. J Mol Liq 207:159–163
14. Baralic I, Andjelkovic M, Djordjevic B, Dikic N, Radivojevic N, Suzin-Zivkovic V, Radojevic-Skodric S, Pejic S (2015) Effect of astaxanthin supplementation on salivary iga, oxidative stress, and inflammation in young soccer players. Evid Based Complement Alternat Med 2015:1–10
15. Becerra MO, Contreras LM, Lo MH, DiazJM HGC (2020) Lutein as a functional food ingredient: Stability and bioavailability. J Funct Foods 66:1–12
16. BerryA JD, Humbelin M, Jore JPM, Hoste B, Cleenwerck VM, Bretzel W, Mayer AF, Lopez-Ulibarri R, Shanmugam B, Swings J, Pasamontes L (2003) Paracoccus zeaxanthinifaciens sp. nov., a zeaxanthin-producing bacterium. Int J Syst Evol Microbiol 53:231–238
17. Biswas P, Wu CY (2005) Nanoparticles and the environment. J Air Waste Manag Assoc 55:708–746
18. Bohn T, Desmarchelier C, El SN, Keijer J, Van Schothorst E, Ruhl R, Borel P (2019) β-Carotene in the human body: metabolic bioactivation pathways - from digestion to tissue distribution and excretion. Proc Nutr Soc 78:68–87
19. Boon CS, McClements DJ, Weiss J, Decker EA (2010) Factors influencing the chemical stability of carotenoids in foods. Crit Rev Food Sci Nutr 50:515–532
20. Bouvier F, Harlingue A, Backhaus RA, Kumagai MH, Camara B (2000) Identification of neoxanthin synthase as a carotenoid cyclase paralog. JBIC J Biol Inorg Chem 267:6346–6352
21. Caballero B, Finglas PM, Toldra F (2016) Encyclopedia of food and health. Academic Press, Oxford, UK, pp 670–675
22. Carle R, Schweiggert R (2016) Handbook on natural pigments in food and beverages: industrial applications for improving food color. Woodhead Publishing, Sawston, Ciriminna
23. Cerda-Olmedo E (2001) Phycomyces and the biology of light and color. FEMS Microbiol Rev 25:503–512
24. Cheng Q (2006) Structural diversity and functional novelty of new carotenoid biosynthesis genes. J Ind Microbiol Biotechnol 33:552–559
25. Ciriminna R, Fidalgo A, Meneguzzo F, Ilharco LM, Mario Pagliaro M (2016) Lycopene: emerging production methods and applications of a valued carotenoid. ACS Sustainable Chem Eng 3:643–650

26. Colak H, Karakose E (2017) Green synthesis and characterization of nanostructured ZnO thin films using *Citrus aurantifolia* (lemon) peel extract by spin-coating method. J Alloys Compd 690:658–662
27. Coyne T, Ibiebele TI, Baade PD, McClintock CS, Shaw JE (2009) Metabolic syndrome and serum carotenoids: findings of a cross-sectional study in Queensland, Australia. Br J Nutr 102:1668–1677
28. Czeczuga B, Upre K (1990) Carotenoids in lichens of the *Lobaria* and *Peltigera* genera from India. Bull Bot Surv India 32:80–85
29. Esatbeyoglu T, Rimbach G (2017) Canthaxanthin: from molecule to function. Mol Nutr Food Res 61(6):1600469
30. Fordham IM, Clevidence BA, Wiley ER, Zimmerman RH (2001) Fruit of autumn olive: a rich source of lycopene. Hort Sci 36:1136–1137
31. Frigard NU, Maresca JA, Yunker CE, Jones AD, Bryant BA (2004) Genetic manipulation of carotenoids biosynthesis in green sulfur bacterium *Chlorobium tepidum*. J Bacteriol 186:5210–5220
32. Galasso C, Corinaldesi C, Sansone C (2017) Carotenoids from marine organisms: biological functions and industrial applications. Antioxidants 23:96
33. Gao Y, Xu D, Ren D, Zeng K, Wu X (2020) Green synthesis of zinc oxide nanoparticles using *Citrus sinensis* peel extract and application to strawberry preservation: a comparison study. LWT 136:109297
34. Gervasi T, Santini A, Daliu P, Salem AZM, Gervasi C, Pellizzeri V, Cicero N (2020) Astaxanthin production by *Xanthophyllomyces* dendrorhous growing on a low cost substrate. Agrofor Syst 94:1229–1234
35. Giossi C, Cartaxana P, Cruz S (2020) Photoprotective role of neoxanthin in plants and algae. Molecules 11:1–16
36. Goldsmith TH, Krinsky NI (1960) The epoxide nature of the carotenoid, neoxanthin. Nature 188:491–493
37. Gonnelli C, Giordano C, Fontani U, Salvatici MC, Ristori S (2017) Green synthesis of gold nanoparticles from extracts of *Cucurbita pepo* L. leaves: insights on the role of plant ageing. Lecture notes in bioengineering, pp 155–164
38. Gonzalez S, Astner S, An W, Goukassian D, Pathak MA (2003) Dietary lutein/zeaxanthin decreases ultraviolet B-induced epidermal hyperproliferation and acute inflammation in hairless mice. J Invest Dermatol 121:399–405
39. Guo L, Xu X, Zhao J, Hu S, Xu L, Kuang H, Xu C (2022) Multiple of 15 triazine herbicides by gold nanoparticles based-paper sensor. Nano Res 15:5483–5491
40. Haider AJ, AL-Anbari RH, Kadhim GR, Salame CT (2017) Exploring potential environmental applications of TiO_2 nanoparticles. Energy Procedia 119:332–345
41. Hasan S (2015) A review on nanoparticles: their synthesis and types. Res J Recent Sci 4:1–3
42. Hayashi M, Ishibashi T, Kuwahara D, Hirasawa K (2021) Commercial production of astaxanthin with *Paracoccus carotinifaciens*. Adv Exp Med Biol 1261:11–20
43. Heber D, Lu QY (2002) Overview of mechanisms of action of lycopene. Exp Biol Med (Maywood) 227:920–923
44. Hernandez-Diaz JA, Garza-Garcia JJ, León-Morales JM, Zamudio-Ojeda A, Arratia-Quijada J, Velazquez-Juarez G, Lopez-Velazquez JC, Garcia-Morales S (2021) Antibacterial activity of biosynthesized selenium nanoparticles using extracts of *Calendula officinalis* against potentially clinical bacterial strains. Molecules 26:5929
45. Hu J, Nagarajan D, Zhang Q, Chang JS, Lee DJ (2018) Heterotrophic cultivation of microalgae for pigment production: a review. Biotechnol Adv 36:54–67
46. Hu ZC, Zheng YG, Wang Z, Shen YC (2006) pH control strategy in astaxanthin fermentation bioprocess by *Xanthophyllomyces dendrorhous*. Enzyme Microb Technol 39:586–590
47. Humbelin M, Thomas A, Lin J, Jore J, Berry A (2002) Genetics of isoprenoidbiosynthesis in *Paracoccus zeaxanthinifaciens*. Gene 297:129–139
48. Husen A (2020) Introduction and techniques in nanomaterials formulation. In: Husen A, Jawaid M (eds) Nanomaterials for agriculture and forestry applications. Elsevier Inc. 50 Hampshire St., 5th Floor, Cambridge, MA 02139, USA, pp 1–14

49. Husen A, Siddiqi KS (2022) Advances in smart nanomaterials and their applications. Elsevier Inc., 50 Hampshire St., 5th Floor, Cambridge, MA 02139, USA
50. Husen A (2022) Engineered nanomaterials for sustainable agricultural production, soil improvement and stress management. Elsevier Inc., 50 Hampshire St., 5th Floor, Cambridge, MA 02139, USA
51. Husen A, Iqbal M (2019) Nanomaterials and plant potential. Springer International Publishing AG, Gewerbestrasse 11, 6330 Cham, Switzerland. https://doi.org/10.1007/978-3-030-05569-1
52. Husen A, Jawaid M (2020) Nanomaterials for agriculture and forestry applications. Elsevier Inc., 50 Hampshire St., 5th Floor, Cambridge, MA 02139, USA. https://doi.org/10.1016/C2018-0-02349-X
53. Hussain I, Singh NB, Singh A, Singh H, Singh SC (2016) Green synthesis of nanoparticles and its potential application. Biotechnol Lett 38:545–360
54. Hwang JS, Han SG, Lee CH, Seo HG (2018) Lutein suppresses hyperglycemia-induced premature senescence of retinal pigment epithelial cells by upregulating SIRT1. J Food Biochem 42:1–7
55. Indrayani I, Egeland ES, Moheimani NR, Borowitzka MA (2022) Carotenoid production of *Botryococcus braunii* CCAP 807/2 under different growth conditions. J Appl Phycol 34:1177–1188
56. Iqbal Z, Javad S, Naz S, Shah AA, Shah AN, Paray BA, Gulnaz A, Abdelsalam NR (2022) Elicitation of the in vitro cultures of selected varieties of *Vigna radiata* L. With zinc oxide and copper oxide nanoparticles for enhanced phytochemicals production. Front Plant Sci 13:908532
57. Jian Y, Zhang C, Wang Y, Li Z, Chen J, Zhou W, Huang W, Jiang M, Zheng H, Li M, Miao H, Zhang F, Li H, Wang Q, Sun B (2021) Characterization of the role of the neoxanthin synthase gene *BoaNXS* in carotenoid biosynthesis in chinese kale. Genes (Basel) 12:1–13
58. Jin-Chul K, Madhusudhan A, Husen A (2021) Smart nanomaterials in biomedical applications. Springer Nature Switzerland AG, Gewerbestrasse 11, 6330 Cham, Switzerland
59. Joost U, Sutka A, Oja M, Smits K, Dobelin N, Loot A, Nommiste E (2018) Reversible photodoping of TiO_2 nanoparticles for photochromic applications. Chem Mater 30:8968–8974
60. Jose IG, Miguel P, Beatriz F, Astrid K, Ulo N (2017) Emissions of carotenoid cleavage products upon heat shock and mechanical wounding from a foliose lichen. Environ Exp Bot 133:87–97
61. Kaczor A, Barańska M, Czamara K (2016) Carotenoids: overview of nomenclature, structures, occurrence and functions. Wiley, Hoboken, pp 1–13
62. Karakurt Y, Süleyman H, Keskin Cimen F, Tasli G, Ucak T, Icel E, Kurt N (2019) The effects of lutein on optic nerve injury induced by ethambutol and isoniazid: an experimental study. Cutan Ocul Toxicol 38:136–140
63. Katherine LSV, Edgar CC, Jerry WK, Luke RH, Julie CD (2008) Extraction conditions affecting supercritical fluid extraction (SFE) of lycopene from watermelon. Bioresour Technol 99:7835–7841
64. Kaulmann A, Bohn T (2014) Carotenoids, inflammation, and oxidative stress-implications of cellular signaling pathways and relation to chronic disease prevention. Nutr Res 34:907–929
65. Kaushik N, Thakkar MS, Snehit S, Mhatre MS, Rasesh Y, Parikh MS (2010) Biological synthesis of metallic nanoparticles. Nanomed Nanotechnol Biol Med 6:257–262
66. Kelman D, Ben-Amotz A, Berman-Frank I (2009) Carotenoids provide the major antioxidant defense in the globally significant N_2-fixing marine Cyanobacterium, *Trichodesmium*. Environ Microbiol 11:1897–1908
67. Khalid M, Bilal M, Iqbal HM, Huang D (2018) Biosynthesis and biomedical perspectives of carotenoids with special reference to human health-related applications. Biocatal Agri Biotechnol 17:399–407
68. Khaw YS, Yusoff FM, Tan HT, Noor Mazli NAI, Nazarudin MF, Shaharuddin NA, Omar AR (2021) The critical studies of fucoxanthin research trends from 1928 to June 2021: a bibliometric review. Mar Drugs 19:1–25

69. Kholany M, Coutinho JA, Ventura SP (2022) Carotenoid production from microalgae: the Portuguese scenario. Molecules 27:2540
70. Kim DY, Vijayan D, Praveenkumar R, Han JI, Lee K, Park JY, Oh YK et al (2016) Cell-wall disruption and lipid/astaxanthin extraction from microalgae: chlorella and Haematococcus. Bioresour Technol 199:300–310
71. Kim JH, Na HJ, Kim CK, Kim JY, Ha KS, Lee H, Chung HT, Kwon HJ, Kwon YG, Kim YM (2008) The non-provitamin A carotenoid, lutein, inhibits NF-kappaB-dependent gene expression through redox-based regulation of the phosphatidylinositol 3-kinase/PTEN/Akt and NF-kappaB-inducing kinase pathways: role of H(2)O(2) in NF-kappaB activation. Free Radic Biol Med 45:885–896
72. Kirti K, Amita S, Priti S, Kumar AM, Jyoti S (2014) Colorful world of microbes: carotenoids and their applications. Adv Biol 2014:1–13
73. Koo SH (2013) Nonalcoholic fatty liver disease: molecular mechanisms for the hepatic steatosis. Clin Mol Hepatol 19:210–215
74. Koo SY, Hwang JH, Yang SH, Um JI, Hong KW, Kang K, Pan CH, Hwang KT, Kim SM (2019) Anti-obesity effect of standardized extract of microalga *Phaeodactylum tricornutum* containing fucoxanthin. Mar Drugs 17:1–15
75. Korani SM, Korani T, Sathyapalan SA (2019) Therapeutic effects of Crocin in autoimmune diseases: a review. BioFactors 45:835–843
76. Krinsky NI, Johnson EJ (2005) Carotenoid actions and their relation to health and disease. Mol Asp Med 26:459–516
77. Krubasik P, Sandmann G (2000) Molecular evolution of lycopene cyclases involved in the formation of carotenoids with ionone end groups. Biochem Soc Trans 28:806–810
78. Kuila T, Bose S, Khanra P, Mishra AK, Kim NH, Lee JH (2012) A green approach for the reduction of graphene oxide by wild carrot root. Carbon 50:914–921
79. Kumar RD, Rajendran N, Jayasingam P, Sampathkumar P (2017) Cultivation and chemical composition of microalgae *Chlorella Vulgaris* and its antibacterial activity against human pathogens. J Aquac Mar Biol 5:65–71
80. Kumar T, Rajpal A, Arora S, Bhargava R, Hari Prasad KS, Kazmi AA (2016) A comparative study on vermifiltration using epigeic earthworm Eisenia fetida and Eudrilus eugeniae. Desalin Water Treat 57(14):6347–6354
81. Kushwah KS, Verma DK (2021) Biological synthesis of metallic nanoparticles from different plant species. In: PV Pham (ed) 21st century nanostructured materials - physics, chemistry, classification, and emerging applications in industry, biomedicine, and agriculture; Intech Open, London
82. Lakshminarayana R, Raju M, Krishnakantha TP, Baskaran V (2005) determination of major carotenoids in few Indian leafy vegetables by high-performance liquid chromatography. J Agric Food Chem 53:2838–2842
83. Lei Y, Lu J, Luo X, Wu T, Du P, Zhang X, Amine K (2013) Synthesis of porous carbon supported palladium nanoparticle catalysts by atomic layer deposition: application for rechargeable Lithium–O_2 battery. Nano Lett 13:4182–4189
84. Li S, Ding Y, Niu Q, Xu S, Pang L, Ma R, Jing M, Feng G, Tang JX, Zhang Q, Ma X, Yan Y, Zhang J, Wei M, Wang HX, Li F, Guo S (2015) Lutein has a protective effect on hepatotoxicity induced by arsenic via Nrf2 signaling. Biomed Res Int 2015:1–10
85. Lin JJ, Lin WC, Li SD, Lin CY, Hsu SH (2013) Evaluation of the antibacterial activity and biocompatibility for silver nanoparticles immobilized on nano silicate platelets. ACS Appl Mater Interfaces 5(2):433–443
86. Liu D, Gao Y, Kispert LD (2000) Electrochemical properties of natural carotenoids. J Electroanal Chem 488:140–150
87. Liu J, Sun Z, Gerken H, Liu Z, Jiang Y, Chen F (2014) *Chlorella zofingiensis* as an alternative microalgal producer of astaxanthin: biology and industrial potential. Mar Drugs 12:3487–3515
88. Lopez-Vargas E, Ortega-Ortíz H, Cadenas-Pliego G, de Alba RK, Cabrera de la Fuente M, Benavides-Mendoza A, Juarez-Maldonado A (2018) Foliar application of copper nanoparticles increases the fruit quality and the content of bioactive compounds in tomatoes. Appl Sci 8:1020

89. Lu LM, Li HB, Qu F, Zhang XB, Shen GL, Yu RQ (2011) In situ synthesis of palladium nanoparticle–graphene nanohybrids and their application in nonenzymatic glucose biosensors. Biosens Bioelectron 26:3500–3504
90. Maghsoudi S, Taghavi Shahraki B, Rabiee N, Fatahi Y, Bagherzadeh M, Dinarvand R, Ahmadi S, Rabiee M, Tahriri M, Hamblin MR, Tayebi L, Webster TJ (2022) The colorful world of carotenoids: a profound insight on therapeutics and recent trends in nano delivery systems. Crit Rev Food Sci Nutr 62:3658–3697
91. Mahajan A, Arya A, Chundawat TS (2019) Green synthesis of silver nanoparticles using green alga (*Chlorella vulgaris*) and its application for synthesis of quinolines derivatives. Synth commun 49:1926–1937
92. Mahato K, Nagpal S, Shah MA, Srivastava A, Maurya PK, Roy S, Jaiswal A, Singh R, Chandra P (2019) Gold nanoparticle surface engineering strategies and their applications in biomedicine and diagnostics. 3 Biotech 9:1–19
93. Mantzouridou F, Tsimidou MZ (2008) Lycopene formation in *Blakeslea trispora* chemical aspects of a bioprocess. Trends Food Sci Technol 19:363–371
94. Maoka T (2020) Carotenoids as natural functional pigments. J Nat Med 74:1–16
95. Marova I, Haronikova A, Petrik S, Dvorakova T, Breierova E (2012) Production of enriched biomass by red yeasts of *Sporobolomyces* sp. grown on waste substrates. J Microbiol Biotechnol Food Sci 1:534–551
96. Masetto A, Flores-Cotera LB, Díaz C, Langley E, Sanchez S (2001) Application of a complete factorial design for the production of zeaxanthin by *Flavobacterium* sp. J BiosciBioeng 92:55–58
97. Mayer-Miebach E, Behsnilian D, Regier M, Schuchmann HP (2005) Thermal processing of carrots: lycopene stability and isomerisation with regard to antioxidant potential. Food Res Int 38:1103–1108
98. Mendes-Silva T, Andrade R, Ootani M, Mendes P, SaR SM, Souza K, Correia M, Silva M, Oliveira M (2020) Biotechnological potential of carotenoids produced by extremophilic microorganisms and application prospects for the cosmetics industry. Adv Microbiol 10:397–410
99. Mezzomo N, Ferreira S (2016) Carotenoids functionality, sources, and processing by supercritical technology: a review. J Chem 2016:1–16
100. Misra SK, Mukherjee P, Chang HH, Tiwari S, Gryka M, Bhargava R, Pan D (2016) Multi-functionality redefined with colloidal carotene carbon nanoparticles for synchronized chemical imaging, enriched cellular uptake and therapy. Sci Rep 6:29299
101. Moreira MD, Melo MM, Coimbra JM, Dos-Reis KC, Schwan RF, Silva CF (2018) Solid coffee waste as alternative to produce carotenoids with antioxidant and antimicrobial activities. Waste Manage 82:93–99
102. Murillo AG, Hu S, Fernandez ML (2019) Zeaxanthin: metabolism, properties, and antioxidant protection of eyes, heart, liver and skin. Antioxidants (Basel) 8:1–18
103. Murthykumar K, Malaiappan S, Kumar R (2020) Antioxidant and antibacterial effect of lycopene mediated silver nanoparticle against *staphylococcus aureus* and *streptococcus mutansan* in vitro study. Plant Cell Biotechnol Mol Biol 21:90–98
104. Muthuirulappan S, Francis SP (2013) Anti-cancer mechanism and possibility of nano-suspension formulation for a marine algae product fucoxanthin. Asian Pac J Cancer Prev 14:2213–2216
105. Nascimento TCD, Pinheiro PN, Fernandes AS, Caetano PA, Jacob-Lopes E, Zepka LQ (2022) Insights on the bioaccessibility of natural pigments from diatom *Chaetoceros calcitrans*. Molecules 27:1–11
106. Nava OJ, Luque PA, Gómez-Gutiérrez CM, Vilchis-Nestor AR, Castro-Beltrán A, Mota-González ML, Olivas A (2017) Influence of Camellia sinensis extract on Zinc Oxide nanoparticle green synthesis. J Mol Struct 1134:121–125
107. Neville CE, Young IS, Gilchrist SE, McKinley MC, Gibson A, Edgar JD, Woodside JV (2013) Effect of increased fruit and vegetable consumption on physical function and muscle strength in older adults. Age (Dordr) 35:2409–2422

108. Ngamwonglumlert L, Devahastin S, Chiewchan N (2017) Natural colorants: pigment stability and extraction yield enhancement via utilization of appropriatepretreatment and extraction methods. Crit Rev Food Sci Nutr 57:3243–3259
109. Olana MH, Sabir FK, Bekele ET, Gonfa BA (2022) *Citrus sinensis* and *Musa acuminata* peel waste extract mediated synthesis of TiO_2/rGO nanocomposites for photocatalytic degradation of methylene blue under visible light irradiation. Bioinorg Chem Appl 2022:20
110. Osganian SK, MJ Stampfer MJ, Rimm E, Spiegelman D, Manson JE, Willett WC (2003) Dietary carotenoids and risk of coronary artery disease in women. Am J Clin Nutr 77:1390–1399
111. Ouyang B, Li Z, Ji X, Huang J, Zhang H, Jiang C (2019) The protective role of luteinon isoproterenol-induced cardiac failure rat model through improving cardiac morphology, antioxidant status via positively regulating Nrf2/HO-1 signaling pathway. Pharm Biol 57:529–535
112. Palombo P, Fabrizi G, Ruocco V, Ruocco E, Fluhr J, Roberts R, Morganti P (2007) Beneficial long-term effects of combined oral/topical antioxidant treatment with the carotenoids lutein and zeaxanthin on human skin: a double-blind, placebo-controlled study. Skin Pharmacol Physiol 20:199–210
113. Pant J, Goudie MJ, Hopkins SP, Brisbois EJ, Handa H (2017) Tunable nitric oxide release from s-nitroso-n-acetylpenicillamine via catalytic Copper nanoparticles for biomedical applications. ACS Appl Mater Interfaces 9:15254–15264
114. Papaioannou EH, Liakopoulou-Kyriakides M (2010) Substrate contribution on carotenoids production in *Blakeslea trispora* cultivations. Food Bioprod Process 88:306–311
115. Parashar UK, Saxena PS, Srivastava A (2009) Bioinspired synthesis of silver nanoparticles. Dig J Nanomater Biostruct 4:159–166
116. Parial D, Pal RR (2014) Biosynthesis of monodisperse gold nanoparticles by green alga *Rhizoclonium* and associated biochemical changes. J Appl Phycol 27:975–984
117. Patra JK, Baek KH (2014) Green nanobiotechnology: factors affecting synthesis and characterization techniques. J Nanomater 2014:1–12
118. Pechyen C, Ponsanti K, Tangnorawich B, Ngernyuang N (2021) Waste fruit peel – mediated green synthesis of biocompatible gold nanoparticles. J Mater Res Technol 14:2982–2991
119. Perrone S, Tei M, Longini M, Buonocore G (2016) The multiple facets of lutein: a call for further investigation in the perinatal period. Oxid Med Cell Longev 2016:1–8
120. Pollmann H, Breitenbach J, Sandmann G (2017) Engineering of the carotenoid pathway in *Xanthophyllomycesdendrorhous* leading to the synthesis of zeaxanthin. Appl Microbiol Biotechnol 101:103–111
121. Porcu OM, Rodriguez-Amaya DB (2008) Variation in the carotenoid composition of the lycopene-rich Brazilian fruit *Eugenia uniflora* L. Plant Foods Hum Nutr 63:195–199
122. Qin R, Li P, Du M, Ma L, Huang Y, Yin Z, Zhang Y, Chen D, Xu H, Wu X (2021) Spatiotemporal visualization of insecticides and fungicides within fruits and vegetables using gold nanoparticle-immersed paper imprinting mass spectrometryimaging. Nanomaterials 11:1327
123. Rafique M, Sadaf I, Rafique MS, Tahir MB (2017) A review on green synthesis of silver nanoparticles and their applications. Artif Cells Nanomed 45:1272–1291
124. Raja R, Hemaiswarya S, Rengasamy R (2007) Exploitation of *Dunaliella* for β-carotene production. Appl Microbiol Biotechnol 74:517–523
125. Rajesh KM, Ajitha B, Reddy YAK, Suneetha Y, Reddy PS (2018) Assisted green synthesis of copper nanoparticles using *Syzygium aromaticum* bud extract: physical, optical and antimicrobial properties. Optik 154:593–600
126. Rambabu K, Bharath G, Banat F, Show PL (2020) Green synthesis of zinc oxide nanoparticles using *Phoenix dactylifera* waste as bioreductant for effective dye degradation and antibacterial performance in wastewater treatment. J Hazard Mater 402:123560
127. Ramyadevi J, Jeyasubramanian K, Marikani A, Rajakumar G, Rahuman AA (2012) Synthesis and antimicrobial activity of copper nanoparticles. Mater Lett 71:114–116
128. Rao AV (2002) Lycopene, tomatoes, and the prevention of coronary heart disease. Exp Biol Med 227:908–913

129. Rao AV, Agarwal S (2000) Role of antioxidant lycopene in cancer and heart disease. J Am Coll Nutr 19:563–569
130. Rao AV, Ray MR, Rao LG (2006) Lycopene. Adv Food Nutr Res 51:99–164
131. Rapoport A, Guzhova I, Bernetti L, Buzzini P, Kieliszek M, Kot AM (2021) Carotenoids and some other pigments from Fungi and Yeasts. Metabolites 11:92
132. Riley RS, Day ES (2017) Gold nanoparticle-mediated photothermal therapy: applications and opportunities for multimodal cancer treatment. Wiley Interdisc Rev Nanomed Nanobiotechnol 9(4):e1449
133. Rodrigo-Banos M, Garbayo I, Vílchez C, Bonete MJ, Martínez-Espinosa RM (2015) Carotenoids from *Haloarchaea* and their potential in biotechnology. Mar Drugs 13:5508–5532
134. Rodríguez-Saiz M, Fuente J, Barredo J (2010) *Xanthophyllomyces dendrorhous* for the industrial production of astaxanthin. Appl Microbiol Biotechnol 88:645–658
135. Rostamabadi H, Falsafi SR, Jafari SM (2019) Nanoencapsulation of carotenoids within lipid-based nanocarriers. J Control Release 298:38–67
136. Roy A, Bulut O, Some S, Mandal AK, Yilmaz MD (2019) Green synthesis of silver nanoparticles: biomolecule-nanoparticle organizations targeting antimicrobial activity. RSC Adv 9(5):2673–2702
137. Roychoudhury P, Bhattacharya A, Dasgupta A, Pal R (2016) Biogenic synthesis of gold nanoparticle using fractioned cellular components from eukaryotic algae and cyanobacteria. Phycol Res 64:133–140
138. Roychoudhury P, Dabek P, Gloc M, Golubeva A, Dobrucka R, Kurzydłowski K, Witkowski A (2021) Reducing efficiency of fucoxanthin in diatom mediated biofabrication of gold nanoparticles. Materials (Basel) 15:4094
139. Saini RK, Moon SH, Gansuk E, Keum YS (2018) An efficient one-step scheme for the purification of major xanthophyll carotenoids from lettuce, and assessment of their comparative anticancer potential. Food Chem 266:56–65
140. Sajilata MG, Singhal RS, Kamat MY (2008) The carotenoid pigment zeaxanthin-a review. Compr Rev Food Sci Food Saf 7:29–49
141. Salem SS, Fouda A (2020) Green synthesis of metallic nanoparticles and their prospective biotechnological applications: an overview. Biol Trace Elem Res 199:344–370
142. Shafiq T, Uzair M, Iqbal MJ, Zafar M, Hussain SJ, Shah SAA (2021) Green synthesis of metallic nanoparticles and their potential in bio-medical applications. Nano Biomed Eng 13:191–206
143. Sharma G, Nim S, Alle M, Husen A, Kim JC (2022) Nanoparticle-mediated delivery of flavonoids for cancer therapy: prevention and treatment. In: Kim JC, Alle M, Husen A (eds) Smart nanomaterials in biomedical applications. Nanotechnology in the life sciences. Springer, Cham. https://doi.org/10.1007/978-3-030-84262-8_3
144. Sharoni Y, Linnewiel-Hermoni K, Khanin M, Salman H, Veprik A, Danilenko M, Levy J (2012) Carotenoids and apocarotenoids in cellular signalling related to cancer: a review. Mol Nutr Food Res 56:259–269
145. Shegokar R, Mitri K (2012) Carotenoid lutein: a promising candidate for pharmaceutical and nutraceutical applications. J Diet Suppl 9:183–210
146. Shejawal KP, Randive DS, Bhinge SD, Bhutkar MA, Todkar SS, Mulla AS, Jadhav NR (2021) Green synthesis of silver, iron and gold nanoparticles of lycopene extracted from tomato: their characterization and cytotoxicity against COLO320DM, HT29 and hella cell. J Mater Sci Mater Med 32:19
147. Sies H, Stahl W (1995) Vitamins E and C, beta-carotene, and other carotenoids as antioxidants. Am J Clin Nutr 62:1315–1321
148. Sindhu ER, Preethi KC, Kuttan R (2010) Antioxidant activity of carotenoid lutein *in vitro* and *in vivo*. Indian J Exp Biol 48:843–848
149. Singh A, Ahmad S, Ahmad A (2015) Green extraction methods and environmental applications of carotenoids-a review. RSC Adv 5:62358–62393
150. Sommerburg O, Keunen JE, Bird AC, Kuijk FJ (1998) Fruits and vegetables that are sources for lutein and zeaxanthin: the macular pigment in human eyes. Br J Ophthalmol 82:907–910

151. Sowani H, Mohite P, Damale S, Kulkarni M, Zinjarde S (2016) Carotenoid stabilized gold and silver nanoparticles derived from the actinomycete *Gordonia amicalis* HS-11 as effective free radical scavengers. Enzyme Microb Technol 95:164–173
152. Stahl W, Heinrich U, Aust O, Tronnier H, Sies H (2005) Lycopene-rich products and dietary photoprotection. Photochem Photobiol Sci 5:238–242
153. Stephen N, Gayathri R, Niranjana Y, Prasad KY, Das AK, Baskaran V, Ganesan P (2017) Carotenoids: types, sources, and biosynthesis. In plant secondary metabolites. In: Stimulation, extraction, and utilization. Apple Academic Press, New Jersey, pp 1–30
154. Takaichi S (2011) Carotenoids in algae: distributions, biosyntheses and functions. Mar drugs 9:1101–1118
155. Tan D, Yu X, Chen M, Chen J, Xu J (2017) Lutein protects against severe traumatic brain injury through anti-inflammation and antioxidative effects via ICAM-1/Nrf-2. Mol Med Rep 16:4235–4240
156. Tanskul S, Khoonchumnan S, Watanasit S, Oda K (2013) Application of a new red carotenoid pigment-producing bacterium, *Enterobacter* sp. P_{41}, as a feed supplement for chicken. Afr J Biotechnol 12:64–69
157. Thi TUD, Nguyen TT, Thi YD, Thi KHT, Phan BT, Pham KN (2020) Green synthesis of ZnO nanoparticles using orange fruit peel extract for antibacterial activities. RSC Adv 10:23899–23907
158. Thiyagarasaiyar K, Goh BH, Jeon YJ, Yow YY (2020) Algae metabolites in cosmeceutical: an overview of current applications and challenges. Mar Drugs 18:323
159. Thomas SE, Harrison EH (2016) Mechanisms of selective delivery of xanthophylls to retinal pigment epithelial cells by human lipoproteins. J Lipid Res 57:1865–1878
160. Tinoi J, Rakariyatham N, Deming RL (2005) Simplex optimization of carotenoid production by *Rhodotorula glutinis* using hydrolyzed mung bean waste flour as substrate. Process Biochem 40:2551–2557
161. Torres N, Guevara-Cruz M, Velázquez-Villegas LA, Tovar AR (2015) Nutrition and atherosclerosis. Arch Med Res 46:408–426
162. Venil CK, Malathi M, Velmurugan P, Devi PR (2020) Green synthesis of silver nanoparticles using canthaxanthin from *Dietzia maris* AURCCBT01 and their cytotoxic properties against human keratinocyte cell line. J Appl Microbiol 130:1730–1744
163. Wackenroder H (1831) Ueber das Oleum radicis Dauci aetherum, das Carotin, den Carotenzucker und den officinellen succus Dauci; so wie auch über das Mannit, welches in dem Möhrensafte durch eine besondere Art der Gahrung gebildet wird. Geigers Magazin Pharmazie 33:144–172
164. Willstatter R, Page HJ (1914) Chlorophyll. XXIV. The pigments of the brown algae. Justus Liebigs Ann Chem 404:237–271
165. Wu L, Lin D (2019) 25-Molecular aspects of carotenoid metabolizing enzymes and implications for ophthalmology. In: Handbook of nutrition, diet, and the eye. Academic Press, Cambridge, MA, USA, pp 415–424
166. Xavier AAO, Perez-Galvez A (2016) Carotenoids as a source of antioxidants in the diet. In: Stange C (ed) Carotenoids in nature, vol 79. Springer, Cham, pp 359–375
167. Yamada YMA, Watanabe T, Ohno A, Uozumi Y (2012) Development of polymeric palladium-nanoparticle membrane-installed microflow devices and their application in hydrodehalogenation. Chemsuschem 5:293–299
168. YilmazB SK, Bilen H, Bahcecioglu IH, Bilir B, Ashraf S, Kucuk O (2015) Carotenoids and non-alcoholic fatty liver disease. Hepatobiliary Surg Nutr 4:161–171
169. You ZQ, Wu Q, Zhou XM, Zhang XS, Yuan B, Wen LL, Xu WD, Cui S, Tang XI, Zhang X (2019) Receptor-mediated delivery of astaxanthin-loaded nanoparticles to neurons: An enhanced potential for subarachnoid hemorrhage treatment. Front Neurosci 13:1–13
170. Zaid RM, Chong FC, Teo EYL, Ng EP, Chong KF (2015) Reduction of graphene oxide nanosheets by natural beta carotene and its potential use as supercapacitor electrode. Arab J Chem 8(4):560–569

171. Zakynthinos G, Varzakas T (2016) Carotenoids: from plants to food industry. Curr Res Nutr Food Sci 4:38–51
172. Zhang Hu, Tang Y, Zhang, Zhang Y, Shuofeng Z, Jing Q, Wang X, Ran K, Chunchao H, Zhenquan L (2015) Fucoxanthin: a promising medicinal and nutritional ingredient. Evid-Based Complement Alter Med 2015:1–10
173. Zhang L, Mazouzi Y, Salmain M, Liedberg B, Souhir Boujday S (2020) Antibody-gold nanoparticle bioconjugates for biosensors: synthesis, characterization and selected applications. Biosens Bioelectron 165:1–41
174. Zhang WL, Zhao YN, Shi ZZ, Cong D, Bai YS (2018) Lutein inhibits cell growth and activates apoptosis via the PI3K/AKT/mTOR signaling pathway in A549 human non-small-cell lung cancer cells. J Environ Pathol Toxicol Oncol 37:341–350

Terpenoids in Nanomaterials: Synthesis, Characterization, and Their Application

Kratika Singh, Ambreen Bano, Rolee Sharma, and Swati Sharma

Abstract Natural compounds have been recently explored for their potential to generate nanoparticle (NPs) out of which secondary metabolites, like terpenoid-rich essential oils, are found to be the most promising ones. Terpenoid metabolites are responsible for performing a wide variety of essential functions in plants. Moreover, they also exhibit a wide spectrum of biological activities like antiviral, antibacterial, antimalarial, anticancer activity as well as anti-inflammatory. Terpene-based drugs such as Taxol and Artemisinin are already well known for their bioactivity. Tea tree oil has long been used in traditional medicine as a powerful antimicrobial agent against a variety of pathogenic bacteria. These terpenoids can be directly adsorbed on the surface of NPs through interactions with carbonyl groups or electrons as well as they can be used for synthesis of nanosystems as nanogels/nanocolloids and can also be entrapped or adsorbed in any nanosystem. This chapter elaborates on the types and characteristics of biosynthesized terpenoids-based NPs and their potential pharmacological applications.

Keywords Nanoparticle · Terpenoids · Characterization · Pharmacological efficacy

1 Introduction

Plant extracts have long been recognized for their extensive clinical, biological, and therapeutic applications [1–7] and are currently also being deployed for nanomaterial synthesis, specially metallic nanoparticle (NPs). Metal NPs, like nickel, iron, gold, silver, and copper NPs, have a vast array of applications in science, including physics, medicine, chemistry, forestry, agriculture as well as engineering [8–16]. Metal NPs

K. Singh · A. Bano · S. Sharma (✉)
IIRC-3, Plant-Microbe Interaction and Molecular Immunology Laboratory, Faculty of Sciences, Department of Biosciences, Integral University, Lucknow, UP, India
e-mail: ssharma@iul.ac.in

R. Sharma
Department of Life Sciences and Biotechnology, C.S.J.M. University, Kanpur 208024, India

A. Husen (ed.), *Secondary Metabolites Based Green Synthesis of Nanomaterials and Their Applications*, Smart Nanomaterials Technology,
https://doi.org/10.1007/978-981-99-0927-8_5

therapeutic deployment has been reported against a variety of fungal, bacterial, and viral pathogens [17]. In phytonanotechnology various extracts from diverse plant parts like leaf extract, roots, stem, seeds, flower, and so on are being used to synthesize NPs [18].

Traditional methods for producing these NPs, including decomposition and erosion, have drawbacks such as producing a defective surface, increasing manufacturing costs, consuming a lot of energy, and ultimately producing a small amount [19]. Chemical synthesis method, like the sol–gel process and chemical reduction, employs hazardous toxic and chemicals reducing agents that remains adsorbed on synthesized NPs [20]. Comparatively greener synthesis methods result in lower cost and higher purity NPs [21]. Thus, phytonanotechnology does not need expensive laboratory equipment and is scalable for larger-scale synthesis [22]. Plant extract-based NP biosynthesis is much stable and eradicates the need for cell maintenance in microorganism-based synthesis [23]. Furthermore, the use of chemically synthesized NPs in clinical as well as medical applications is limited due to calcination at high temperature and pressure, as well as chemical contamination [24]. Henceforth, plant extracts are being used as stabilizing/reducing agents or capping agents aiding in the production of NPs.

Interaction of plants along with NPs has resulted in improvements in many research fields, including medicine, food, and agriculture [25, 26]. Nanotechnology, on the other side, has typically been used in the discipline of plant sciences, where NPs have been used as fertilizer pesticides, growth regulators, antimicrobials, and biosensors, as well as plant NPs production and nanobionics [27–30]. To achieve greater success, it is necessary to understand the various characteristics of the interaction between NPs and plants [28, 31, 32].

Plant Metabolites Importance in NPs synthesis

The plant kingdom produces a broad range of metabolites with diverse pharmacological and biological properties [33]. Plants use primary metabolites like fatty acids, carbohydrates, amino acids, nucleic, and other components to grow and develop [34–37]. Plants produce secondary metabolites in response to a variety of abiotic and biotic stressors [34, 38, 39]. Figure 1 depicts an overview of the major metabolites classes found in plants.

In recent years, the primary as well as secondary metabolites have been constantly considered in the redox reaction toward eco-friendly NPs preparation [40, 41]. Proteins, organic acids, amino acids, and vitamins, and SMs such as terpenoids, flavonoids, polyphenols, alkaloids, polysaccharides, and heterocyclic compounds, all play important roles in the metal and metal oxide NPs synthesis and stabilization of various shapes and sizes, as previously stated [42–45]. It has been considered that the secondary metabolites suitable for the NPs preparation are found in numerous taxa of the plant kingdom, mainly dicot plant species [46].

Flavonoids and terpenoids are the most common plant metabolites which are being used for the production of NPs, as shown in Table 1. In the polyphenol complexes flavonoids are the large group that includes various classes that can decrease metal ions into NPs, such as anthocyanins chalcones, flavonols, and

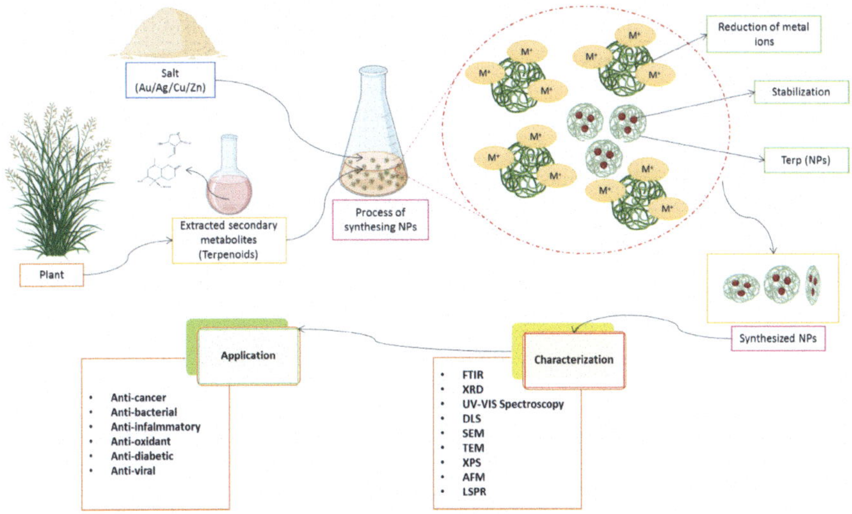

Fig. 1 The mechanism of biosynthesis of metallic NPs

isoflavonoids [47]. Many plant species contain anti-oxidative compounds (reductants) of various structural types, such as terpenoids flavonoids, coumarins, phenylpropanoids, anthraquinone, and xanthone, that are the ideal candidates for producing natural NPs [48, 49]. As a result, there is enormous potential for novel chemical structures and their actions to play a significant role in future drug discoveries for numerous diseases treatment. For example, plant extracts that have flavonoids such as rosmarinic acid and luteolin may decrease Ag + ions during the production of silver NPs (AgNP) [50]. A comparable study discovered that the flavonoids –OH group (myricetin and quercetin) can be oxidized to carbonyl groups during metal ion bioreduction [51]. It was also inferred by Naeem et al. [52] that *Punica granatum* L. contains phenolic acids like caffeic acid, gallic acid, protocatechuic acid, and ellagic acid which according to them is reducing agents of AuNPs biosynthesis.

Origin, Chemical Structure, and Function of Terpenes/Terpenoids

Terpenoid occur mostly in higher plants but they are also present in several bacteria, fungi, and even in some invertebrates as secondary metabolites. Terpenoids are found in all parts of plants and serve several essential functions. Terpenes are the major and structurally diverse secondary metabolites (Fig. 2). They function as signaling molecule, aiding plants in responding to temperature changes, and warding off competitors and pathogens. They protect plants from variety of biotic and abiotic stresses in this way [53]. A complex of volatile organic compound mixtures are generated by plants as secondary metabolites [54]. Smaller molecules, like sesquiterpenoids or monoterpenoids, are frequently found in essential oils, whereas larger molecules like triterpenoids are primarily found in resin and balsam [55]. Hydrocarbons like terpenes, terpenoids, and sesquiterpenes are found in essential

Table 1 Common plant metabolites which are being used for the production of NPs

NPs	Source plant (family) common name	Bark phytoconstituents responsible for the reduction	Type size of particle (nm)	Shape
Ag	*Albizia chevalieri* Harms. (Fabaceae)	Alkaloids, flavonoids, phenols, and terpenoids	~30	Spherical
Au	*Cassia fistula* L. (Fabaceae) Golden tree	Reducing sugars and terpenoids, secondary metabolites, such as lupeol, β-sitosterol, and hexacosanol	55.2–98.4	
Ag	*Elaeodendron croceum* Thunb. DC. (Celastraceae)—Saffron wood	Amino acids, proteins, polysaccharides, alkaloids, polyphenols, terpenoids or triterpenes, tannins, saponins, and vitamins	12.6–41.4	Spherical
Pd	*Eucommia ulmoides* Oliv. (Eucommiaceae)—Hardy rubber tree	Polyphenols, phytosterol, flavonoids, alkaloids, triterpenoids, aminoacids, and proteins	12.6	Spherical and quasi-spherical with FCC
Ag	*Ficus benghalensis* var. krishnae (Moraceae)—Krishna butter cup	Phenols, flavonoids, tannins, terpenoids, proteins, alkaloids, saponins, and vitamins	15–28	Spherical
Ag	*Ficus benghalensis* (Moraceae) Banyan tree *Azadirachta indica* Neem tree (Meliaceae) *A. Juss* (Meliaceae) Indian lilac	Flavonoids, terpenoids, and phenols	40–50	Spherical
Ag	*Holarrhena antidysenterica* L. (Aponycaceae) Wall. Tellicherry bark or conessi	Terpenoids, alkaloids, flavonoids, and phenols	32	Spherical

(continued)

Table 1 (continued)

NPs	Source plant (family) common name	Bark phytoconstituents responsible for the reduction	Type size of particle (nm)	Shape
Ag	*Melia azedarach* L. (Meliaceae)—Indian lilacelliptica	Triterpenoids, flavonoids, glycosides steroids, and carbohydrates	4–30	Spherical,
Ag–Au	*Melia azedarach* L. (Meliaceae)—Indian lilacelliptica	Triterpenoids, flavonoids, glycosides steroids, and carbohydrates	15–80	Hexagonal
Ag	*Moringa oleifera* Lam. (Moringaceae)—Moringa	Terpenoids, flavonoids, and polysaccharides	40	Spherical, pentagon
Cu	*Terminalia arjuna* Wigh and Arn (Combretaceae)—Arjuna tree	Polyphenols (flavonoids), terpenoids, ketones, aldehydes	~23	Spherical
Cu–Ag	–	Polyphenols, flavonoids, terpenoids, and reducing sugars	~20–30	Spherical

Note Ag—silver nanoparticles; Au—gold nanoparticles; Ag–Au—combination between silver and gold nanoparticles; Cu—copper nanoparticles; Pd—palladium nanoparticles; FCC—face-centered-cubic structure

oils, along with oxygenated compounds such as esters, ethers, alcohols, aldehydes, ketones, phenols, lactones, and phenol ethers [56]. Units of isoprenoid combine to form sesquiterpenoids and monoterpenoids [57], which serve as building blocks for new metabolite like plant sterols, rubber, carotenoid, hormones, turpentine, and chlorophyll's phytol tail [58].

Terpene occurs naturally as organic chemicals obtained from 5 carbon isoprene units that have been modified and assembled in various ways. Terpene synthase gene expression can be induced by a diversity of abiotic and biotic factors resulting in various combinations of terpenes. Terpenoids are terpenes that have been oxygenated. Terpenoids account for 90% of essential oils and are the natural products' biggest family found in plants. Terpenoids found in plant essential oils, like monoterpenoids and sesquiterpenoids, have been identified as an important contributor to NPs synthesis [59, 60]. Terpenoids are valued in naturopathic/ medicine for their analgesic, antibacterial, antifungal, antiviral, anticancer, antiparasitic, and anti-inflammatory properties [61]. Some fungal terpenoid nanomaterials also reported as xylaranic acid silver nanoparticle system (AgNPs) were obtained from *Xylaria primorskensis* and were found to improve solubility and bioavailability, antimicrobial properties as compared to xylaranic acid [62].

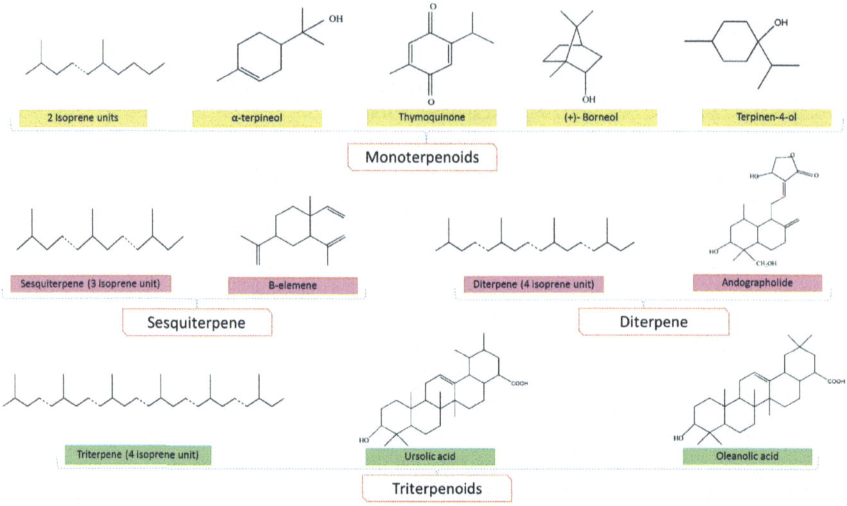

Fig. 2 Chemical structures of different types of terpenes including monoterpene, sesquiterpene, diterpene, and triterpene

Types of Terpenoids

Terpenoids are a type of organic polymer and are composed of five carbons as an initial unit Isoprene, and are classified [63] as monoterpenes (C10), sesquiterpenes (C15), diterpenes (C20), triterpenes (C30), tetraterpenes (C40), or polyterpenes as represented in Fig. 2. Some important terpenoids include farnesene, limonene, lycopene, linalool, myrcene, taxadiene, pinene, nerolidol, β-caryophyllene, caryophyllene oxide, phytol, and astaxanthin. These metabolites can conjugate or bind with NPs and hence can be used for drug discovery and compound purification [64, 65].

Plant Extract-Based Nano-materials (NMs)

Nanomaterials comprise unique physiochemical properties, and they give scaffolds for functionalization with secondary metabolites. Controlled release of biomolecules has been shown by certain nanomaterials like gold, polymeric, and magnetic [66–68]. The synthesis of NPs using plants is called as green synthesis of nanoparticles. Stem, root, flower, seed, and many other parts of plants are used for the synthesis. A vast variety of secondary metabolites are produced by plants with different functional groups which results in the generation of NPs [69].

Numerous studies have shown that essential oils are important in the synthesis of NPs. Several studies indicate that secondary metabolites found in plant-derived essential oils are the primary AgNP synthesizers [70]. Essential oils extracted from various species of plant, including Myristica fragrans [71], Cocos nucifera [72], Ricinus communis [73], Anacardium occidentale [74], and a few vegetables [75], have been used directly to synthesize NPs and other metal NPs. The low molecular weight aliphatic and aromatic constituents, such as terpenoid, play the most important role in the synthesis of AgNPs [76].

The leaf broth of *Azadiracta indica* contains reducing sugar/terpenoids which is used for the biosynthesis of pure metallic silver nano-particles [77]. Polyols like terpenoids, flavones, and polysaccharides are considered to be pivotal in the reduction of silver and chloroaurate ions by extract of *Cinnamomum camphora* leaves [78]. It has been proposed that phytoconstituents have important roles in metal salt reduction, and additionally, act as capping and stabilizing agents for NP synthesis (Table 1).

Nanoformulations and Terpenoids

Phytonanotechnology using terpenoids has now been used extensively as it has an edge over conventional methods as terpenes are reported [79, 80] to be encapsulated in nanostructure giving the following benefits:

- The bioactivity of NPs is increased,
- The complications with the plant extracts have been reduced,
- Persistent release of biomolecules,
- Improved biodistribution and bioabsorption,
- Enhanced solubility,
- Low dose requirement.

Like in 2018, for encapsulation of thymol nanostructured lipid carriers (NLC) have been employed which presents various properties like antimicrobial, antioxidant, and antiseptic properties [81]. By using the cutaneous acute inflammation model, the anti-inflammatory activity of NLC has been evaluated, and it has been reported that the thymol-NLC-gel showed higher edema inhibition as compared to the free thymol. Likewise, it is also suggested by a report that higher dose of non-encapsulated lupeol have been shown less inflammatory effect as comparison to lowest dose of lupeol-loaded nanosystem [82], impact of lycopene also explored on skin edema and inflammation [83]. Encapsulated terpenes like thymoquinone have been found to exhibit neurological properties [84]. Nanoencapsulated derivatives of *Ginkgo-biloba* have demonstrated anti-inflammatory effects in Parkinson's disease and arthritis models [85, 86]. All these investigations indicated the promising role of nanostructured terpenes systems.

Types of Terpenoid-Based Nanomaterials

In phytonanotechnology, Terpenoids can play a vital role and can be nanoformulated like lycopene NPs or lycopene/β-cyclodextrin inclusion complex using supercritical antisolvent precipitation for commercial use as drugs, nutraceuticals, or fragrances [87] and can also be used for preparation of NPs and nanocomposites as

- Reducing agents: metal salt reduction can be used for preparation of NPs
- Capping and Stabilizing agents for protection or formation of NPs
- Stabilizing or crosslinking agents in polymeric systems
- In making colloidal nanoparticle systems as dispersed in oil in water with Tween 80

- Applied intermixed with Nanofibers, nanocomposites, hydrogels, and nanofilm systems: the applications are via entrapment in scaffolds and coatings
- Encapsulated in lipid-based nanosystems or drug delivery systems as vesicular nanocarriers, solid lipid nanocarriers, and lipid nanocarriers

The metallic plant nanomaterials are differentiated into two subclasses (1) Metallic NPs (2) Metal Oxide NPs.

1.1 Metallic Nanoparticles

Various metabolites have been produced by metallic NPs which have been employed in several plant species, and metallic NPs have been reported to possess activities like mass propagation, genetic manipulation, elimination of microbial content [88]. The characteristics of the MNPs affects the production of metabolites (e.g., the used concentration, the time of exposure and artificial source, size), and it is also affected by type of plant culture.

Silver nanoparticles (AgNPs) are certainly the most exploited MNPs as elicitors of plant metabolites in in vitro cultures of various plant species [38, 89, 90]. For instance, it is reported that for the conversion of silver ions into NPs the terpenoids present in germanium leaf takes part actively [91]. In *Cinnamomum zeylanisum* extracts eugenol is found as a main terpenoid which is an important component for the bioreduction of HAuCl4 and $AgNO_3$ metal salts into their respective metal NPs. In FTIR results eugenol –OH groups disappear along with the formation of Au and Ag NPs. The other functional groups like alkyne, carbonyl, etc., appeared after the formation of Au NPs [92]. Also reported that on exposure to AgNPs after ROS generation resulted in an improvement of essential oil components, for example, geraniol, citronellyl formate, and E-caryophyllene in *Pelargonium graveolens* seedlings [93]. A possible interaction with plant growth regulators has also been proposed as they also affect the production of secondary metabolites. Various combinations of MNPs have been reported effective for metabolites production in plants species [94, 95]. It has been reported that a more complex effect was reported for different combinations of Ag and Au NPs used as elicitors of essential oil production in cell suspension culture of *Lavandula angustifolia*, with a reduced accumulation of low molecular weight components (for example trans-pinocarveol and 1,8-cineole), however, a concomitant increase in high molecular weight compounds (for instance, cadalene) was noticed [96].

1.2 Metal Oxide Nanoparticles

The mechanism for the preparation of metal oxide NPs from plants is exactly not known but in general there are three phases are involved: (i) the activation phase (bioreduction of metal ions or salts along with nucleation process of the reduced metal ions), (ii) the growth phase (spontaneous combination of tiny particles with greater ones) via a process acknowledged as Ostwald ripening, and (iii) the last one is termination phase (defining the final shape of the NPs) [97, 98].

Zinc oxide nanoparticles (ZnONPs) have some unique properties like band gap and high surface area to volume ratio properties due to which they have huge applications and impact on plant metabolites [99–102]. They can be grown by the green synthesis of *Dysphania ambrosioides* (local name: "epazote") that is presently worn extensively in traditional Mexican cuisine and is rich in organic compounds as flavonoids and terpenes which may support the NPs synthesis [103]. It has been reported that ZnO-NPs and terpenes have broad antimicrobial activity against microorganisms in oral biofilms [104–108]. The ZnONPs were reported to increase the production of metabolites in plants species and culture systems. The total flavonoid contents were accelerated in callus culture of *Echinacea purpurea* [109]. Similarly, the phenolic content was also increased by the ZnONPs and anthocyanin (3.28 mg g1 FW) contents in potato plants when applied at 300 and 500 ppm concentration in media, respectively [110].

The iron oxide nanoparticles (Fe_3O_4 NPs) are easily available for production of secondary metabolites NPs production because of their various properties like simple and cost-effective as they as a range of applications in different yields of science [111–113]. For example, Fe_3O_4NPs can be more effective than other MONPs such as ZnONPs to elicit production of metabolites, for example, hairy root culture of *Cichorium intybus*. Indeed, results showed that contrary to ZnONPs, Fe_2O_3NPs application proved to be proficient elicitor treatment to enhance both growth and production of phenolic and flavonoid compounds in hairy root culture of *C. intybus*. [113].

Brown algae *Bifurcaria. bifurcate* terpene containing extracts were used for the synthesis of copper oxide NPs dimension of 5–45 nm, and they were shown as assortment of Cu(I) and Cu(II) oxides with crystalline nature. Because of incompletely filled d orbitals, they are the essential class of semiconductors [114–117].

In a study, terpenoids were used to prepare titanium dioxide nanoparticles (TiO_2NPs). TiO_2NPs are one of the most released NPs in the environment especially because of their wide use as a UV filters in sunscreens. Therefore, plants are more prone to TiO_2NPs, and they can be directly uptake from the environment [118], mainly caused by ionic Ti [119, 120]. Furthermore, the Authors showed that drought stress-induced oxidative damages can be overpassed by foliar application of TiO_2NPs at appropriate concentrations.

Terpenoid Loaded Nanosystems

The Terpenoid-based nanosystems (Table 2) as oil in water nanoemulsions made by dispersion are very common while Solid lipid NPs like zein-based or Polymeric/co-polymeric NPs are also being explored as PNIPAAM–PEG-based for injectables. Encapsulated polymeric particles are also being studied in recent years.

Table 2 Application of terpenoid nanosystems

Nanoparticle	Terpenoid used	Application	References
Polymeric nanoparticles (methyl methacrylate and Triethylene Glycol Dimethacrylate copolymers)	D-Limonene loaded (10% w/w of dry particles)	Enhanced antimicrobial properties for potential application in food packaging	[121]
Zein nanoparticles	Limonene loaded	Antimicrobial activity against *P. syringae*	[122]
Thermosensitive PNIPAAM–PEG-based co-polymeric nanoparticles	Lycopene co-polymeric nanoparticle-encapsulated formulations of commercial and extracted lycopene	In vitro anticancer activity and chemopreventive effect on murine skin inflammation and tumorigenesis	[83]
Encapsulated in the Eudragit RL100 nanoparticles	Lycopene	Passive targeting to the tumor site passive targeting to the tumor site and cytotoxicity in prostate cancer, Lycopene/β-cyclodextrin inclusion complex using supercritical antisolvent precipitation	[123]
Sustained-release injectable mPEG–PLGA–mPEG nanoparticles	Co-encapsulation of four terpenes components (ginkgolides A, B, C, and bilobalide) from Ginkgo biloba extract	Herbal medicines	[124]
Citral and linalool nanoemulsions mixed with 5% Tween 80	5% citral	Delayed ripening and antibacterial activity against *Listeria monocytogenes*	[125]
Citral-in-water nanoemulsions	Citral	Bactericidal properties against six food-associated bacteria	[126]
Chitosan/citral oil in water nanoemulsions	Chitosan/citral	Antimicrobial activity against plant pathogens *Erwinia carotovora, Aspergillus niger*, and *Rhizopus stolonifer*	[127]
Solid lipid nanoparticles	10% limonene	Enhanced dermal delivery of all-trans-retinoic acids	[128]

2 Properties

2.1 Optical Properties

The reaction of plant extracts with silver salt which occurs due to redox outburst results in the change in color. The oxidation into terpenoids, whereas other metabolites act as reducing agent by the plant apparatus reduces Ag^{+} ions into Ag^{0} [129]. Due to surface plasmon resonance (SPR), Au and AgNPs show strong absorption of electromagnetic wave in the evident range. There are various properties which affect the SPR like shape, size, inter-particle distances [130, 131]. However, Terpene-based AgNPs SPR depends primarily on their essential features such as stability, surface-to-volume ratio, electronic and optical properties [132]. There is difference in the properties of Terpene-based AgNPs and chemically synthesized ones. The important characteristic features include efficient stability with higher catalytic, anticancer, antimicrobial, antifungal, antiprotozoal, and larvicidal activities. The morphology of terpene-based AgNPs was found to be convenient and having fine size spreading properties by Vilas et al. [133]. For the morphological analysis of Terpene-based NPs, various techniques are used like SEM, TEM, HRTEM, AFM, and FESEM [132]. The absorption of radiation occurs at wavelengths of 400–490 nm (λmax due to the transition of electrons [134]. The separated peaks formed by λmax values represent that TerpAgNPs formed in the solution have special sizes and shapes and are in the form of aggregates. Higher temperature and pH may produce multiple SPR bands therefore representing anisotropic phenomenon that leads to various shapes of TerpAgNPs however, in some cases, the concentration of salt and extract may also lead to hyperchromic shift [134–137].

Stability

In the optical properties, the stability with time in the nanoparticle solutions is not an observable attribute [77]. At low concentration of the solution, the metal colloidal particles are self-dispersed and steady for a few days or weeks [138]. Electrostatic/charge stabilization or polymeric stabilization was used for the study of the stability of Terpene-based AgNPs [139]. In the synthesis by chemical methods, one of the most common troubles is aggregation of AgNPs, therefore, to stop down aggregation some stabilizers like polyvinyl pyrrolidone, ammonia, citrate, gelatin, cellulose, and starch can be added [140, 141]. However, plant terpenoids are also used to coat NPs or embed them in organic matrix preventing their aggregation and making them stabile for a prolonged period of time [77, 142].

Characterization

NPs' physicochemical properties influence their behavior, safety, biodistribution, and efficacy. As a result, characterization of terpenoid-based NPs is critical in order to assess the functional properties of the synthesized particles. Like other NPs terpenoid-based particles are also characterized by using XRD, UV–vis spectroscopy, SEM, TEM, FTIR, XPS, DLS, and AFM are used for characterization. Recently, Rauf et al. [143] characterized AuNPs and AgNPs using *Mentha longifolia* leaves extracts (MLE) having huge amount of terpenoids by using UV–visible spectroscopy), AFM, FTIR, and TEM techniques. The shape and size of some terpenoid-based NPs are already presented in Table 1.

UV–Visible Spectroscopy

It is a highly efficient as well as accurate technique for the primary characterization of synthesized NPs, as well as used for monitoring the stability and synthesis of NPs. NPs have distinct optical properties that cause them to interact intensely with light precise wavelengths [144]. Furthermore, it is simple, quick, sensitive, easy, selective for various NP types, and needs only a short measurement time, with no calibration required for colloidal suspension particle characterization [145]. The NPs absorption is affected by particle size, chemical environment, and dielectric medium [146]. Almaary et al. [147] used Spectroscopy, TEM, X-ray diffraction, and optical absorption dimensions to characterize Ag-NPs synthesized from Seed-borne fungus *Penicillium duclauxii.*

Dynamic Light Scattering

The nanomaterial's physico-chemical characterization is critical for using radiation scattering techniques to analyze biological activities. DLS can measure the small particles size distribution in suspension or solution on a ranging scale from submicron to one nanometer. Any nanomaterial characterization is required to evaluate its toxic potential. DLS is primarily employed to determine size distributions and particle size in physiological or aqueous solutions. Kavitha et al. [148] used DLS, zeta potential, SEM, XRD, and FTIR for characterization of ZnO NPs biosynthesized using terpenoid (TAP) fractions isolated from leaves of *Andrographis paniculata.*

X-ray Diffraction (XRD)

XRD is a analytical technique that has been used for molecular and crystal structure analysis, particle sizes, qualitative identification of NPs, measuring the degree of crystallinity, quantitative resolution of chemical species, isomorphous substitutions, and other purposes [149].

X-ray Photoelectron Spectroscopy (XPS)

Electron spectroscopy for chemical analysis (ESCA or XPS) is quantitative spectroscopic surface chemical analysis technique XPS which is used to evaluate empirical formula as XPS can identify and characterize specific groups of starburst macromolecules like aromatic rings, P=S, C=O, and C–O [150]. XPS have a distinct role in providing access to semi-quantitative/quantitative, qualitative, and surface speciation information regarding NPs.

Fourier Transform Infrared Spectroscopy (FTIR)

FTIR gives reproducibility, accuracy, and a good signal-to-noise ratio. It allows for the detection of small absorbance changes on the order of 103, allowing for difference spectroscopy [151]. FTIR spectroscopy is frequently used in academic and industrial research to determine that in the NPs synthesis biomolecules are involved or not. Moreover, FTIR has been applied to the investigation of nano-scaled materials, like the confirmation of functional molecules covalently grafted onto graphene, carbon nanotubes, silver, and gold NPs, or interactions between substrate and enzyme during the catalytic process.

Scanning Electron Microscopy (SEM)

SEM is a high-resolution surface imaging technique able of resolving different nanomaterial shapes, size distributions, particle sizes, and the surface morphology of synthesized particles at the nano and microscales [152]. Generally, SEM is used for the study of particle morphology [153]. Energy-dispersive X-ray spectroscopy (EDX) and SEM can be used together to elucidate chemical composition analysis like in study of terpenoid-based ZnO NPs by Chérif et al. [154].

Transmission Electron Microscopy (TEM)

TEM is a precious, widely used for the characterization of nanomaterials, for quantitative measurements of particle size, size distribution, and observing their morphology.

Localized Surface Plasmon Resonance (LSPR)

In a metallic NPs the coherent, collective spatial oscillation of conduction electrons that can be directly excited by near-visible light is known as LSPR which is used for determining nanoparticle dimensions, shapes, along with its compositions of the localized surface plasmon resonance (LSPR) condition is influence by the particle size and shape, electronic properties of the NPs, dielectric environment, or temperature.

Atomic Force Microscopy (AFM)

AFM is commonly used to study nanomaterials aggregation, dispersion, as well as their shape, size, structure, and sorption; three available scanning modes are non-contact mode, contact mode, and intermittent sample contact mode. It is also used to characterize nanomaterial interactions with supported lipid bilayers in real time that current electron microscopy (EM) techniques cannot do.

Applications

Terpenoid-based NPs/nanocomposite/nanosystems represent an encouraging approach as novel drugs as well as new drug delivery systems.

3 Pharmacological Application of Terpenoid-Mediated Nanoparticles

Anticancer Activity

During the past years, the study of inorganic NPs for biomedical applications has evolved into a rapidly expanding and interesting field of study. AuNPs' have unique physicochemical properties making them an excellent source for pharmacological applications. AuNPs terpenoid-based green synthesis employs plant extracts having medicinal importance, which may persist on the NPs surface and act as carriers in this condition. The AuNPs use also reduces the possibility of adverse effects as well as limits the damage to noncancerous cells [155]. Patil et al. [156] synthesized terpenoid-based AuNPs and found them to be a novel cancer therapy agent with aggregation, dose as well as size-dependent cytotoxic activity against various cancer cells.

Combating multidrug resistance mechanisms is a complex issue in the combat against cancer cells and pathogenic bacterial, and smart application of metal oxide NPs (MONPs: Cu_2O, CuO, TiO_2, Fe_3O_4, and ZnO NPs), metal nanoparticles (MNPs: Au, Ag, Pt, Cu, and Pd NPs), and metal nanocomposites can be an appealing alternative to traditional drugs (Fig. 3).

Madhavan et al. [157] reported that the *Pedalium murex* linn methanolic leaves extract has anticancer activity besides the human lung cancer cell line A549. Chittasupho and Athikomkulchai [158] used MTT assay to determine *the Combretum quadrangulare* leaf extract with NPs on lung cancer cells cytotoxic effect and induced apoptosis in the cell line A549. Hence, NPs synthesized from leaf extract of *Combretum quadrangulare* Kurz could be a favorable candidate for the lung cancer

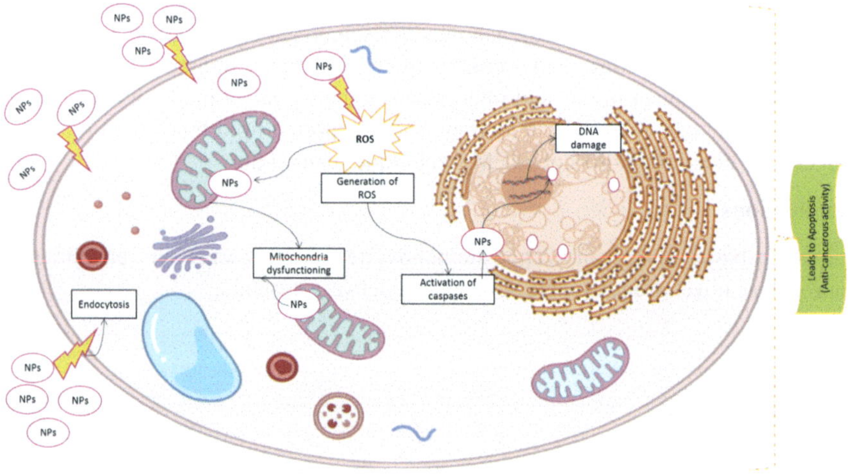

Fig. 3 Anti-cancerous activity of NPs

therapy drugs development. Balasubramanian et al. [159] reported that *Jasminum auriculatum* leaf extract is used to stabilize and reduce Au NPs. The biogenic Au NPs cytotoxicity indicated that the NPs has a significant dose-dependent inhibitory effect in the proliferation of the human cervical cancer cell line with the IC_{50} value of 104 μg/ml.

4 Antimicrobial Activity

Antibacterial Activity

Biogenic gold NPs derived from *Jasminum auriculatum* leaf extract demonstrated significant antimicrobial activity against human pathogenic fungus (*Lecanicillium lecanii, Aspergillus fumigatus, Candida albicans,* and *Trichoderma viride*) and bacteria (*Staphylococcus aureus, Escherichia coli, Klebsiella pneumonia,* and *Streptococcus pyogenes*) [159].

AgNPs synthesized by various methods were used in the in vitro determination of bactericidal activity against the bacterial species listed in Table 1. Antibacterial activity of AgNPs' is determined by the concentration of NPs exposed to bacteria and the type of bacteria [160]. The antibacterial activity of AgNPs' possible mechanisms is depicted diagrammatically in Fig. 2. The mechanism underlying AgNPs' antibacterial activity is complex and poorly understood. Smaller NPs have greater antibacterial activity because they have higher surface exposure to bacterial membrane [161]. Ag^+ positive charge interacts with the negative charge on the cell wall of bacteria, causing an increase in cell permeability or leakage, and changes in cell wall morphology resulting in cell death [162].

Długosz et al. [163, 164] used the propolis-derived selenium NPs activity against *Pseudomonas aeruginosa, Escherichia coli, Candida albicans,* and *Staphylococcus aureus* was confirmed. The MBC of the microorganisms were 25, 25, 100, 100, and mg/l for *E. coli, C. albicans P. aeruginosa,* and *S. aureus,* respectively.

Sani et al. [165] synthesized CuO-NPs and Ag-NPs from Carica papaya aqueous leaf extract and evaluated the antibacterial effects of plant-derived bimetallic NPs on multidrug-resistant bacterial strains multidrug-resistant (MDR) bacteria, including *Staphylococcus aureus, Pseudomonas aeruginosa,* and *Klebsiella pneumonia.* This research reveals the synergistic effect of BNPs (Ag-NPS/CuO-NPs) in combating antibiotic resistance in MDR bacteria (Fig. 4).

Antiviral Activity

Viruses can enter hosts quickly and expand their colonies rapidly. NSPs derived from terpenoid-rich plant extracts were found to be effective antiviral agents and can be used to treat and control viral pathogens. Suriyakalaa et al. [166] investigated the anti-HIV efficacy of biosynthesized NSPs and discovered an effective anti-HIV activity at the onset of reverse transcription. Metal NPs prevent viruses from entering the host, making them an effective antiviral agent [167].

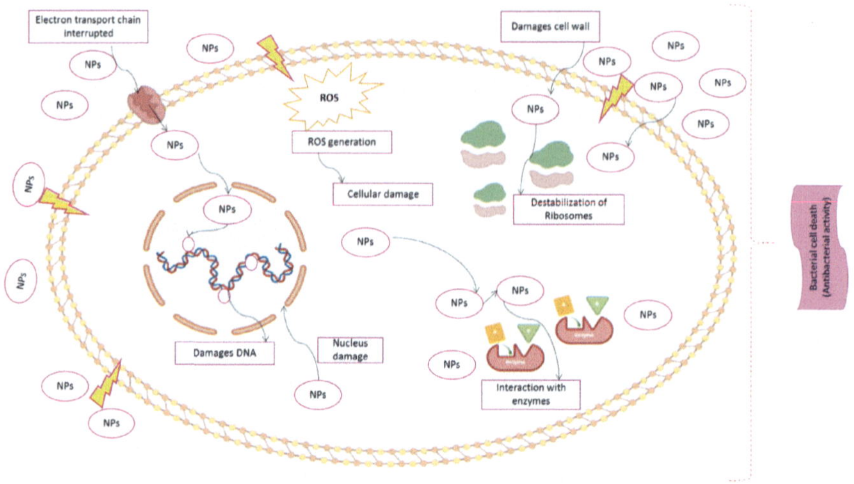

Fig. 4 Antibacterial activity caused by NPs

In a recent study, Długosz et al. [163, 164] reported that Selenium sulfide NPs were synthesized in natural deep eutectic solvents from terpenoid-rich extracts of spices such as cayenne pepper, black pepper, curcumin, and cinnamon. They concluded that the obtained systems exhibited higher antiviral activity against Human influenza virus A/H1N1 and Betacoronavirus 1 along with biocidal activity against strains of *C. albicans, E. coli, P. aeruginosa* and *S. aureus*. Selenium sulfide NPs suspension stabilized by spice extracts was also found to be effective against influenza viruses and B-coronavirus.

Anti-inflammatory Activity

Along with antibacterial properties, Terpenoids have excellent skin absorption and penetration properties hence terpenoid-based NPs are being studied the applications of NPs and nanocomposites in lipid-based systems as scaffolds as well as coatings for wound dressings like nanofibers, hydrogels, and films [168].

The anti-inflammatory activity is an important mechanism of wound healing. It is a biological channel process that produces cytokines and interleukins, which are then produced by macrophages, T-lymphocytes, and B-lymphocytes [169]. Anti-inflammatory mediators such as cytokines, IL-2, and IL-1, for example, are produced in primary immune organs [170]. These mediators are involved in a variety of biochemical pathways, making them a significant disease control agent. Because of their progressive effect on tissue regeneration and thus boosting wound healing, platinum and gold NPs have been shown to have anti-inflammatory properties [171], making them natural anti-inflammation agents. The results showed that the EO of *C. pseudomontana* and NGPs solutions had significant anti-inflammatory activity [172].

Antioxidant Activity

Plants are known to contain phenolic, flavonoids, triterpenoids, as well as coumarins compounds, which appear to have significant role in the synthesis and stabilization of AgNPs. Because of the capped phenolic compounds, the NPs have antioxidant activity that can be used to combat the harmful effects of free radicals [173].

It was observed that antioxidant and antibacterial activity of silver NPs may be attributed in the identified phenolic compounds in *Euphorbia wallichii* leaf extract [174]. Rajan et al. [175] prepared *Lantana camara* terpene-rich extract and reported that high antioxidant activity of NPs is expected because of favored adsorption of the antioxidant material from the extract onto the surface of the NPs.

Biosynthesized silver NPs using *Brachychiton populneus* extract containing terpenoids were reported to have god antioxidant (DPPH assay), anti-inflammatory (albumin denaturation assay), antidiabetic (alpha amylase assay) as well as cytotoxic potential (MTT assay) against HEK293 and U87 cell lines.

Anti-diabetic Activity

Diabetes Mellitus is a type of metabolic dysfunction in which blood sugar levels are uncontrolled. Some foods and a well-balanced diet, as well as synthetic insulin drugs, can help preclude diabetes at certain levels, but treating diabetes completely is a difficult task. However, nanomaterials biosynthesized could be used as an alternative drug to treat diabetes mellitus [176].

Crystalline nano-suspension of lycopene were engineered for management of oxidative stress in diabetes and were found to be effective in animal experiments by Mishra and Kumari [176].

In a recent study conducted by Al-Radadi [177] reported that Au-NPs with a particle size of 25.31 nm were synthesized using *Hylocereus polyrhizus* extract which is rich in terpenoids acts as reducing and stabilizing agent and demonstrates a potential s an antioxidant and anti-cancer agent. Au-NPs green production with *H. polyrhizus* fruit extract can be useful for medical applications due to its blood biocompatibility and physiological stability. The NPs also demonstrated dose-dependent cytotoxic activity against many cancer cell lines.

In a streptozotocin (STZ)/high-fat diet (HFD) induced diabetes in rats, AuNPs were synthesized using the leaf extract of *Dittrichia viscosa* with major secondary metabolites including, terpenoids and its treatment was found to reduce blood glucose levels, gene expression as well as hepatic PEPCK activity (P 0.05). This study reported that AuNPs synthesized from leaf extract of *D. viscosa* can reduce hyperglycemia in rats with STZ/HFD-induced diabetes, possibly by inhibiting hepatic gluconeogenesis by inhibiting the activity and expression of the hepatic PEPCK gene [178].

Azadirachta indica is highly effective and medicinally valuable traditional medicinal plant rich in terpenoids. *A. indica* aqueous kernel extracts were evaluated for their ability to synthesize Ag-NPs and also their anti-diabetic and anti-inflammatory activity in vitro by Chi et al. [179].

Metal NPs like nickel, iron, gold, silver, and copper have a vast array of applications in different disciplines of science [180]. Metal NPs therapeutic deployment has been reported against a variety of fungal, bacterial, and viral pathogens [17]. As a result, plant extracts when used as stabilizing/reducing agents or capping agents may aid in production of NPs without expensive laboratory equipment, as well as possible adaptation for larger-scale synthesis [22]. Using plant extracts green synthesis protocols offer many advantages over traditional chemical and physical methods, such as ease of access, using harmless solvents and creating less toxic substances, safety by having metabolites vast range at disposal, lower energy consumption and cost-effectiveness, and most importantly environmental protection.

All these findings indicate the highly promising potential of nanostructured terpenes systems for inflammatory diseases. Terpenoids are valued in naturopathic/ medicine for their analgesic, antibacterial, antifungal, antiviral, anticancer, antiparasitic, and anti-inflammatory properties [61]. Numerous studies have shown that essential oils are important in the synthesis of NPs. Several studies indicate that secondary metabolites found in plant-derived essential oils are the primary AgNP synthesizers [70]. Essential oils extracted from various species of plant, including *Myristica fragrans* [71], *Cocos nucifera* [72], *Ricinus communis* [73], *Anacardium occidentale* [74], and a few vegetables [75], have been used directly to synthezise NPs and other metal NPs. The low molecular weight aliphatic and aromatic constituents, such as terpenoid, play the most important role in the synthesis of AgNPs [76].

Terpenoid Loaded Nanosystems

The antimicrobial activity of terpenes and terpenoids has been long recognized, recently Terpenoid loaded Nanosystems are also being studied for different purposes as food packaging sustained fragrance release for commercial products [181]. The applications of their nanoemulsions are listed in Table 2.

Menthol-Based Nanoparticles for Drug Delivery

Many Menthol-based NPs have also been studied for drug delivery. Menthol-modified BSA NPs based on co-delivery of albendazole (Abz) and nano-silver have been obtained by Liang et al. [182] and were proven to have superior anti-glioma efficacy. Menthol-loaded PLGA Micro and Nanospheres were prepared using the multiple emulsion/solvent evaporation technique by Holz et al. [183].

5 Conclusion

Terpenes and terpenoids have made a special place in nanophytotechnology. Terpenoid NPs/nanoemulsions as well as terpenoid-based nanoparticle systems are sustainable as well as cost-effective. They have great promise for application as novel drugs and also novel drug delivery methods as antimicrobial bandages. Although some detailed investigations are still required to develop a comprehensive and clear concept. The future research is needed on their impact on humans as well as the environment.

References

1. Ayalew H, Tewelde E, Abebe B, Alebachew Y, Tadesse S (2022) Endemic medicinal plants of Ethiopia: Ethnomedicinal uses, biological activities and chemical constituents. J Ethnopharmacol 293:115307
2. Husen A (2021) Traditional herbal therapy for the human immune system. Taylor & Francis Group, LLC, 6000 Broken Sound Pkwy, NW
3. Husen A (2022a) Exploring poisonous plants: medicinal values, toxicity response and other potential uses. Taylor & Francis Group, LLC, 6000 Broken Sound Pkwy, NW
4. Husen A (2022b) Environmental pollution and medicinal plants. Taylor & Francis Group, LLC, 6000 Broken Sound Pkwy, NW
5. Husen A (2022c) Herbs, shrubs and trees of potential medicinal benefits. Taylor & Francis Group, LLC, 6000 Broken Sound Pkwy. NW
6. Husen A, Bachheti RK, Bachheti A (2021) Non-timber forest products (Food, Healthcare and Industrial Applications). Springer Nature Switzerland AG, Gewerbestrasse 11, 6330 Cham, Switzerland
7. Twaij BM, Hasan MN (2022) Bioactive secondary metabolites from plant sources: types, synthesis, and their therapeutic uses. Int J Plant Biol 13(1):4–14
8. Husen A, Iqbal M (2019a) Nanomaterials and plant potential: an overview. In: Husen A, Iqbal M (eds) Nanomaterials and plant potential. Springer International Publishing AG, Gewerbestrasse 11, 6330 Cham, pp 3–29 (https://doi.org/10.1007/978-3-030-05569-1_1
9. Husen A, Siddiqi KS (2014) Phytosynthesis of nanoparticles: concept, controversy and application. Nanoscale Res Lett 9(229):1–24
10. Husen A, Siddiqi KS (2014) Carbon and fullerene nanomaterials in plant system. J Nanobiotechnology 12(1):1–16
11. Husen A, Iqbal M (2019b) Nanomaterials and plant potential. Springer International Publishing AG, Gewerbestrasse 11, 6330 Cham, Switzerland. https://doi.org/10.1007/978-3-030-05569-1
12. Husen, A, Jawaid, M (2020) Nanomaterials for agriculture and forestry applications. Elsevier Inc., 50 Hampshire St., 5th Floor, Cambridge, MA 02139, USA. https://doi.org/10.1016/C2018-0-02349-X
13. Kumar A, Choudhary A, Kaur H, Guha S, Mehta S, Husen A (2022) Potential applications of engineered nanoparticles in plant disease management: a critical update. Chemosphere 295:133798
14. Kumar A, Choudhary A, Kaur H, Mehta S, Husen A (2021) Metal-based nanoparticles, sensors and their multifaceted application in food packaging. J Nanobiotechnology 19(256):1–25. https://doi.org/10.1186/s12951-021-00996-0
15. Sharma P, Pandey V, Sharma MMM, Patra A, Singh B, Mehta S, Husen A (2021) A review on biosensors and nanosensors application in agroecosystems. Nano Res Lett 16:136. https://doi.org/10.1186/s11671-021-03593-0
16. Sharma P, Bano A, Singh SP, Atkinson JD, Lam SS, Iqbal HM, Tong YW (2022) Nanomaterials as highly efficient photocatalysts used for bioenergy and biohydrogen production from waste toward a sustainable environment. Fuel 329:125408
17. Islam S, Thangadurai D, Adetunji CO, Nwankwo W, Kadiri O, Makinde S, … Adetunji JB (2020) Nanomaterials and nanocoatings for alternative antimicrobial therapy. In: Handbook of nanomaterials and nanocomposites for energy and environmental applications, pp 1–17
18. Chandraker SK, Kumar R (2022) Biogenic biocompatible silver nanoparticles: a promising antibacterial agent. Biotechnol Genet Eng Rev 1–35

19. Habibullah G, Viktorova J, Ruml T (2021) current strategies for noble metal nanoparticle synthesis. Nanoscale Res Lett 16(1):1–12
20. Perveen R, Shujaat S, Qureshi Z, Nawaz S, Khan MI, Iqbal M (2020) Green versus sol-gel synthesis of ZnO nanoparticles and antimicrobial activity evaluation against panel of pathogens. J Market Res 9(4):7817–7827
21. Subhapriya S, Gomathipriya P (2018) Green synthesis of titanium dioxide (TiO_2) nanoparticles by Trigonella foenum-graecum extract and its antimicrobial properties. Microb Pathog 116:215–220
22. Kyomuhimbo HD, Michira IN, Mwaura FB, Derese S, Feleni U, Iwuoha EI (2019) Silver–zinc oxide nanocomposite antiseptic from the extract of bidens pilosa. SN Appl Sci 1(7):1–17
23. Yazdanian M, Rostamzadeh P, Rahbar M, Alam M, Abbasi K, Tahmasebi E, Tebyaniyan H, Ranjbar R, Seifalian A, & Yazdanian A (2022) The potential application of green-synthesized metal nanoparticles in dentistry: a comprehensive review. Bioinorg Chem Appl
24. Naidi SN, Harunsani MH, Tan AL, Khan MM (2021) Green-synthesized CeO_2 nanoparticles for photocatalytic, antimicrobial, antioxidant and cytotoxicity activities. J Mater Chem B 9(28):5599–5620
25. Ali SS, Al-Tohamy R, Koutra E, Moawad MS, Kornaros M, Mustafa AM, … Sun J (2021) Nanobiotechnological advancements in agriculture and food industry: applications, nanotoxicity, and future perspectives. Sci Total Environ 792:148359
26. Mishra V, Mishra RK, Dikshit A, Pandey AC (2014) Interactions of nanoparticles with plants: an emerging prospective in the agriculture industry. In: Emerging technologies and management of crop stress tolerance. Academic Press, pp 159–180
27. Hassanzadeh P, Atyabi F, Dinarvand R (2022) Nanobionics: from plant empowering to the infectious disease treatment. J Control Release 349:890–901
28. Hu J, Xianyu Y (2021) When nano meets plants: a review on the interplay between nanoparticles and plants. Nano Today 38:101143
29. Shang Y, Hasan MK, Ahammed GJ, Li M, Yin H, Zhou J (2019) Applications of nanotechnology in plant growth and crop protection: a review. Molecules 24(14):2558
30. Sharma G, Nim S, Alle M, Husen A, Kim JC (2022) Nanoparticle-mediated delivery of flavonoids for cancer therapy: prevention and treatment. In: Kim JC, Alle M, Husen A (eds) Smart nanomaterials in biomedical applications. nanotechnology in the life sciences. Springer, Cham. https://doi.org/10.1007/978-3-030-84262-8_3
31. Elnahal AS, El-Saadony MT, Saad AM, Desoky ESM, El-Tahan AM, Rady MM, El-Tarabily KA (2022) The use of microbial inoculants for biological control, plant growth promotion, and sustainable agriculture: a review. Eur J Plant Pathol 1–34
32. Bano A, Gupta A, Sharma S, & Sharma R (2020) Recent developments in nanocarrier-based nutraceuticals for therapeutic purposes. Biogenic Nano-Particles and their Use in Agro-ecosystems, 371–391
33. Wink M (2015) Modes of action of herbal medicines and plant secondary metabolites. Medicines 2(3):251–286
34. Bano A, Gupta A, Rai S, Fatima T, Sharma S, Pathak N (2021) Mechanistic role of reactive oxygen species and its regulation via the antioxidant system under environmental stress. Plant Stress Physiol—Perspect Agricult 1–18
35. Kandar CC (2021) Secondary metabolites from plant sources. In: Bioactive natural products for pharmaceutical applications. Springer, Cham, pp 329–377
36. Terletskaya NV, Korbozova NK, Grazhdannikov AE, Seitimova GA, Meduntseva ND, Kudrina NO (2022). Accumulation of Secondary Metabolites of Rhodiola semenovii Boriss. In: Situ in the dynamics of growth and development. Metabolites 12(7):622
37. Zhao Y, Zhang Q, Li J, Yan X, He H, Gao X, Jia G (2021) High temperature in the root zone repressed flowering in Lilium× formolongi by disturbing the photoperiodic pathway and reconfiguring hormones and primary metabolism. Environ Exp Bot 192:104644
38. Gupta A, Mishra R, Rai S, Bano A, Pathak N, Fujita M, … Hasanuzzaman M (2022) Mechanistic insights of plant growth promoting bacteria mediated drought and salt stress tolerance in plants for sustainable agriculture. Int J Mol Sci 23(7):3741

39. Yeshi K, Crayn D, Ritmejerytė E, Wangchuk P (2022) Plant secondary metabolites produced in response to abiotic stresses has potential application in pharmaceutical product development. Molecules 27(1):313
40. Hassanisaadi M, Bonjar AHS, Rahdar A, Varma RS, Ajalli N, Pandey S (2022) Eco-friendly biosynthesis of silver nanoparticles using Aloysia citrodora leaf extract and evaluations of their bioactivities. Mater Today Commun 104183
41. Pungle R, Nile SH, Makwana N, Singh R, Singh RP, Kharat AS (2022) Green synthesis of silver nanoparticles using the Tridax procumbens plant extract and screening of its antimicrobial and anticancer activities. Oxid Med Cell Longevity
42. Jeevanandam J, Kiew SF, Boakye-Ansah S, Lau SY, Barhoum A, Danquah MK, Rodrigues J (2022) Green approaches for the synthesis of metal and metal oxide nanoparticles using microbial and plant extracts. Nanoscale 14(7):2534–2571
43. Nitnavare R, Bhattacharya J, Thongmee S, Ghosh S (2022) Utilizing photosynthetic microbes in nanobiotechnology: applications and perspectives. Sci Total Environ 156457
44. Shreyash N, Bajpai S, Khan MA, Vijay Y, Tiwary SK, Sonker M (2021) Green synthesis of nanoparticles and their biomedical applications: a review. ACS Appl Nano Mater 4(11):11428–11457
45. Vijayaraghavan K, Ashokkumar T (2017) Plant-mediated biosynthesis of metallic nanoparticles: a review of literature, factors affecting synthesis, characterization techniques and applications. J Environ Chem Eng 5(5):4866–4883
46. Darroudi M, Yazdi MET, Amiri MS (2020) Plant-mediated biosynthesis of nanoparticles. In: 21st century nanoscience—a handbook. CRC Press, pp 1–1
47. Nagajyothi PC, Cha SJ, Yang IJ, Sreekanth TVM, Kim KJ, Shin HM (2015) Antioxidant and anti-inflammatory activities of zinc oxide nanoparticles synthesized using Polygala tenuifolia root extract. J Photochem Photobiol, B 146:10–17
48. Hano C, Abbasi BH (2021) Plant-based green synthesis of nanoparticles: production, characterization and applications. Biomolecules 12(1):31
49. Shen N, Wang T, Gan Q, Liu S, Wang L, Jin B (2022) Plant flavonoids: classification, distribution, biosynthesis, and antioxidant activity. Food Chem 132531
50. Reddy NV, Li H, Hou T, Bethu MS, Ren Z, Zhang Z (2021) Phytosynthesis of silver nanoparticles using Perilla frutescens leaf extract: characterization and evaluation of antibacterial, antioxidant, and anticancer activities. Int J Nanomed 16:15
51. Ghoreishi SM, Behpour M, Khayatkashani M (2011) Green synthesis of silver and gold nanoparticles using Rosa damascena and its primary application in electrochemistry. Physica E 44(1):97–104
52. Naeem GA, Muslim RF, Rabeea MA, Owaid MN, Abd-Alghafour NM (2020) Punica granatum L. mesocarp-assisted rapid fabrication of gold nanoparticles and characterization of nano-crystals. Environ Nanotechnol Monitor Manag 14:100390
53. Gershenzon J, Dudareva N (2007) The function of terpene natural products in the natural world. Nat Chem Biol 3(7):408–414
54. Boncan DAT, Tsang SS, Li C, Lee IH, Lam HM, Chan TF, Hui JH (2020) Terpenes and terpenoids in plants: interactions with environment and insects. Int J Mol Sci 21(19):7382
55. Yang W, Chen X, Li Y, Guo S, Wang Z, Yu X (2020) Advances in pharmacological activities of terpenoids. Nat Prod Commun 15(3):1934578X20903555
56. Edris AE (2007) Pharmaceutical and therapeutic potentials of essential oils and their individual volatile constituents: a review. Phytother Res: Int J Devoted Pharmacol Toxicol Eval Nat Prod Derivat 21(4):308–323
57. Sangwan NS, Farooqi AHA, Shabih F, Sangwan RS (2001) Regulation of essential oil production in plants. Plant Growth Regul 34(1):3–21
58. Zwenger S, Basu C (2008) Plant terpenoids: applications and future potentials. Biotechnol Mol Biol Rev 3(1):1
59. Khan MA, Khan T, Nadhman A (2016) Applications of plant terpenoids in the synthesis of colloidal silver nanoparticles. Adv Coll Interf Sci 234:132–141

60. Omar ZA, Abduljabar RS, Sajadi SM, Mahmud SA, Yahya RO (2022) Recent progress in eco-synthesis of essential oil-based nanoparticles and their possible mechanisms. Ind Crops Prod 187:115322
61. Cox-Georgian D, Ramadoss N, Dona C, Basu C (2019) Therapeutic and medicinal uses of terpenes. In: Medicinal plants. Springer, Cham pp 333–359
62. Adnan M, Patel M, Reddy MN, Alshammari E (2018) Formulation, evaluation and bioactive potential of Xylaria primorskensis terpenoid nanoparticles from its major compound xylaranic acid. Sci Rep 8(1):1–12
63. Boureau L, Pribat A, Mortain-Bertrand A, Bert LS, Rolin D, Teyssier E, Gallusci P (2013) Metabolic engineering of isoprenoid biosynthesis. Natural products. Phytochemistry, botany and metabolism of alkaloids, phenolics and terpenes. Springer, Berlin, Germany, pp. 2814–2851
64. Hernández ÁP, Micaelo A, Piñol R, García-Vaquero ML, Aramayona JJ, Criado JJ, … Fuentes M (2022) Comprehensive and systematic characterization of multi-functionalized cisplatin nano-conjugate: from the chemistry and proteomic biocompatibility to the animal model. J Nanobiotechnol 20(1):1–19
65. Marslin G, Siram K, Maqbool Q, Selvakesavan RK, Kruszka D, Kachlicki P, Franklin G (2018) Secondary metabolites in the green synthesis of metallic nanoparticles. Materials 11(6):940
66. Bruchez M Jr, Moronne M, Gin P, Weiss S, Alivisatos AP (1998) Semiconductor nanocrystals as fluorescent biological labels. Science 281(5385):2013–2016
67. Perrault SD, Walkey C, Jennings T, Fischer HC, Chan WC (2009) Mediating tumor targeting efficiency of nanoparticles through design. Nano Lett 9(5):1909–1915
68. Tang Z, Wang Y, Podsiadlo P, Kotov NA (2006) Biomedical applications of layer-by-layer assembly: from biomimetics to tissue engineering. Adv Mater 18(24):3203–3224
69. Pirtarighat S, Ghannadnia M, Baghshahi S (2019) Green synthesis of silver nanoparticles using the plant extract of Salvia spinosa grown in vitro and their antibacterial activity assessment. J Nanostruct Chem 9(1):1–9
70. Abbas HS, Mahmoud AM, Wahed RA, Elsantawy MAA, Hamdy NM, Ismail SE, Nabil MA (2021) Prospects of using bioactive compounds in nanomaterials surface decoration and their biomedical purposes. Int Nano Lett 1–14
71. Faisal S, Jan H, Shah SA, Shah S, Khan A, Akbar MT, … Syed S (2021) Green synthesis of zinc oxide (ZnO) nanoparticles using aqueous fruit extracts of Myristica fragrans: their characterizations and biological and environmental applications. ACS Omega 6(14):9709–9722
72. Asdagh A, Karimi Sani I, Pirsa S, Amiri S, Shariatifar N, Eghbaljoo–Gharehgheshlaghi H, … Taniyan A (2021) Production and characterization of nanocomposite film based on whey protein isolated/copper oxide nanoparticles containing coconut essential oil and paprika extract. J Polym Environ 29(1):335–349
73. Waris M, Nasir S, Abbas S, Azeem M, Ahmad B, Khan NA, … Mahboob S (2020) Evaluation of larvicidal efficacy of Ricinus communis (Castor) and synthesized green silver nanoparticles against Aedes aegypti L. Saudi J Biol Sci 27(9):2403–2409
74. Vilas V, Philip D, Mathew J (2016) Biosynthesis of Au and Au/Ag alloy nanoparticles using Coleus aromaticus essential oil and evaluation of their catalytic, antibacterial and antiradical activities. J Mol Liq 221:179–189
75. Ni ZJ, Wang X, Shen Y, Thakur K, Han J, Zhang JG, … Wei ZJ (2021) Recent updates on the chemistry, bioactivities, mode of action, and industrial applications of plant essential oils. Trends Food Sci Technol 110:78–89
76. Bakkali F, Averbeck S, Averbeck D, Idaomar M (2008) Biological effects of essential oils—a review. Food Chem Toxicol 46(2):446–475
77. Shankar SS, Rai A, Ahmad A, Sastry M (2004) Rapid synthesis of Au, Ag, and bimetallic Au core–Ag shell nanoparticles using Neem (Azadirachta indica) leaf broth. J Colloid Interf Sci 275:496–502
78. Huang J, Li Q, Sun D, Lu Y, Su Y, Yang X, … Chen C (2007) Biosynthesis of silver and gold nanoparticles by novel sundried Cinnamomum camphora leaf. Nanotechnology 18(10):105104

79. Bonifácio B, da Silva PB, Ramos M, Negri K, Maria Bauab T, Chorilli M (2014) Int J Nanomed 9:1–5
80. Fonseca-Santos B, Gremião MPD, Chorilli M (2015) Nanotechnology-based drug delivery systems for the treatment of Alzheimer's disease. Int J Nanomed 10:4981
81. Pivetta TP, Simões S, Araújo MM, Carvalho T, Arruda C, Marcato PD (2018) Development of nanoparticles from natural lipids for topical delivery of thymol: Investigation of its anti-inflammatory properties. Colloids Surf, B 164:281–290
82. Cháirez-Ramírez MH, Sánchez-Burgos JA, Gomes C, Moreno-Jiménez MR, González-Laredo RF, Bernad-Bernad MJ, … Rocha-Guzmán NE (2015) Morphological and release characterization of nanoparticles formulated with poly (dl-lactide-co-glycolide) (PLGA) and lupeol: In vitro permeability and modulator effect on NF-κB in Caco-2 cell system stimulated with TNF-α. Food Chem Toxicol 85:2–9
83. Bano S, Ahmed F, Khan F, Chaudhary SC, Samim M (2020) Targeted delivery of thermoresponsive polymeric nanoparticle-encapsulated lycopene: in vitro anticancer activity and chemopreventive effect on murine skin inflammation and tumorigenesis. RSC Adv 10(28):16637–16649
84. Ramachandran S, Thangarajan S (2016) A novel therapeutic application of solid lipid nanoparticles encapsulated thymoquinone (TQ-SLNs) on 3-nitroproponic acid induced Huntington's disease-like symptoms in wistar rats. Chem Biol Interact 256:25–36
85. Han Y (2005) Ginkgo Terpene component has an anti-inflammatory effect on Candida Albicans-caused arthritic inflammation. Int Immunopharmacol 5:1049–1056
86. Zhao Y, Xiong S, Liu P, Liu W, Wang Q, Liu Y, … Chen T (2020) Polymeric nanoparticles-based brain delivery with improved therapeutic efficacy of ginkgolide B in parkinson's disease. Int J Nanomed 15:10453
87. Nerome H, Machmudah S, Fukuzato R, Higashiura T, Youn YS, Lee YW, Goto M (2013) Nanoparticle formation of lycopene/β-cyclodextrin inclusion complex using supercritical antisolvent precipitation. J Supercritical Fluids 83:97–103
88. Gonçalves S, Mansinhos I, Romano A (2023) Role of nanoparticles on modulation of plant secondary metabolism. In: Engineered nanomaterials for sustainable agricultural production, soil improv stress manag. Academic Press, pp 447–473
89. Chung IM, Rajakumar G, Thiruvengadam M (2018) Effect of silver nanoparticles on phenolic compounds production and biological activities in hairy root cultures of Cucumis anguria. Acta Biol Hung 69(1):97–109
90. Yasur J, Rani PU (2013) Environmental effects of nanosilver: impact on castor seed germination, seedling growth, and plant physiology. Environ Sci Pollut Res 20(12):8636–8648
91. Shankar SS, Ahmad A, Pasricha R, Sastry M (2003) Bioreduction of chloroaurate ions by geranium leaves and its endophytic fungus yields gold nanoparticles of different shapes. J Mater Chem 13(7):1822–1826
92. Singh AK, Talat M, Singh DP, Srivastava ON (2010) Biosynthesis of gold and silver nanoparticles by natural precursor clove and their functionalization with amine group. J Nanopart Res 12:1667–1675
93. Logeswari P, Silambarasan S, Abraham J (2015) Synthesis of silver nanoparticles using plants extract and analysis of their antimicrobial property. J Saudi Chem Soc 19:311–317
94. Fazal H, Abbasi BH, Ahmad N, Ali M (2016) Elicitation of medicinally important antioxidant secondary metabolites with silver and gold nanoparticles in callus cultures of Prunella vulgaris L. Appl Biochem Biotechnol 180(6):1076–1092
95. Singh R, Singh DP, Gupta P, Jain P, Mishra T, Kumar A, … Shirke PA (2019) Nanoparticles alter the withanolide biosynthesis and carbohydrate metabolism in Withania somnifera (Dunal). Indus Crops Prod 127:94–109
96. Wesołowska A, Jadczak P, Kulpa D, Przewodowski W (2019) Gas chromatography-mass spectrometry (GC-MS) analysis of essential oils from AgNPs and AuNPs elicited Lavandula angustifolia in vitro cultures. Molecules 24(3):606
97. Glusker JP, Katz AK, Bock CW (1999) Metal ions in biological systems. Rigaku J. 16:8–17

98. Si S, Mandal TK (2007) Tryptophan-based peptides to synthesize gold and silver nanoparticles: a mechanistic and kinetic study. Chemistry–A Euro J 13(11):3160–3168
99. Chamani E, Karimi Ghalehtaki S, Mohebodini M, Ghanbari A (2015) The effect of Zinc oxide nano particles and Humic acid on morphological characters and secondary metabolite production in *Lilium ledebourii* Bioss. Iran J Genet Plant Breed 4(2):11–19
100. Asl KR, Hosseini B, Shara A, Palazon J (2019) Eng Life Sci 19:73–89
101. Kołodziejczak-Radzimska A, Jesionowski T (2014) Zinc oxide—from synthesis to application: a review. Materials 7(4):2833–2881
102. Oloumi H, Soltaninejad R, Baghizadeh A (2015) The comparative effects of nano and bulk size particles of CuO and ZnO on glycyrrhizin and phenolic compounds contents in Glycyrrhiza glabra L. seedlings. Indian J Plant Physiol 20(2):157–161
103. Álvarez-Chimal R, García-Pérez VI, Álvarez-Pérez MA, Arenas-Alatorre JÁ (2021) Green synthesis of ZnO nanoparticles using a Dysphania ambrosioides extract. Structural characterization and antibacterial properties. Mater Sci Eng: C 118:111540
104. Catalán A, Pacheco JG, Martínez A, Mondaca MA (2008) In vitro and in vivo activity of Melaleuca alternifolia mixed with tissue conditioner on Candida albicans. Oral Surg, Oral Med, Oral Pathol, Oral Radiol, Endodontol 105(3):327–332
105. Hernández-Sierra JF, Ruiz F, Pena DCC, Martínez-Gutiérrez F, Martínez AE, de Jesús Pozos A, … Castañón GM (2008) The antimicrobial sensitivity of Streptococcus mutans to nanoparticles of silver, zinc oxide, and gold. Nanomed: Nanotechnol Biol Med 4(3):237–240
106. Jones N, Ray B, Ranjit KT, Manna AC (2008) Antibacterial activity of ZnO nanoparticle suspensions on a broad spectrum of microorganisms. FEMS Microbiol Lett 279(1):71–76
107. Martínez A, Rojas N, García L, González F, Domínguez M, Catalán A (2014) In vitro activity of terpenes against Candida albicans and ultrastructural alterations. Oral Surg Oral Med Oral Pathol Oral Radiol 118(5):553–559
108. Xu JS, Li YAO, Cao XUE, Cui Y (2013) The effect of eugenol on the cariogenic properties of Streptococcus mutans and dental caries development in rats. Exp Ther Med 5(6):1667–1670
109. Karimi N, Behbahani M, Dini G, Razmjou A (2018) Enhancing the secondary metabolite and anticancer activity of Echinacea purpurea callus extracts by treatment with biosynthesized ZnO nanoparticles. Adv Nat Sci: Nanosci Nanotechnol 9(4):045009
110. Raigond P, Raigond B, Kaundal B, Singh B, Joshi A, Dutt S (2017) Effect of zinc nanoparticles on antioxidative system of potato plants. J Environ Biol 38(3):435
111. Ahamed TES, Ahamed ESS (2018) Synergy prospect low gamma irradiation doses incorporating elicitation with iron nanoparticles to hyper production biomass yield and bioactive secondary metabolites for cress, medicinal plant. J Plant Sci 6(5):157–163
112. Sharaf E, Khayam Nekoei S, Fotokian MH, Davoodi D, Hadavand Mirzaei H, Hasanloo T (2013) J Measur Phys Behav 2:177–184
113. Mohebodini M, Fathi R, Mehri N (2017) Optimization of hairy root induction in chicory (Cichorium intybus L.) and effects of nanoparticles on secondary metabolites accumulation. Iran J Genet Plant Breed 6(2):60–68
114. Abboud Y, Saffaj T, Chagraoui A, El Bouari A, Abboud Y, Saffaj T, Chagraoui A, El Bouari A, Brouzi K, Tanane O, Ihssane B (2014) Biosynthesis, characterization and antimicrobial activity of copper oxide nanoparticles (CONPs) produced using brown alga extract (Bifurcaria bifurcata). Appl Nanosci 4(5):571–576
115. Ahmadi SJ, Outokesh M, Hosseinpour M, Mousavand T (2011) A simple granulation technique for preparing high-porosity nano copper oxide (II) catalyst beads. Particuology 9(5):480–485
116. Jani AMM, Losic D, Voelcker NH (2013) Nanoporous anodic aluminium oxide: advances in surface engineering and emerging applications. Prog Mater Sci 58(5):636–704
117. Ramgir N, Datta N, Kaur M, Kailasaganapathi S, Debnath AK, Aswal DK, Gupta SK (2013) Metal oxide nanowires for chemiresistive gas sensors: issues, challenges and prospects. Colloids Surf, A 439:101–116
118. Keller AA, McFerran S, Lazareva A, Suh S (2013) Global life cycle releases of engineered nanomaterials. J Nanopart Res 15(6):1–17

119. Lei Z, Mingyu S, Xiao W, Chao L, Chunxiang Q, Liang C, ... Fashui H (2008) Antioxidant stress is promoted by nano-anatase in spinach chloroplasts under UV-B radiation. Biol Trace Element Res 121(1):69–79
120. Vittori Antisari L, Carbone S, Gatti A, Vianello G, Nannipieri P (2015) Uptake and translocation of metals and nutrients in tomato grown in soil polluted with metal oxide (CeO_2, Fe_3O_4, SnO_2, TiO_2) or metallic (Ag Co, Ni) engineered nanoparticles. Environ Sci Pollut Res 22(3):1841–1853
121. Andriotis EG, Papi RM, Paraskevopoulou A, Achilias DS (2021) Synthesis of d-limonene loaded polymeric nanoparticles with enhanced antimicrobial properties for potential application in food packaging. Nanomaterials 11(1):191
122. Bidyarani N, Kumar U (2019) Study of limonene loaded Zein nanoparticles for sustainable agriculture. In: Adv Spectrosc: Mol Mater. Springer, Singapore, pp 257–267
123. Goswami A, Patel N, Bhatt V, Raval M, Kundariya M, Sheth N (2022) Lycopene loaded polymeric nanoparticles for prostate cancer treatment: formulation, optimization using Box-behnken design and cytotoxicity studies. J Drug Deliv Sci Technol 67:102930
124. Han L, Fu Y, Cole AJ, Liu J, Wang J (2012) Co-encapsulation and sustained-release of four components in ginkgo terpenes from injectable PELGE nanoparticles. Fitoterapia 83(4):721–731
125. Prakash A, Vadivel V (2020) Citral and linalool nanoemulsions: impact of synergism and ripening inhibitors on the stability and antibacterial activity against Listeria monocytogenes. J Food Sci Technol 57(4):1495–1504
126. Lu WC, Huang DW, Wang CC, Yeh CH, Tsai JC, Huang YT, Li PH (2018) Preparation, characterization, and antimicrobial activity of nanoemulsions incorporating citral essential oil. J Food Drug Anal 26(1):82–89
127. Marei GIK, Rabea EI, Badawy ME (2018) Preparation and characterizations of chitosan/citral nanoemulsions and their antimicrobial activity. Appl Food Biotechnol 5(2):69–78
128. Charoenputtakun P, Pamornpathomkul B, Opanasopit P, Rojanarata T, Ngawhirunpat T (2014) Terpene composited lipid nanoparticles for enhanced dermal delivery of all-trans-retinoic acids. Biol Pharmaceut Bull b14-00015
129. Gengan R, Anand K, Phulukdaree A, Chuturgoon A (2013) A549 lung cell line activity of biosynthesized silver nanoparticles using Albizia adianthifolia leaf. Colloids Surf, B 105:87–91
130. Link S, El-Sayed MA (2003) Optical properties and ultrafast dynamics of metallic nanocrystals. Ann Rev Phys Chem 54:331–366
131. Pimprikar PS, Joshi SS, Kumar AR, Zinjarde SS, Kulkarni SK (2009) Influence of biomass and gold salt concentration on nanoparticle synthesis by the tropical marine yeast Yarrowia lipolytica NCIM 3589. Colloids Surf, B 74:309–316
132. Shedbalkar U, Singh R, Wadhwani S, Gaidhani S, Chopade BA (2014) Microbial synthesis of gold nanoparticles: current status and future prospects. Adv Colloid Interface Sci 209:40–48
133. Vilas V, Philip D, Mathew J (2014) Catalytically and biologically active silver nanoparticles synthesized using essential oil. Spectrochim Acta Part A Mol Biomol Spectrosc 132:743–750
134. Okafor F, Janen A, Kukhtareva T, Edwards V, Curley M (2013) Green synthesis of silver nanoparticles, their characterization, application and antibacterial activity. Int J Environ Res Public Health 10:5221
135. Jagtap UB, Bapat VA (2013) Green synthesis of silver nanoparticles using Artocarpus heterophyllus Lam. seed extract and its antibacterial activity. Indus Crops Prod 46:132–137
136. Krishnaraj C, Ramachandran R, Mohan K, Kalaichelvan PT (2012) Optimization for rapid synthesis of silver nanoparticles and its effect on phytopathogenic fungi. Spectrochim Acta A Mol Biomol Spectrosc. 93:95–99
137. Kumar R, Roopan SM, Prabhakarn A, Khanna VG, Chakroborty S (2012) Agricultural waste Annona squamosa peel extract: biosynthesis of silver nanoparticles. Spectrochim Acta Part A Mol Biomol Spectrosc 90:173–176
138. AbdelHamid AA, Al-Ghobashy MA, Fawzy M, Mohamed MB, Abdel-Mottaleb MMSA (2013) Phytosynthesis of Au, Ag, and Au–Ag Bimetallic nanoparticles using aqueous extract of Sago Pondweed (Potamogeton pectinatus L.). ACS Sustain Chem Eng 1:1520–1529

139. Sperling R-A, Parak W (2010) Surface modification, functionalization and bioconjugation of colloidal inorganic nanoparticles. Philos Trans Royal Soc Lond A: Math Phys Eng Sci 368:1333–1383
140. Khan Z, Singh T, Hussain JI, Obaid AY, Al-Thabaiti SA, El-Mossalamy EH (2013) Starchdirected green synthesis, characterization and morphology of silver nanoparticles. Colloids Surf, B 102:578–584
141. Singh R, Shedbalkar UU, Wadhwani SA, Chopade BA (2015) Bacteriagenic silver nanoparticles: synthesis, mechanism, and applications. Appl Microbiol Biotechnol 99:4579–4593
142. Rashid M, Sabir S (2014) Biosynthesis of self-dispersed silver colloidal particles using the aqueous extract of P. peruviana for sensing dl-Alanine. ISRN Nanotechnol 2014:7
143. Rauf A, Ahmad T, Khan A, Maryam, Uddin G, Ahmad B, … Al-Harrasi A (2021) Green synthesis and biomedicinal applications of silver and gold nanoparticles functionalized with methanolic extract of Mentha longifolia. Artif Cells, Nanomed Biotechnol 49(1):194–203
144. Tanna JA, Chaudhary RG, Juneja HD, Gandhare NV, Rai AR (2015) Histidine-capped ZnO nanoparticles: an efficient synthesis, spectral characterization and effective antibacterial activity. BioNanoScience 5(3):123–134
145. Brar SK, Verma M (2011) Measurement of nanoparticles by light-scattering techniques. TrAC, Trends Anal Chem 30(1):4–17
146. Sharma V, Chotia C, Ganesan V, Okram GS (2017) Influence of particle size and dielectric environment on the dispersion behaviour and surface plasmon in nickel nanoparticles. Phys Chem Chem Phys 19(21):14096–14106
147. Almaary KS, Sayed SR, Abd-Elkader OH, Dawoud TM, El Orabi NF, Elgorban AM (2020) Complete green synthesis of silver-nanoparticles applying seed-borne Penicillium duclauxii. Saudi J Biol Sci 27(5):1333–1339
148. Kavitha S, Dhamodaran M, Prasad R, Ganesan M (2017) Synthesis and characterisation of zinc oxide nanoparticles using terpenoid fractions of Andrographis paniculata leaves. Int Nano Lett 7(2):141–147
149. Das R, Ali E, Abd Hamid SB (2014) Current applications of x-ray powder diffraction-a review. Rev Adv Mater Sci 38(2)
150. Patil RB, Chougale AD (2021) Analytical methods for the identification and characterization of silver nanoparticles: a brief review. Mater Today: Proc 47:5520–5532
151. Kong J, Yu S (2007) Fourier transform infrared spectroscopic analysis of protein secondary structures. Acta Biochim Biophys Sin 39(8):549–559
152. Sharma NK, Vishwakarma J, Rai S, Alomar TS, AlMasoud N, Bhattarai A (2022a) Green route synthesis and characterization techniques of silver nanoparticles and their biological adeptness. ACS Omega
153. Fissan H, Ristig S, Kaminski H, Asbach C, Epple M (2014) Comparison of different characterization methods for nanoparticle dispersions before and after aerosolization. Anal Methods 6(18):7324–7334
154. Chérif I, Dkhil YO, Smaoui S, Elhadef K, Ferhi M, Ammar S (2022) X-ray diffraction analysis by modified Scherrer, Williamson–Hall and Size–Strain Plot Methods of ZnO nanocrystals synthesized by oxalate route: a potential antimicrobial candidate against foodborne pathogens. J Clust Sci 1–16
155. Liu XY, Wang JQ, Ashby CR Jr, Zeng L, Fan YF, Chen ZS (2021) Gold nanoparticles: synthesis, physiochemical properties and therapeutic applications in cancer. Drug Discov Today 26(5):1284–1292
156. Patil MP, Jin X, Simeon NC, Palma J, Kim D, Ngabire D, … Kim GD (2018) Anticancer activity of Sasa borealis leaf extract-mediated gold nanoparticles. Artif Cells, Nanomed, Biotechnol 46(1):82–88
157. Madhavan SA, Vinotha P, Uma V, Mahadevi M (2020) Anticancer activity of Pedalium murex linn methanolic leaves extract against A549 human lung cancer cell line. Asian J Adv Res 33–40
158. Chittasupho C, Athikomkulchai S (2018) Nanoparticles of Combretum quadrangulare leaf extract induce cytotoxicity, apoptosis, cell cycle arrest and anti-migration in lung cancer cells. J Drug Deliv Sci Technol 45:378–387

159. Balasubramanian S, Kala SMJ, Pushparaj TL (2020) Biogenic synthesis of gold nanoparticles using Jasminum auriculatum leaf extract and their catalytic, antimicrobial and anticancer activities. J Drug Deliv Sci Technol 57:101620
160. Panáček A, Smékalová M, Večeřová R, Bogdanová K, Röderová M, Kolář M, ... Kvítek L (2016) Silver nanoparticles strongly enhance and restore bactericidal activity of inactive antibiotics against multiresistant Enterobacteriaceae. Colloids Surf B: Biointerf 142:392–399
161. Tang S, Zheng J (2018) Antibacterial activity of silver nanoparticles: structural effects. Adv Healthc Mater 7(13):1701503
162. Patil SV, Borase HP, Patil CD, Salunke BK (2012) Biosynthesis of silver nanoparticles using latex from few euphorbian plants and their antimicrobial potential. Appl Biochem Biotechnol 167(4):776–790
163. Długosz O, Chmielowiec-Korzeniowska A, Drabik A, Tymczyna L, Banach M (2022a) Bioactive selenium nanoparticles synthesized from propolis extract and quercetin based on natural deep eutectic solvents (NDES). J Clust Sci 1–12
164. Długosz O, Ochnik M, Sochocka M, Franz D, Orzechowska B, Anna CK, ... Banach M (2022b) Antimicrobial and antiviral activity of selenium sulphide nanoparticles synthesised in extracts from spices in natural deep eutectic solvents (NDES). Sustain Mater Technol 32:e00433
165. Sani I, Ukwuani-Kwaja AN, Abdulkadir D (2022) Antibacterial activities of plant-derived metallic nanoparticles on some selected multidrug-resistant clinical isolates. Asian J Biol Sci 15(1):15–26
166. Suriyakalaa U, Antony JJ, Suganya S, Siva D, Sukirtha R, Kamalakkannan S, ... Achiraman S (2013) Hepatocurative activity of biosynthesized silver nanoparticles fabricated using Andrographis paniculata. Colloids Surf B: Biointerf 102:189–194
167. Sun RWY, Chen R, Chung NPY, Ho CM, Lin CLS, Che CM (2005) Silver nanoparticles fabricated in Hepes buffer exhibit cytoprotective activities toward HIV-1 infected cells. Chem Commun 40:5059–5061
168. Lasoń E (2020) Topical administration of terpenes encapsulated in nanostructured lipid-based systems. Molecules 25(23):5758
169. Jacob SJP, Finub JS, Narayanan A (2012) Synthesis of silver nanoparticles using Piper longum leaf extracts and its cytotoxic activity against Hep-2 cell line. Colloids Surf, B 91:212–214
170. Satyavani K, Ramanathan T, Gurudeeban S (2011) Green synthesis of silver nanoparticles by using stem derived callus extract of bitter apple (Citrullus colocynthis). Dig J Nanomater Biostruct 6(3):1019–1024
171. Gurunathan S, Lee KJ, Kalishwaralal K, Sheikpranbabu S, Vaidyanathan R, Eom SH (2009) Antiangiogenic properties of silver nanoparticles. Biomaterials 30(31):6341–6350
172. Muniyappan N, Nagarajan NS (2014) Green synthesis of gold nanoparticles using Curcuma pseudomontana essential oil, its biological activity and cytotoxicity against human ductal breast carcinoma cells T47D. J Environ Chem Eng 2(4):2037–2044
173. Mittal AK, Kaler A, Banerjee UC (2012) Free radical scavenging and antioxidant activity of silver nanoparticles synthesized from flower extract of rhododendron dauricum. Nano Biomed Eng 4(3)
174. Arif M, Ullah R, Ahmad M, Ali A, Ullah Z, Ali M, ... Sher H (2022) Green synthesis of silver nanoparticles using Euphorbia wallichii leaf extract: its antibacterial action against citrus canker causal agent and antioxidant potential. Molecules 27(11):3525
175. Rajan R, Chandran K, Harper SL, Yun SI, Kalaichelvan PT (2015) Plant extract synthesized silver nanoparticles: an ongoing source of novel biocompatible materials. Ind Crops Prod 70:356–373
176. Mishra SB, Kumari N (2021) Engineering of crystalline nano-suspension of lycopene for potential management of oxidative stress–linked diabetes in experimental animals. BioNanoScience 11(2):345–354
177. Al-Radadi NS (2022) Biogenic proficient synthesis of (Au-NPs) via aqueous extract of Red Dragon Pulp and seed oil: characterization, antioxidant, cytotoxic properties, anti-diabetic anti-inflammatory, anti-Alzheimer and their anti-proliferative potential against cancer cell lines. Saudi J Biol Sci 29(4):2836–2855

178. Ayyoub S, Al-Trad B, Aljabali AA, Alshaer W, Al Zoubi M, Omari S, … Tambuwala MM (2022) Biosynthesis of gold nanoparticles using leaf extract of Dittrichia viscosa and in vivo assessment of its anti-diabetic efficacy. Drug Deliv Transl Res 1–7
179. Chi NTL, Narayanan M, Chinnathambi A, Govindasamy C, Subramani B, Brindhadevi K, … Pikulkaew S (2022) Fabrication, characterization, anti-inflammatory, and anti-diabetic activity of silver nanoparticles synthesized from Azadirachta indica kernel aqueous extract. Environ Res 208:112684
180. Husen A, Siddiqi KS (2022) Advances in smart nanomaterials and their applications. Elsevier Inc., 50 Hampshire St., 5th Floor, Cambridge, MA 02139, USA
181. Manfredini N, Ilare J, Invernizzi M, Polvara E, Contreras Mejia D, Sironi S, … Sponchioni M (2020) Polymer nanoparticles for the release of fragrances: how the physicochemical properties influence the adsorption on textile and the delivery of limonene. Indus Eng Chem Res 59(28):12766–12773
182. Liang J, Zhu Y, Gao C, Ling C, Qin J, Wang Q, Wang J (2019) Menthol-modified BSA nanoparticles for glioma targeting therapy using an energy restriction strategy. NPG Asia Mater 11(1), 38
183. Holz JP, Bottene MK, Jahno VD, Einloft S, Ligabue R (2018) Menthol-loaded PLGA micro and nanospheres: Synthesis, characterization and degradation in artificial saliva. Mater Res 21
184. Bano A, Gupta A, Rai S, Sharma S, Pathak N (2021) Elucidation of bioactive potential of two commonly grown north Indian Psidium guajava viz., Lalit and Shweta against pathogenic foodborne and MDR bacteria. Biointerf Res Appl Chem 11:14090–14102

Lignin and Their Role in Nanomaterials Synthesis and Applications

Surendra Pratap Singh

Abstract About 50 million metric tons per annum of lignin are produced as a by-product of the paper industry. It is anticipated that the production of chemicals and renewable fuels from lignocellulosic biomass as a sustainable alternative to petroleum in the future. Black liquor is considered a low value-added substance. It is used as a chemical recovery furnace fuel for generation of power. Industrial lignin is used as active component of cement, drilling fluids, and adhesives. Lignin is suitable additive for thermoplastic to increase mechanical, physical chemical and UV-tolerance properties because of its aromatic nature. It can participate in radical-mediated cross-linking reactions, due to abundance of functional groups available for chemical or derivatization reactions. Lignin can be used alone or in conjunction with other polymers to form the basis for a variety of nanomaterials. In this chapter, the extraction of lignin, manufacturing and applications of various types of lignin-based nanomaterials are summarized.

Keywords Biomass · Lignocellulosic · Lignin · Nanomaterials · Nano particles · Nanocomposite

1 Introduction

The gradual depletion of fossil fuel resources and the continuous growth of the global population has sparked interest in discovering an alternative, clean and globally available resource. Lignocellulosic biomass is the second most abundant material. It might replace many petroleum-based products. ‘Trees are the giants of the kingdom Plantae, due to the fact of their large age, size, and the special rigid wooden structure of the stems and branches. Trees are perennial, seed-bearing plants (Spermatophytae), which are classified commonly into two categories known as hardwoods (angiosperms or dicotyledonous angiosperms) and softwoods (gymnosperms). Hardwood trees produce covered seeds within flowers, while Softwoods are also referred

S. P. Singh (✉)
Pulp and Paper Research Institute, Jaykaypur, Rayagada, Odisha, India
e-mail: spsinghbh@gmail.com

A. Husen (ed.), *Secondary Metabolites Based Green Synthesis of Nanomaterials and Their Applications*, Smart Nanomaterials Technology,
https://doi.org/10.1007/978-981-99-0927-8_6

to as conifers (coniferous woods) since they have seeds that are produced in cones and not covered [44]. However, these general names are not a measure of "hardness" because there is considerable overlap in the range of average specific gravities of hardwoods and softwoods. Some hardwoods are relatively soft and some softwoods are quite hard. Other classification is based on the retention of needle- or scale-like leaves by most softwood species, as opposed to the annual leaf shedding by most hardwoods. Thus, major commercial soft woods and hardwoods are generally called "evergreen" trees (i.e., they retain new leaves for several years) and "deciduous" trees (i.e., they commonly shed their broad or blade-like leaves each autumn at the end of the tree's growing season) [29]. However, many hardwood species growing in tropical conditions are nonetheless "evergreen" and do not shed their leaves. The bark, stem, top, leaves and needles, branches, and roots are the main structural parts of tree. Only the debarked stem wood is commonly used for furniture and pulping, even though all of these components are useful as feedstocks for conversion processes that are frequently used with renewable natural resources.

In general, wood is not a uniform material, with respect to its anatomical, physical, and chemical properties, and is degradable, for example, by fungi, microorganisms, and heating. It is made up of different kinds of more or less specialized cells, performing the necessary functions of water transport (roughly half of the mass of a living tree consists of water), metabolism and mechanical support [7]. The wood anatomy is characteristic to many hardwood species of trees, i.e., the types of woody cells, their percentages and arrangements differ between the different types of tree [57].

Wood cells are chemically heterogeneous and built up of a polymeric matrix of structural components: polysaccharides, such as cellulose, hemicelluloses and lignin (shown in Fig. 1). These macromolecular substances are not uniformly distributed within the wood cell wall, and their relative concentration varies between different parts of the tree. Non-structural components, such as extractives and inorganic substances, which represent only a minor fraction, are mostly composed of low-molar-mass compounds mainly deposited outside the cell wall [51]. In addition, trace amounts of nitrogen-containing compounds, such as proteins and alkaloids, are present in the wood cell wall. Both softwoods and hardwoods are widely distributed on earth, ranging from tropical to arctic regions [43]. The number of known softwood species (about 1 000) is relatively low when compared to that of hardwood species (about 35 000). However, only a minor part of these wood species is currently utilised commercially because of the more extensive exploitation of tropical forests.

When examining the macroscopic structure, or morphology, of wood, it is evident that there are differences between hardwoods and softwoods, between different species, and also between different wood tissues within a single tree. Wood is composed of elongated cells that are mainly oriented in the longitudinal direction of the stem. The cells are connected to each other with openings, called pits, which allow water and nutrients to be conducted in the tree. In softwoods, the cells are mainly fibrous in form and are therefore termed fibres (tracheids). In hardwoods, there is a broader variety of different specialized cell types, such as fibres, vessels (pores), and parenchyma cells. In a mature tree, the vast majority of both hardwoods

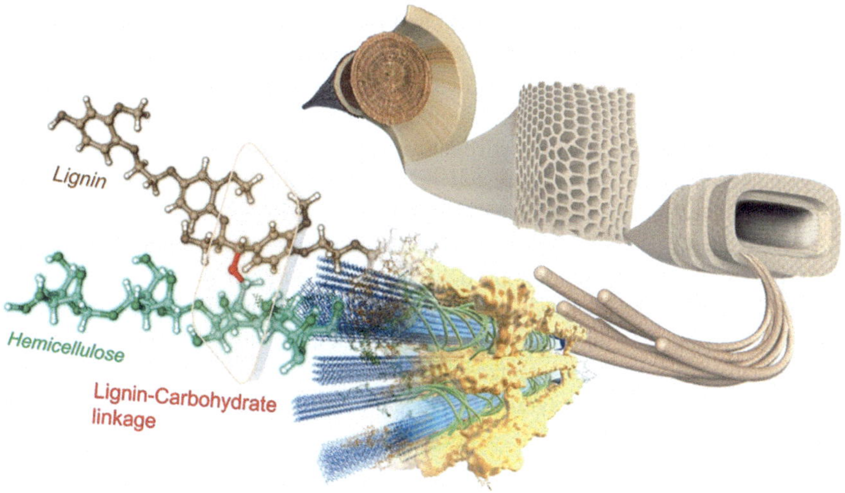

Fig. 1 A three-dimensional view of the lignin–carbohydrate complex in the wood cell wall (adopted from [47]

and softwoods cells are hollow and dead, thus the tissue formed is mainly composed of cell walls and voids, the latter being the lumina, or hollow interiors of the cells [32].

The cross-section of a tree stem shows the macroscopic structure of wood (xylem) and bark. In addition to the characteristic features shown, some softwoods also contain both vertical and horizontal resin canals [34]. Non-wood fibres may have distinctly different properties compared with hardwood or softwood fibres (Table 1). Different aspects include arabinogalactan and rhamnoarabino galactan. Pectin substances form a heterogeneous group, including galacturonans, galactans, and arabinan. For example, spruce wood pectins have a backbone of (1→4)-linked partly methyl-esterified a-D-galacturonic acid units interspersed with (1→2)-linked a-L-rhamnose units.

Wood is the main raw material in the timber, paper industry and consequently, an important by-product of this industry is lignin [34]. Although lignin currently has

Table 1 Average chemical composition of ligno-cellulosic biomass

Raw material	Cellulose	Hemi-celluloses	Pectin	Lignin
Spruce	43	29	–	27
Flax bast	70–80	12–18	1–3	3
Hemp core	34–40	20–25	42–4	20–25
Flax shive	35–40	25–30	2–4	25–30
Aspen	53	31	–	16
Hemp bast	60–70	10–15	1–3	4

direct applications in the pharmaceutical and polymer industry, the main drawback is that it can only be incorporated in small quantities, as its mechanical properties and thermal degradation must be considered. Due to chemical modification of lignin seems to be the better way to use lignin as a starting material for the chemical synthesis new materials, such as hydrogels or nano composites materials [17]. Thus, the scientists are now more interested in the preparation of lignin nanoparticles (LNPs) and investigating their potential applications [9].

It is expected that LNPs will play a vital role in promoting lignin valorization, much like synthetic polymer nanoparticles are in the polymer industry. The conversion of lignin into valuable products was published in a number of research articles. However, there aren't many thorough reviews that concentrate on the synthesis of lignin nanoparticles and their application together. The present chapter summarized all the information regarding synthesis LNPs and potential applications in numerous areas.

2 Structure and Properties of Lignin

Lignin is one of the primary components of plant cell walls. It is a strong, non-fibrous chemical that plays a key function in increasing a plant's strength and stiffness by supporting the cell wall's polysaccharide structure. Lignin is a natural phenolic polymer with a high molecular weight, complex composition, and structure. The biosynthesis of lignin significantly contributes to tissue/organ development, plant development. responses to a variety of biotic and abiotic stressors and resistance to lodging. These biosynthesised through an enzyme-catalyzed dehydrogenative polymerization of three basic precursors: Sinapyl, coniferyl, and p-coumaryl alcohols, precursors of syringyl (S) lignin, guaiacyl (G) and p-hydroxyl phenol (H) units, respectively (shown in Fig. 2). Lignin is chemically combined with the carbohydrates, or that lignin is high-molecular and forms a three-dimensional network in the wood [65]. The polymerization. process results in the formation of a randomly branched and cross-linked structure, for which frequent carbon-to-carbon linkages between phenyl propane units is characteristic. Figure 3 illustrates the general type of polymeric structure. In addition, lignins contain minor amounts of various aromatic acids (e.g., vanillic, ferulic, p-coumaric, syringic and p-hydroxybenzoic) in ester like combination [20].

By means of appropriate color reactions, lignins have been shown to be present in all vascular plants, with the curious exception of tropical tree ferns (*Dicksonia*). In mature plants, stems, roots, bark, leaves, fruit shells, and seed-hairs are all lignified to varying degrees. Lignins are absent in nonvascular algae, fungi, mushrooms, lichens and liverworts, and their presence is doubtful in mosses [39].

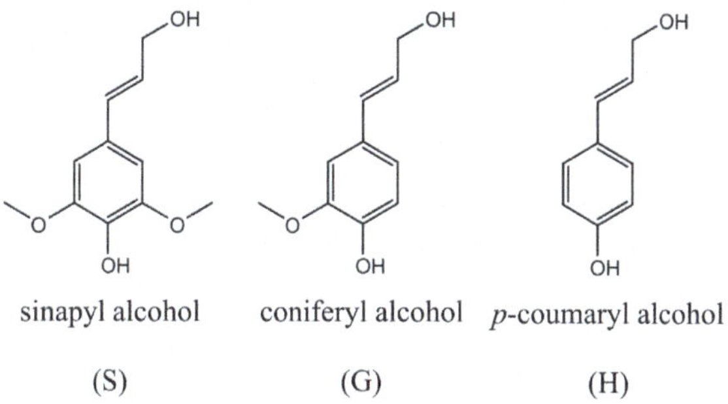

Fig. 2 The three monolignols considered as the building blocks of lignin (adopted from [12]

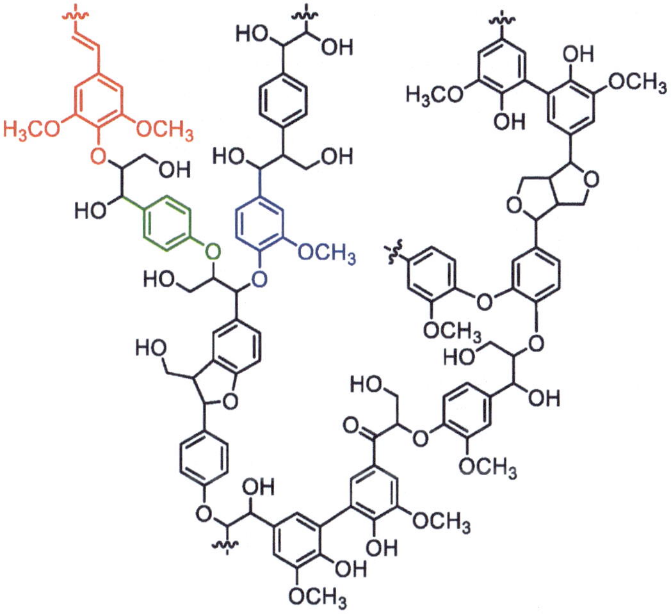

Fig. 3 Presentation of schematic diagram of macromolecular structure of lignin (adopted from [31]

3 Classification of Lignin

Lignins are broadly classified in the following groups:

(1) "Guaiacyl" lignins are present in conifers, and a number of cryptogams, including ferns (Pteridophyta, club mosses (Lycopodium), and horsetails (Equisetum), forming a homogeneous group. The amount of guaiacyl propane units generally varies roughly between 80 and 96%, with the exception of lignins in the compression wood of conifers (~70%) [59]. The rest of the units consist probably of p-hydroxyphenyl propane units, with very small amounts of syringyl

propane moieties present. Vanillic and ferulic ester groups are found only in trace quantities in conifer lignins, while cryptogam lignins contain somewhat larger amounts of these groups, together with esters of protocatechuic, p-hydroxybenzoic, and syringic acids. Lignin contents vary widely even within the same species e.g., 24 to 34% in normal conifer woods, 35 to 40% in compression wood of conifers, and approximately 15 to 30% in mature fern plants [24].

(2) "Guaiacyl-syringyl" all angiosperms contain lignins., both arborescent and herbaceous, some exceptional conifers (*Gentales, Tetraclinis articulata, Podocarpus amarus*), and in the unusual cryptogam *Salaginella.* The *p*-hydroxyphenyl propane units are present in trace quantities only, and the syringyl propane-guaiacyl propane ratio varies even within the same genus. This variance is reflected in the ratio of C6C1-aldehydes isolated in nitrobenzene oxidation, although it should be noted that the syringaldehyde-vanillin ratio (S/V) is always larger than the ratio of the parent moieties in lignin. Guaiacyl-syringyl lignins may be divided in the following subgroups:

a. Temperate zone hardwood lignins. Lignin content 0.16 to 24%, S/V = 1 to 5, usually 3, Klason lignin MeO = 17 to 22%. The amount of ester groups minor, except in *Populus* sp. that contain p-hydroxybenzoate groups absent in other hardwoods.
b. Tropical hardwood lignins. Lignin content 25 to 33%, S/V clearly lower than in the previous group. Klason lignin MeO = 1 5 to 1 8.5%.
c. Guaiacyl-syringyl lignins in conifers. Lignin content 23 to 32%, S/V = 1 to 3.
d. Herbaceous dicotolydon lignins have been characterized only superficially.
e. Grass lignins (present in cereal straws, bamboo, etc.) Lignin content 17 to 23%,S/V = 0.5 to 1.0.

These lignins contain sign, amounts (7 to 12%) of ester groups, in which p-coumaric and ferulic acids are the main constituents. The former acid is converted to p-hydoxyl benzaldehyde in nitrobenzene oxidation. The paper industry, which uses mechanical, chemical and enzymatic processes, is the principal source for lignin. Different process methods produce different kinds of lignins, such as alkali lignin, which is primarily obtained from the kraft process, lignosulphonates obtained by the sulphite process, organosolv lignin obtained by the organosolv process, and hydrolytic lignin obtained by enzymatic hydrolysis [14].

4 Extraction of Lignin

These lignins contain sign, amounts (7 to 12%) of ester groups, in *p*-coumaric and ferulic acids are the main constituents.

The former acid is converted to *p*-hydoxyl benzaldehyde in nitrobenzene oxidation.

The black liquor is the main byproduct of the pulp and paper industry. It is a primary resource of lignin. There are various pulping processes in the paper industry such as chemical, mechanical, enzymatic, organosolv etc. Alkali lignin, which is primarily obtained from the sulphate process, organosolv lignin obtained by the organosolv process, lignosulphonates, obtained by the sulphite process and hydrolytic lignin, obtained by enzymatic hydrolysis, are just a few examples of the various lignin types that are produced by various processes [17].

The chemical pulping process is based on the principle of liberating fibres from the wood matrix by dissolving maximum lignin from the middle lamella. The wood chip dimensions-particularly thickness are of major importance in this context. As cooking proceeds reactive ions must diffuse into the chips. If the diffusion distance is too long and the rate of diffusion too slow, the chemicals are completely consumed before they can reach the chip centers, which results in non-uniform cooking and delignification [22]. Thus, there is a critical balance between the rate of ion transport, chip thickness, and the rate of chemical reaction. The diffusion rate is controlled by the concentration gradient between the free liquor outside the chips and liquid in the chips. Delignification inside thick chips will be more non-uniform than in thin chips. Cooking non-uniformity can be reduced and perhaps eliminated by proper chip manufacturing and screening, perfect impregnation, and low enough cooking temperature. Delignification can be carried out in alkaline, neutral and acidic conditions. The process variables affecting the end result of pulping can be divided into two main groups: those related to raw material properties, and those related to process conditions. Raw material properties cannot be easily controlled, but should be taken into account when setting operating conditions. For example, the proportions between the wood species and the basic properties of wood can only be controlled by good management of supply logistics, but chip quality and particle size distribution are within the control of the pulp mill itself [37]. Process parameters such as chemical charge (grams of chemicals per grams of dry wood), cooking liquor composition, cooking time, and temperature are adjusted by process control measures [8].

4.1 Kraft Process

The kraft pulping process is the dominant method for manufacturing bleachable and unbleached paper pulp. Wood is debarked and chipped in small pieces.The chips are screened in to eliminate fines, pin chips and over-sized chips. The "accepted" chips are either fed to the digester or to an impregnation vessel. The chips are steamed with direct steam to eliminate as much of the air in the chips as possible. The digester is then filled with warm (80 °C–100 °C) cooking liquor to submerge the chips. In the kraft process wood chips are cooked in a solution of sodium hydroxide (NaOH) and sodium sulphide (Na^+. Due to the hydroxide ion (OH^-) in the alkaline environment the bonds in the lignin macromolecules are broken and lignin is therefore fragmented into smaller segments. The cooking liquor is a mixture of caustic soda and sodium sulphite (i.e., regenerated cooking liquor) and spent black liquor from a previous cook [53].

The main active components of the kraft cooking liquor are the $OH^-$and HS^-ions. The digester contents are heated to 150 °C–170 °C temperature, either by direct steam or by indirect heating in a liquor/ steam heat exchanger by extracting liquor from the digester, circulating it through a heat exchanger to heat it and reintroducing it into the digester. The liquor circulation will help to even out the temperature and chemical concentration gradients within the digester, making the delignification more homogeneous [15]. The cooking temperature and time are maintained through H factor calculation for the desired degree of delignification. After pulping, the digested pulp is discharged into a blow tank with the help of pumps and the digester pressure. The heat released is recovered in a blow heat recovery system. Volatile compounds formed during heating and cooking are continuously purged from the digester to control the cooking pressure [21]. The gases go to a condenser system for recovery of volatile compounds such as turpentine. The sodium salts of these fragments are soluble in the cooking liquor, and therefore easy to remove with the black liquor.

Compared to the sulphite process, the sulphide in the pulping liquor in the kraft process increases the rate of delignification considerably and produces a stronger pulp at higher yield. However, the pulp is much darker than soda pulp and sulphite pulp and harder to beat and bleach. The kraft process is highly effective and its main advantage is the recovery and reuse of pulping chemicals. Furthermore, kraft pulping causes emissions of malodorous gases, principally organic sulphides [42], which nowadays are mostly removed by gas scrubbers in the exhaust pipe. Several developments of the recovery boiler for the regeneration of kraft spent chemicals also the energy economy of the kraft process has been continuously improved. Commonly, kraft lignin is used in the manufacture of epoxy resins and polyurethanes [3].

4.2 Sulphite Process

The sulphite pulping originates its name from the use of a bisulphite solution as the delignication medium of the woody biomass. The cation used is usually sodium, magnesium, calcium, or ammonium. The sulphite process, which can be applied over a wide pH range, are mainly applied in the production of specialty grade pulps (i.e. dissolving or rayon grade pulp). The active pulping chemicals in the sulphite cooking liquor involve bisulphite (HSO^-) with a suitable cation (normally *Ca, Mg, Na, K or* NH_4^+). The pH of the delignifying medium can vary between acidic, neutral and alkaline, depending on the cation used. There are various equilibrium reactions between the bisulphite, the cation and water. The first commercial sulphite pulping process was based on a magnesium based sulphite process, but magnesium could not be recovered and reused at the time and was too expensive for large-scale commercial operation [40].

The solubilities of various sulphite solutions determine which cation to use. Calcium requires pH ~2 to stay in solution, while magnesium allows operation up to pH ~4. The sodium and ammonium sulphite solutions can be strongly alkaline without precipitation. Acid processes are those in which the pH is 2–3, bisulphite processes operate in the pH range of 3–5, neutral sulphite processes cover the pH range 6–9, and alkaline sulphite processes operate above pH ~11. It is difficult to give a general description of sulphite pulping because it covers such a wide range of pH levels and cations used. The classical calcium acid sulphite process is carried out as follows:

The cooking acid is made by reacting limestone with sulphur dioxide gas and water in a counter-currently operated acid tower filled with limestone. Sulphur dioxide gas is passed upward through the tower filled with limestone, and water is trickled down the tower. The contact of $SO_2$with limestone produces a calcium bisulphite solution which flows out of the bottom section of the acid tower. This raw acid is strengthened by *S02* gas recovered from the previous cooks and by make-up gas.

Carefully debarked wood is chipped, screened, and fed into the digesters as in the kraft pulping process. In the digester, the cooking liquor thoroughly penetrates the chips. The digester contents are heated by circulating the liquor through a strainer, a circulation system with indirect steam heating, and back to the digester. Excess acid is drained from the digester at roughly 100 °C. The heating by liquor circulation is then continued to the selected cooking temperature (125–140 °C in acid cooking). The cook is then interrupted by degassing the digester; SO_2 is piped to pressurised acid accumulators for use in subsequent cooks. Degassing is generally carried out in two phases to different high- and low-pressure acid accumulators. The spent liquor is often recovered by displacing it directly from the digester into an evaporation plant, while at the same time the digester contents are cooled. The pulp is washed, screened, bleached, and dried. The spent liquor can be used for manufacturing by-products, such as ethanol, proteins, yeast, vanillin, and lignosulphonates. The rest of the spent liquor is evaporated and incinerated in an anoxidative recovery boiler.

Many low-molecular-mass aromatic compounds apparently are formed during sulphite delignification. The monomeric compounds typically include 4-hydroxybenzoic acid, vanillin, vanillic acid, acetovanillone, dihydroconiferyl alcohol, syringol, syringaldehyde, syringic acid and acetosyringone. Some dimeric compounds are also obtained.

Unbleached sulphite pulp has a higher initial brightness, and it is also easier to bleach. The carbohydrate yield is higher at a given kappa number for sulphite pulps [60]. Odour problems are smaller, as are the investment costs. The sulphite process offers the flexibility of producing dissolving grade pulp with high α-cellulose contents, because it can be used over the whole pH range.9 Furthermore, lignosulphonates obtained from sulphite pulping have much broader use as a by-product than lignin from the kraft process because of their solubility in water.Calcium was the base favored in later installations, causing the process to be named calcium sulphite pulping [16].

4.3 Organosolv Process

Both kraft and sulphite pulping have some serious drawbacks, such as air and water pollution and high investment costs. However, with organosolv methods it might be possible to avoid some of these problems, since they allow using both sulphur- and chlorine-free conditions during pulping and bleaching. Organosolv (solvent-basedor solvolysis) pulping is a collective name for processes using an organic solvent, although many use an aqueous solution instead of inorganic species solvated in water as in kraft and sulphite pulping. Generally, the most commonly used solvents can be divided into alcohols, organic acids and others. Of these, especially ethanol pulping has the greatest number of references in the literature, probably due to the cheap price of ethanol compared to other organic solvents. Other important organic solvents are methanol, formic acid and acetic acid48. The group "others" include various phenols, amines, glycols, nitrobenzene, dioxane, dimethylsulphoxide, sulpholane, and liquid carbon dioxide. The primary function of the organic solvent in organosolv cooking is to make the lignin more soluble in the pulping liquor [30].

Organosolv pulping methods can be divided into six categories according to their cooking chemistry as,

I. Methods involving thermal autohydrolysis that use the hydrolysing effect of organic acids cleaved from the wood during cooking
II. Acid catalysed methods using acidic materials to cause hydrolysis
III. Methods using phenols and acid catalysts (This could also be part of the previous category.)
IV. Alkaline organosolv cooking methods
V. Sulphite and sulphide cooking in organic solvents
VI. Cooking using oxidation of lignin in an organic solvent.

Some organosolv processes are Alcell, Organoceil, alkali-sulphite-anthraquinone-methanol (ASAM), alcohol reinforced kraft cooking, Acetosolv, Acetocell, Formacell, Milox, Lignol, Lignofibre (LGF), tetrahydrofurfuryl alcohol (THFA) and processes use neutral alkali-earth metalsalts (NAEM).Organosolv pulping with tetrahydrofurfuryl alcohol (THFA) has been used to pulp rice straw58. Between 0.15 and 0.5% of catalytic hydrochloric acid *(HCl)* was added to 80–95% solutions of THFA. The optimal conditions of THFA/f/C/ cooking were found to be: 95% THFA, 0.50% *HCl*, temperature 120 °C, and cooking time240 min. The pulp yield was higher, but the strength properties are poorer than those of corresponding kraft pulp. The process reportedly offered high potential for using the lignin and hexosans isolated from the waste liquor for biofuels, and for lactic acid and polylactide (PLA), and possibly also for biomaterial production [6].

4.4 Neutral Sulphite Semi-Chemical (NSSC) Process

Semichemical pulping generally uses sulphite chemicals at close to neutral or neutral pH. The most common process is neutral sulphite semi-chemical (NSSC) pulping [19]. Sodium and ammonium are the only possible bases due to the pH requirement. The process is mainly applied to hardwood species or sawdust to produce special pulps for board manufacturing; the most common application is for the corrugating medium. Fibre length and strength are not particularly important quality features in this process, but fibre stiffness is required. The pulping process begins with a chemical phase similar to chemical pulping, using chip pre-steaming to eliminate air from the chips and to enhance penetration of chemicals into them [58]. Homogeneous penetration is particularly important here, since the active chemical doses applied are so low that the chemicals are depleted quickly. Chipping techniques are set to produce short and thin chips to improve impregnation. This is feasible because fibre length is not a major quality factor in these types of pulp. The idea of chemical pretreatment at elevated temperature is to break just enough bonds the wood matrix to enable fibres to be liberated without extensive damage, while applying moderate mechanical energy during defibration [13].

The chemical reactions are intended to achieve limited delignification through the combined effects of sulphonation and hydrolysis (sulphitolysis). High reaction temperatures (160–190 °C) are used to accelerate sulphonation. The residual lignin content is left at a level of 15–20% of pulp. Part of the middle lamella lignin which is left on newly exposed fibre surfaces after defibration will be dispersed into the liquid phase in subsequent pulp refining and washing operations [63]. More hemicellulose than lignin is dissolved by hydrolysis in semichemical pulping. Near-neutral pH is used in cooking to minimise carbohydrate losses, and the cooking liquor has a high buffering capacity (bicarbonate-carbonate) to compensate for pH drops caused by the formation of free acids through decomposition of hemicelluloses. Cellulose remains basically unchanged in semichemical pulping. The softened chips are refined under high pressure and temperature to form a raw pulp, which is screened and washed. Spent liquor is recovered and regenerated into new cooking chemicals either in a separate regeneration system or in cross recovery with a sulphite or kraft pulp mill [33].

4.5 Enzymatic Hydrolysis Process

Additionally, lignin received from enzymatic hydrolysis of the agricultural residues or wood ligno-cellulosic carbohydrates (LCC) by using cellulolytic enzymes leave lignin as the residual substance. In enzymatic hydrolysis process, the LCC is subjected to successive enzymatic treatment for the completion of hydrolysis and dissolution of the carbohydrates of the hemicellulose and cellulose, increasing the insoluble recovered lignin quantity. Generally, recovered lignin samples contain 7–8% carbohydrates, 65–80% lignin, and other impurity. Proteins are also generated during the enzymatic hydrolysis process. The lignin obtained from this treatment is used as an emulsifier, dispersant, and binder [67].

The biodegradability of the lignin, which is a very desirable quality for lignin-based products, is another intriguing property of the substance in addition to its abundant availability and low cost. In nature, a number of bacteria and some fungi contribute to the breakdown of lignin. It is well known that fungi break down lignin more effectively than bacteria, whose fragmentation is slower and less extensive. Lignin degrading microbes are formed by unique extracellular enzyme systems, which split lignin through radical-based oxidative reactions. The microbes produce a variety of combinations of enzymes with multiple isoforms and isoenzymes by reacting to different environmental stimuli like temperature, nutrient availability and oxygen concentration. These enzymes enable efficient decomposition and degradation of lignin into the LCC. However, lignin breakdown by organisms other than fungi has not been extensively researched. But lignin-degrading bacteria are reported. The lignin-degrading bacterium are classified into three main class as α-proteobacteria, γ-proteobacteria and actinomycetes.

5 Role of Lignin in Synthesis of Nanomaterials

In order for valorisation, the lignin has been efficiently converted into nanofibers, nanocapsules, nanotubes, nanoparticles, and other nano-structured materials. LNPs significantly improve the mechanical and thermal features of their corresponding polymer nanocomposites. The presence of reactive functional groups on the LNP's surface, enables surface modification reactions which can contribute to the interfacial adhesion between the matrix material and nano-fillers. Various types of lignin-based nanostructured materials can be obtained using different methods (shown in Fig. 4), such as acid precipitation, solvent exchange, interfacial cross linking, polymerization, and CO_2 antisolvent, ultra-sonication and so on. The resultant LNPs may have a pre defined size, shapes and dimensional distribution.

In materials science, colloidal and nano particles (herein, known as microparticles) are used to achieve desired properties such as thermal stability, material strength, colour and light stability etc. Normally, nanoparticle isolation is less energy demanding than chemical-synthesis routes. This is a direct result of nanoparticles possessing surface energies that are comparatively smaller than typical reaction enthalpies (e.g., reduction, dissolution). Polymeric nanoparticles are playing a key role in a wide spectrum of areas. Lignin superstructures are comparable to other biomolecules, where these can also be transformed into useful nanoparticles. Technical or industrial lignins have been proposed for the development of value-added nanostructured materials that preserve the colloidal composition and, structure features of the macromolecule. They are of high interest for encapsulation, protection and controlled release. Few functions are already bestowed to lignin by nature, to

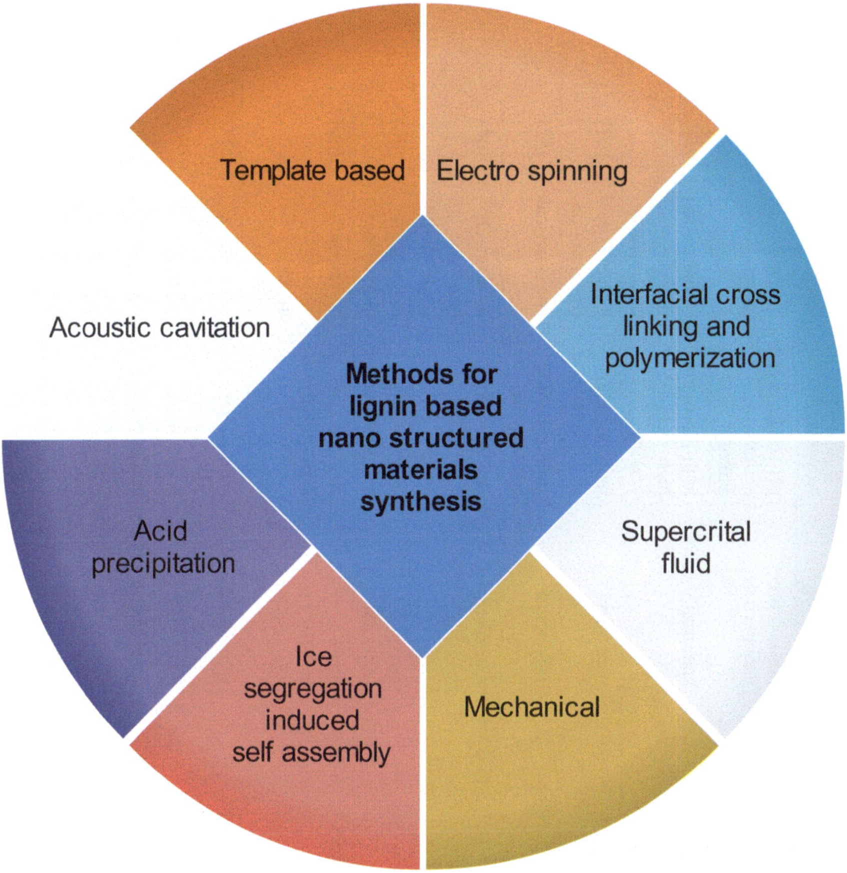

Fig. 4 Methods for synthesis of micro- and nano-particles of lignin

fight against (bio)degradation (microbial, UV, heat), oxidation, and dehydration [2]. The main lignin-derived ingredients, and useful products have been highlighted in Table 2.

6 Application of Lignin Nanomaterials

In general, nanomaterials have shown numerous applications [26–28, 35, 36, 54, 56], further exploration of lignin, including its use and applications, have been ongoing for a long time. The biomedical field (gene therapy, pharmaceutical excipients, controlled drug release, etc.) as well as food packaging are using lignin nanomaterials. They are also being used in microfluidics, UV protection as antioxidants, and food packaging [11]. Many researchers have described the prospect of lignin as a high-value added

Table 2 Preparation of lignin nanostructured materials

Methods	Lignin Source	Solvent	Diameter(nm)	Morphology	Advantage	Disadvantage	Key references
Acid Solution as Anti-solvent	low-sulphonated lignin	ethylene glycol	84 ± 5	porous core nanoparticle	pH-stable, higher antimicrobial activity	irregular morphology	[50]
Aerosol flow reactor	alkali, kraft and organosolvlignins	H_2O, DMF	30–2000	spherical	high-throughput high yield scalable	complicated size fractionation	[1]
rapid freezing, ice sublimation	kraft lignin	H_2O	< 100	nanofibers	reproducible, scalable method	high energy consumption	[61]
electro-spinning	Alcell lignin	ethanol	400–1000	submicron fibres	simple method, stable in air	non-nanoscale	[52]
electro-spinning	kraft lignin	H_2O	61 ± 3, 509 ± 34	nanofibers	simple method	no research on compatibility	[64]
Acid Solution as Anti-solvent	kraft lignin	ethylene glycol	45–250	quasi-spherical	stable within a broad pH range	irregular morphology	[49]
Acid Solution as Anti-solvent	pristine lignin	ethylene glycol	48.85 ± 16.38	quasi-spherical	uniform size distribution	no research on stability in acidic medium	[66]
Acid Solution as Anti-solvent	low-sulphonated lignin	ethylene glycol	50–250	spherical structure	increased thermal stability and crystallinity	no research on stability in medium	[25]
Acid Solution as Anti-solvent	low-sulphonated lignin	ethylene glycol	40–200	aggregate structure	biodegradable biocompatible	stable only at pH (1–9)	[18]

(continued)

Table 2 (continued)

Methods	Lignin Source	Solvent	Diameter(nm)	Morphology	Advantage	Disadvantage	Key references
Acid Solution as Anti-solvent	low-sulphonated lignin	NaOH solution ($pH = 11.44$)	85.9	aggregate structure	biodegradable biocompatible	stable only at $pH < 5$	[18]
Acid Solution as Anti-solvent	Organosolv wheat straw	H_2O/ ethanol mixture	100–463	irregular	high yield, valourization of wheat straw	uncontrollable size,complex process	[5]
Supercritical CO_2 as Anti-solvent	kraft lignin	DMF	38	Quasi-spherical	uniform size, highly monodisperse	toxic organic solvent	[45]
Supercritical CO_2 as Anti-solvent	organosolv lignin	acetone	144 ± 30	spherical	uniform dispersion	toxic solvent complex process	[38]
Ultrasonication and Homogenization	wheat straw lignin, sarkanda grass lignin	H_2O	100	spherical	simple physical method, no organic solvents	no research on stability	[23]
Ultrasonication and Homogenization	dioxane soluble fragment of alkali lignin	DMSO	80–200	solid/ hollow spherical colloids	novel method, structural tunability	toxic organic solvent	[41]
Ultrasonication and Homogenization	kraft lignin	alkali water solution	300–1100	micro/nano capsules spherical	uniform size, no organic solvents	non-nanoscale	[62]

(continued)

Table 2 (continued)

Methods	Lignin Source	Solvent	Diameter(nm)	Morphology	Advantage	Disadvantage	Key references
Ultrasonication and Homogenization	alkali lignin	H_2O	200	nanoparticle dispersion	simple method, no organic solvents	no-isolated nanoparticles	[10]
Ultrasonication and Homogenization	kraft lignin	H_2O	< 100	irregular	simple mechanical treatment	not uniform size,irregular morphology	[46]
Ultrasonication and Homogenization	organosolv lignin residues	ethanol/water	200	colloidal spheres	simple method, uniform size	no research on stability	[48]

product [55]. The industrial or technical lignin has numerous prospective industrial and pharmaceutical applications due to its high adaptability and low toxicity. There are many high-value lignin prospectives, including cost effectiveness, bulk accessibility and the rising demand for bio-based and renewable materials. Lignin can be utilized efficiently for tanning or dye agent dispersal. The active phenolic groups present in lignin, such as amino and organosulphur, help effectively scatter dye compound homogenously in aqueous solutions. The utilization of lignin has been restricted to a small number of low-value uses, as shown in Fig. 5.

LNps are used in many value-added applications, such as controlled drug release, green antibacterial agents and wound-healing for cosmetic and medical products. LNPs could also be used in daily items like textiles and packaging, but also for more technical purposes like anchors, nanoreactors for biocatalysts, adhesives and additives for composite. Due to their surface chemistry and structure, LNPs are effective antibacterial and antioxidative agents. It has more antibacterial action when compared to lignin because of the lower molecular weight and higher phenolic content. The

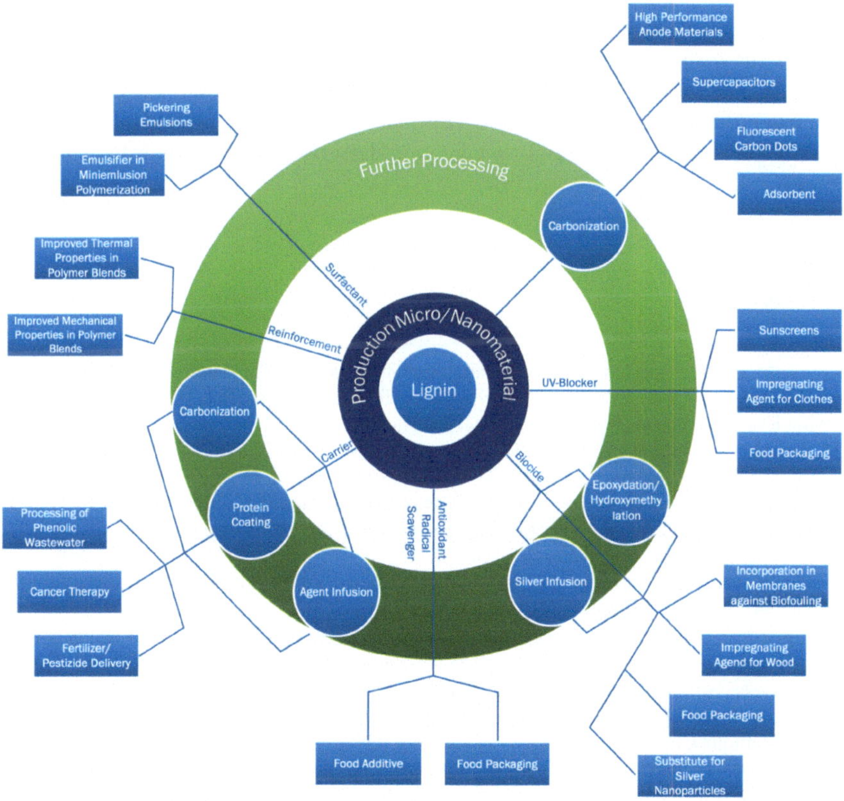

Fig. 5 An overview of lignin micro- and nanoparticles for different applications (adopted from [4]

lignin's polyphenols attack and lyse the cell wall, allowing the cell membrane to bleed inside. The phenomena is supported by the LNP's antioxidant properties.

7 Future Perspective and Conclusions

The main by-product of the pulp and paper industries is lignin, whose abundance has greatly raised interest in lignin-based products. The valourization of materials that were earlier considered production waste by creating new materials based on them has created a potential new economic opportunity in an industry that is becoming more aware of environmental issues. The fabrication of thermoset and polyurethanes has received the majority of recent attention in the integration of lignin-based materials. Even Nevertheless, a growing number of research teams are creating lignin-based nanomaterials each year, despite the fact that the study of these materials continues to be of high importance in the field. A critical study has been done of the production and use of LNPs as a green reinforcer for different nanomaterials. All of the listed preparation techniques are simple ones that enabled the quick synthesis of LNPs while using less ingredients. This encourages a greener environment, which is now highly valued. On the other hand, the applications that have been identified span from fundamentally improved LNPs qualities including antioxidant, antibacterial, UV protectant, biomaterials, medication delivery systems, and water filtration system.

References

1. Ago M, Huan S, Borghei M, Raula J, Kauppinen EI, Rojas OJ (2016) High-throughput synthesis of lignin particles (~30 nm to~ 2μm) via aerosol flow reactor: Size fractionation and utilization in pickering emulsions. ACS Appl Mater Interfaces 8:23302–23310
2. Ago M, Tardy BL, Wang L, Guo J, Khakalo A, Rojas J (2017) Supramolecular assemblies of lignin into nano-and microparticles. MRS Bull 42:371–378
3. Aguayo MG, Fernández Pérez A, Reyes G, Oviedo C, Gacitúa W, Gonzalez R, Uyarte O (2018) Isolation and characterization of cellulose nanocrystals from rejected fibers originated in the kraft pulping process. Polymers 10:1145
4. Beisl S, Friedl A, Miltner A (2017) Lignin from micro-to nanosize: applications. Int J Mol Sci 18:2367
5. Beisl S, Loidolt P, Miltner A, Harasek M, Friedl A (2018) Production of micro-and nanoscale lignin from wheat straw using different precipitation setups. Molecules 23:633
6. Brosse N, Hussin MH, Rahim AA (2017) Organosolv processes. Biorefineries 153–176.
7. Cai W, ChengYK THH, Tai HC, Lo SF (2022) Identification and characterization of wood from antique Chinese guqin zithers. J Cult Herit 53:72–79
8. Camarero S, Ibarra D, Martinez AT, Romero J, Gutiérrez A, José C (2007) Paper pulp delignification using laccase and natural mediators. Enzyme Microb Technol 40:1264–1271
9. Chauhan PS (2020) Lignin nanoparticles: Eco-friendly and versatile tool for new era. Bioresour Technol Rep 9:100374

10. Chen Y, Zheng K, Niu L, Zhang Y, Liu Y, Wang C, Chu F (2019) Highly mechanical properties nanocomposite hydrogels with biorenewable lignin nanoparticles. Int J Biol Macromol 128:414–420
11. Collins MN, Nechifor M, Tanasă F, Zănoagă M, McLoughlin A, Stróżyk MA, Culebras M, Teacă CA (2019) Valorization of lignin in polymer and composite systems for advanced engineering applications–a review. Int J Biol Macromol 131:828–849
12. Costa CAE, Vega-Aguilar SA, Rodrigues AE (2021) Added-value chemicals from lignin oxidation. Molecules 26:4602
13. Do VT, Tanifuji K, Jin G, Ohi H (2020) Neutral sulfite semi-chemical pulping of moso-bamboo. Japan tappi journal 74:379–387
14. Ejaz U, Sohail M (2021) Lignin: a renewable chemical feedstock. Handbook of Smart Materials, Technologies, and Devices: Applications of Industry 4.0. Springer, pp 1–15.
15. Fearon O, Kuitunen S, Ruuttunen KS, Alopaeus V, Vuorinen T (2020) Detailed modeling of kraft pulping chemistry. Delignification. Ind. Eng. Chem. Res 59:12977–12985
16. Ferreira JC, Evtuguin DV, Prates A (2020) Effect of cellulose structure on reactivity of eucalyptus acid sulphite dissolving pulp. Cellulose 27:4763–4772
17. FigueiredoPK L, Hirvonen JT, Kostiainen MA, Santos HA (2018) Properties and chemical modifications of lignin: towards lignin-based nanomaterials for biomedical applications. Prog Mater Sci 93:233–269
18. Frangville C, Rutkevičius M, Richter AP, Velev OD, Stoyanov SD, Paunov VN (2012) Fabrication of environmentally biodegradable lignin nanoparticles. Chem Phys Chem 13:4235–4243
19. Galeano SF, Dillard BM (1972) Process modifications for air pollution control in neutral sulfite semi-chemical mills. JAPCA 22:195–199
20. Gellerstedt G, Henriksson G (2008) Lignins: major sources, structure and properties. Monomers, polymers and composites from renewable resources: 201–224.
21. Gierer J (1980) Chemical aspects of kraft pulping. Wood Sci Technol 14:241–266
22. Gierer J (1985) Chemistry of delignification. Wood Sci Technol 19:289–312
23. Gilca IA, Popa VI, Crestini C (2015) Obtaining lignin nanoparticles by sonication. Ultrason Sonochem 23:369–375
24. Glasser WG (2000) Classification of lignin according to chemical and molecular structure. Historical, Biological, and Materials Perspectives ACS Publications, Lignin
25. GuptaAK MS, Nayak S (2014) Synthesis, characterization and application of lignin nanoparticles (LNPs). Mater Focus 3:444–454
26. Husen A, Iqbal M (2019) Nanomaterials and plant potential: an overview. In: Husen A, Iqbal M (eds) Nanomaterials and plant potential, pp 3–29. Springer International Publishing AG, Gewerbestrasse 11, 6330 Cham. https://doi.org/10.1007/978-3-030-05569-1_1
27. Husen A, Siddiqi KS (2014a) Carbon and fullerene nanomaterials in plant system. J Nanobiotechnology 12(1):1–16
28. Husen A, Siddiqi KS (2014b) Phytosynthesis of nanoparticles: concept, controversy and application. Nanoscale Res Lett 9(229):1–24
29. Jain RK (1992) Fuelwood characteristics of certain hardwood and softwood tree species of India. Bioresour Technol 41:129–133
30. Johansson A, Aaltonen O, Ylinen P (1987) Organosolv pulping—methods and pulp properties. Biomass 13:45–65
31. Karunarathna MS, Smith RC (2020) Valorization of lignin as a sustainable component of structural materials and composites: advances from 2011 to 2019. Sustainability 12:734
32. Kataoka Y, Kondo T (1998) FT-IR microscopic analysis of changing cellulose crystalline structure during wood cell wall formation. Macromolecules 31:760–764
33. Khakifirooz A, Samariha A, Kasmani JE (2011) Neutral sulfite semi-chemical pulping of bagasse. World Appl Sci J 13:85–89
34. Koizumin A, Hirai T (2006) Evaluation of the section modulus for tree-stem cross sections of irregular shape. J Wood Sci 52:213–219
35. Kumar A, Choudhary A, Kaur H, Guha S, Mehta S, Husen A (2022) Potential applications of engineered nanoparticles in plant disease management: a critical update. Chemosphere 295:133798

36. Kumar A, Choudhary A, Kaur H, Mehta S, Husen A (2021) Metal-based nanoparticles, sensors and their multifaceted application in food packaging. J Nanobiotechnol 19(256):1–25. https://doi.org/10.1186/s12951-021-00996-0
37. Lora JH, Wayman M (1978) Delignification of hardwoods by autohydrolysis and extraction. Tappi J 61:47–50
38. Lu Q, Zhu M, Zu Y, Liu W, Yang L, Zhang Y, Zhao X, Zhang X, Zhang X, Li W (2012) Comparative antioxidant activity of nanoscale lignin prepared by a supercritical antisolvent (SAS) process with non-nanoscale lignin. Food Chem 135:63–67
39. Lücking R, Nelsen MP (2018) Ediacarans, protolichens, and lichen-derived Penicillium: a critical reassessment of the evolution of lichenization in fungi. Transform Paleobotany:551–590
40. Mendes IS, Caldeira T, Prates A, Evtuguin D (2022) Study of the anions profile throughout the sulphite pulp line. J Wood Chem Technol 42:244–252
41. Mishra P, Wimmer R (2017) Aerosol assisted self-assembly as a route to synthesize solid and hollow spherical lignin colloids and its utilization in layer by layer deposition. Ultrason Sonochem 35:45–50
42. Mongkhonsiri G, Anantpinijwatna A, Charoensuppanimit P, Arpornwichanop A, Gani R, Assabumrungrat S (2021) Novel biorefinery-Integrated-Kraft-pulping network for sustainable development. Chem Eng Process Process Intensif 163:108373
43. Monsigny L, Feghali E, Berthet JC, Cantat T (2020) Feature extraction of wood-hole defects using empirical mode decomposition of ultrasonic signals. NDT and E Int 114:102282
44. Monsigny L, Feghali E, Berthet JC, Cantat T (2018) Efficient reductive depolymerization of hardwood and softwood lignins with Brookhart's iridium (iii) catalyst and hydrosilanes. Green Chem 20:1981–1986
45. Myint AA, Lee HW, Seo B, Son WS, Yoon J, Yoon TJ, Park HJ, Yu J, Yoon J, Lee YW (2016) One pot synthesis of environmentally friendly lignin nanoparticles with compressed liquid carbon dioxide as an antisolvent. Green Chem 18:2129–2146
46. Nair SS, Sharma S, Pu Y, Sun Q, Pan S, Zhu J, Deng Y, Ragauskas AJ (2014) High shear homogenization of lignin to nanolignin and thermal stability of Nanolignin-Polyvinyl alcohol blends. Chemsuschem 7:3513–3520
47. Nishimura H, Kamiya A, Nagata T, Katahira M, Watanabe T (2018) Direct evidence for α ether linkage between lignin and carbohydrates in wood cell walls. Sci Rep 8:1–11
48. Rao X, Liu Y, Zhang Q, Chen W, Liu Y, Yu H (2017) Assembly of organosolv lignin residues into submicron spheres: the effects of granulating in ethanol/water mixtures and homogenization. ACS Omega 2(6):2858–2865
49. Richter AP, Bharti B, Armstrong HB, Brown JS, Plemmons D, Paunov VN, Stoyanov SD, Velev OD (2016) Synthesis and characterization of biodegradable lignin nanoparticles with tunable surface properties. Langmuir 32:6468–6477
50. Richter AP, Brown JS, Bharti B, Wang A, Gangwal S, Houck K, Cohen Hubal EA, Paunov SD, Stoyanov VN, Velev OD (2015) An environmentally benign antimicrobial nanoparticle based on a silver-infused lignin core. Nature Nanotech 10:817–823
51. Richter F, Atreya A, Kotsovinos P, Rein G (2019) The effect of chemical composition on the charring of wood across scales. Proc Combust Inst 37:4053–4061
52. Ruiz-Rosas R, Bedia J, Lallave M, Loscertales I, Barrero A, Rodríguez-Mirasol J, Cordero T (2010) The production of submicron diameter carbon fibers by the electrospinning of lignin. Carbon 48:696–705
53. Sewsynker-Sukai Y, David KEG (2020) Recent developments in the application of kraft pulping alkaline chemicals for lignocellulosic pretreatment: potential beneficiation of green liquor dregs waste. Bioresour technol 306:123225
54. Sharma G, Nim S, Alle M, Husen A, Kim JC. 2022. Nanoparticle-Mediated delivery of flavonoids for cancer therapy: prevention and treatment. In: Kim JC, Alle M, Husen A (eds) Smart nanomaterials in biomedical applications. Nanotechnology in the Life Sciences. Springer, Cham. https://doi.org/10.1007/978-3-030-84262-8_3
55. Sharma S, Sharma A, Mulla SI, Pant D, Sharma T, Kumar A, Sharma S, Sharma A, Mulla SI, Pant D, Sharma T, Kumar A (2020) Lignin as potent industrial biopolymer: an introduction. In:

Sharma S, Kumar A (eds) Lignin: biosynthesis and transformation for industrial applications, pp 1–15. Springer International Publishing, Cham
56. Siddiqi KS, Husen A (2021) Plant Response to Silver Nanoparticles: A Critical Review. Crit Rev Biotechnol 42(7):773–990. https://doi.org/10.1080/07388551.2021.1975091
57. Simonović J, Stevanic J, Djikanović D, Salmén L, Radotić K (2011) Anisotropy of cell wall polymers in branches of hardwood and softwood: a polarized FTIR study. Cellulose 18(6):1433–1440
58. Sitter T, Oveissi F, Fatehi P (2014) A process for producing lignocellulosic flocs from NSSC spent liquor. J Biotechnol 173:19–23
59. Solihat NN, Sari FP, Falah F, Ismayati M, Lubis MAR, Fatriasari W, Santoso EB, Syafii W (2021) Lignin as an active biomaterial: a review. Jurnal Sylva Lestari 9:1–22
60. Sparlinek L, Leitner V, Kamm B (2018) Statistical optimisation of enzymatic detoxification with laccase from Trametes versicolor for spent sulphite liquors using a novel in-situ NMR method. J biotechnol 284:63–67
61. Spender J, Demers AL, Xie X, Cline AE, Earle MA, Ellis LD, Neivandt DJ (2012) Method for production of polymer and carbon nanofibers from water-soluble polymers. Nano Lett 12:3857–3860
62. Tortora M, Cavalieri F, Mosesso P, Ciaffardini F, Melone F, Crestini C (2014) Ultrasound driven assembly of lignin into microcapsules for storage and delivery of hydrophobic molecules. Biomacromol 15:1634–1643
63. Vidaurre GB, Pereira M, Boschetti WTN, Patt R, Colodette JL, Vital BR, de Almeida MNF (2018) NSSC pulping of fast growing trees. NPPRJ 33:404–413
64. Wurm FR, Weiss CK (2014) Nanoparticles from renewable polymers. Front Chem 2:49
65. Xu C, Ferdosian F (2017) Structure and properties of lignin. Conversion of lignin into bio-based chemicals and materials, pp 1–12. Springer
66. Yang W, Kenny JM, Puglia D (2015) Structure and properties of biodegradable wheat gluten bionanocomposites containing lignin nanoparticles. Ind Crops Prod 74:348–356
67. Yu Z, Jameel H, Chang HM, Park S (2011) The effect of delignification of forest biomass on enzymatic hydrolysis. Biores Technol 102:9083–9089

Nanomaterials Synthesis Using Saponins and Their Applications

Apekshakumari Patel, Nimisha Patel, Ahmad Ali, and Hina Alim

Abstract Nanoscience is an exponentially developing field where multidisciplinary principles intersect in formulating notable innovations and discoveries. Synthesizes of such material is achievable through using physical, chemical and biological approach. Here, biosynthesis of the nanomaterial is foster to be the most promising sector in industries, due to its eco-friendly and cost-effective approach. Nanoparticles are the most studied nanomaterial for their extremely small scale, yet diverse applications. Biosynthesis approach involves a one step process including two essential reactions, reduction and capping of the nanoparticle through a bioactive compound extracted from a plant, fungi, bacteria, or algae. Bioactive compounds are viewed with immense potential in various fields of science and technology, applications ranging from pharmacy to industrial usage. One diverse group of compounds that come from such category of compounds is saponins, the most widely distributed compound in the plant kingdom. They are structurally triterpene or steroids with one or more side chains of sugar, fabricating them to be surface active. Conventionally, saponins are of substantial use in food, agriculture and medicine as a consequence of their pharmacological properties including, anti-inflammatory, anticancer, antiviral, antimicrobial and immuno-modulatory effects. This chapter will highlight upon the nanomaterials, especially metal nanoparticles synthesis using saponins, and further applications of achieved nano-products in various sectors.

Keywords Nanomaterial synthesis · Nanoparticles · Nanomaterial application · Active surface

A. Patel · A. Ali · H. Alim (✉)
Department of Life Sciences, University of Mumbai, Vidyanagari Campus, Santacruz (East), Mumbai, Maharashtra 400 098, India
e-mail: hina.alim@lifesciences.mu.ac.in

N. Patel
Department of Life Sciences, J.C. Bose University of Science & Technology, YMCA, Faridabad, Haryana 121006, India

A. Husen (ed.), *Secondary Metabolites Based Green Synthesis of Nanomaterials and Their Applications*, Smart Nanomaterials Technology,
https://doi.org/10.1007/978-981-99-0927-8_7

1 Introduction

Nanomaterial is the most promising material in the field of material science due to the vast range of applications it possesses in various sectors [28–32, 38, 39, 61, 62]. Nanomaterials are defined as any material consisting of one or more of their dimension in the nanoscale, which is in 10^{-9} m, equivalent to the linear arrangement of 10 hydrogen atoms [7]. As to the large surface to area ratio of the nanomaterial, the fabrication of the desired nanomaterial with the required feature can be achieved. There are found ranges of nanomaterials such as sols, colloids, nanocrystals, nanoparticles, nanowires, nanoporous, nanorods, nanofibers, nanoceramics and many more [72]. All of them are classified on the basis of their dimension (Fig. 1). Materials possessing all their dimensions in the nanoscale i.e. all dimensions less than 100 nm, are included in the 0D (zero dimensional) nanomaterials, examples include nanoparticles. The 1D (one dimensional) nanomaterials have two dimensions in nanoscale, example includes nanowire, nanorods and nanofibers. The 2D (two dimensional) nanomaterials have one dimension in nanoscale, examples include nanofilms, nanoplate and nanocoatings. While, the 3D (three dimensional) nanomaterials have no dimension in nanoscale, examples include equiaxed crystallites [2, 46, 72].

Nanomaterials that we currently visualized as the novel field in Science, are so ancient whose exact antiquity is still not derived. However, certain evidence proves the usage of nanomaterial by humans in the ancient times in ranging ways unknowingly, such as the use of clay minerals which can be comparable to natural nanomaterials [11, 18]. In the fourth century A.D., the Romans exhibited the most captivating example of ancient nanoscience with the Lycurgus cup. This cup was made up of dichroic glass, which means in direct light (reflective light condition) it appears pickle green, while when the cup was illuminated internally (transmission light condition)

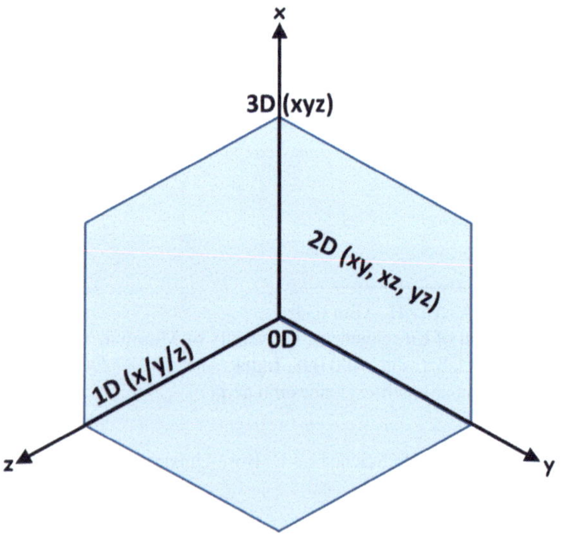

Fig. 1 Dimensions of nanomaterials represented on the Cartesian coordinate system

it appears ruby red in color [25, 41]. Lead sulfate particles (PbS) nanoparticles were known to the ancient Egyptians over 4000 years ago, and these nanoparticles were utilized in an ancient hair-dyeing method [73]. Asbestos nanofibers were used by ancient people to strengthen ceramic mixes some 4500 years ago [18].

Richard Feynman was the Nobel Prize laureate who in 1959 introduced the very first concept of nanotechnology. He quoted the famous line "There's plenty of room at the bottom" in his annual meeting of the American Physical Society [56]. Following the discovery many scientists have developed two broad approaches for the synthesis of nanomaterials. The two main approaches are top-down and bottom-up, both differing in the scale, cost and methodology (Fig. 2) [11].

The top-down approach employs the principle of bulk material breaking down into nanoscale particles. The methods used are the mechanical process, thermal process or optical process. Certain techniques that are recently used in the industrial production of the nanomaterial from the top-down approach include etching, sputtering and lithography. Etching is the chemical removal of surface layers during the manufacturing process. In etching there are two types of etching techniques as wet etching and dry etching [20]. Sputtering method includes using energetic ions (gas or gaseous

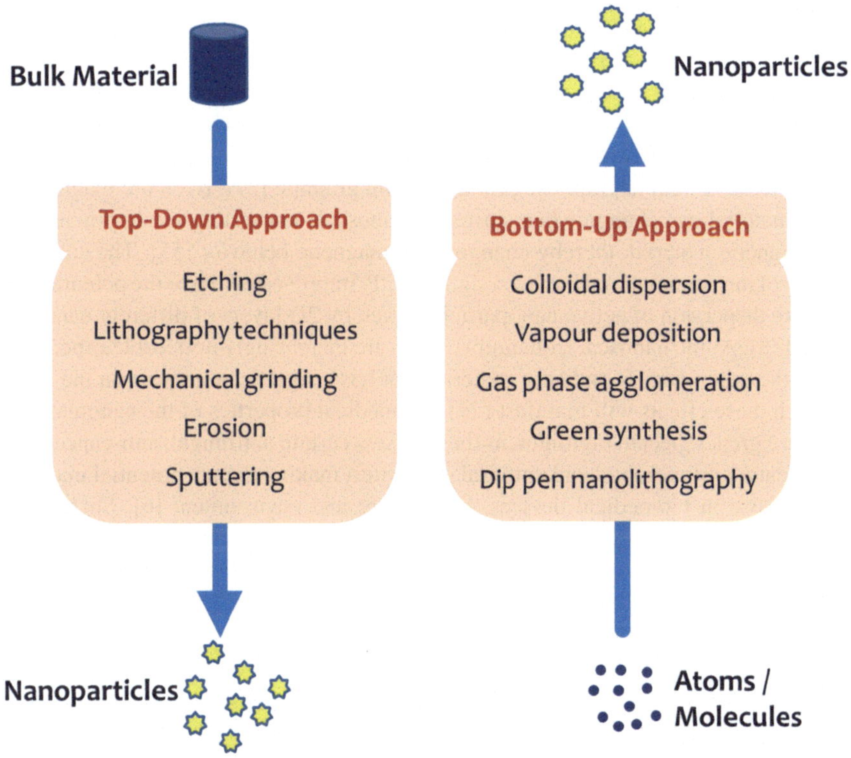

Fig. 2 Approaches for nanoparticle synthesis: Top-Down and Bottom-Up

plasma) bombardment to eject microscopic particles from the surface of the desired material [4, 42]. Lithography is the process of patterning a surface by exposing it to light, electrons or ions, and then depositing the material on that surface to make the desired substance [34].

The bottom-up approach employs the principle of building up atom-by-atom or molecule-by-molecule to achieve a desirable nanoscale structure. Physical and chemical methods involve using highly controlled atom and molecule self-assembly. In a bottom-up process called self-assembly, atoms or molecules arrange themselves into structured nanostructures through chemical-physical interactions. The only method that allows individual atoms, molecules, or clusters to be placed freely one at a time is positional assembly [34]. Green synthesis of nanoparticles is also a bottom-up approach involving two essential reactions—reduction reaction and capping or stabilization reaction. Both these reactions are achieved in a one-step method, which is the mixing of metal salts with the biological extract (plant, fungi, bacteria, and algae) [67].

The following important features can be obtained by adjusting the sizes and morphologies of the nanomaterials, among a variety of other special properties, surface area, mechanical porperties, high thermal and electrical conductivity, magnetism, catalytic property, quantum effects and biomedical properties. In comparison to their bulk counterparts, the surface areas of nanoparticles are often noticeably higher, and this characteristic is shared by all nanomaterials [12]. The mechanical properties exhibited by the nanomaterials are outperformed by their analogues at macroscale [74]. Exceptional thermal and electrical conductivity can be seen at the nanoscale level in comparison to bulk analogues, depending on the nature of the nanomaterial, example graphene produced from graphite [37]. Even the magnetism of the material can change when shifted to nanoscale converting the non-magnetic into magnetic material, thereby changing the magnetic behavior [55]. The catalytic activity of the nanomaterial has been significantly improved owing to the potential for effective dispersion of active nanoparticles given by 2D layers of different nanomaterials [78]. At the nanoscale, quantum effects are increasingly noticeable,especially due to the type of semiconductor material that has a significant impact on the scale at which these effects will manifest [26]. Biomedical properties of the nanomaterial are also a great aspect to highlight, as they possess certain antifungal, anti-cancerous, antioxidant, antimicrobial and antiviral properties, making them a potential material for their use in biomedical devices, health care and environment [6]. Such engineered nanomaterials possess great advantages in attaining sustainable products. Presently, the nanomaterials found in commercial include electronics, scratch free paint cosmetics, remediation, sensors, nanodevices for energy-storage, therapeutics, diagnostic, and environment [60].

The major three procedures to obtain a nanomaterial are physical, chemical and biological approaches. In terms of physical approach, evaporation–condensation and laser ablation emerge as the most important strategies. The large amount of energy required to sustain high temperatures and pressures might raise the manufacturing costs. A broadly applied principle for the nanoparticles formation in chemical and green synthesis is the reduction of metal or metal-oxide into nanoparticles using inorganic and organic reducing agents. Furthermore, immediately capping nanoparticles after creation is required to prevent aggregation of the nanoparticles. Some of the compounds employed in chemical synthesis, however, are non-biodegradable, while others may create harmful by-products [67]. Among them, green synthesis proves to be the most cost-effective, environment friendly, and easily scalable to large-scale nanoparticle manufacturing. The main benefit is that it is a one-step procedure that reduces and caps metal and metal-oxides, resulting in the most stable nanoparticles [59].

In the green synthesis process the essential component is the bioactive compound in the extract, used for the purpose of reduction and capping of nanoparticles. A bioactive compound is a compound that is produced by plants as a secondary metabolite, possessing a defense mechanism against animals upon its ingestion above specific dosage. Beside from the compound's toxicological effects, it also has pharmacological effects on humans and animals [14]. The categorization of bioactive substances is still uneven, with most classifications based on comparable features. The three major classes of bioactives are alkaloids, terpenes and phenolic chemicals [5, 10]. Among them saponins are a broad collection of glycosides discovered in a variety of agriculturally essential plants, particularly legumes. Legumes are mostly known as a staple diet for humans. *Phaseolus vulgaris* beans, soya beans (*Glycine max*), and chickpeas (*Cicer arietinum*) are particularly high in saponins [57]. According to the carbon structure of the aglycones, saponins grouped as steroidal, steroidal alkaloids, and triterpenoidal [68].

2 Saponins and Their Properties

Saponins are found naturally as an amphiphilic chemical in a diverse plant-based meal [22]. Saponins are made up of an aglycone (sapogenin) unit connected to one or more glycon (carbohydrate) chains (Fig. 3). Classification of saponins is shown in Fig. 4. The aglycone unit is made up of a sterol or, more commonly, a triterpene unit [48]. The carbohydrate side chain is frequently linked to sapogenin 3rd carbon, in steroid and triterpenoid saponins [57]. The conjunction of the water-soluble side chain and the nonpolar sapogenin causes a saponin to froth. Saponins are bitter substances that impair the palatability of cattle diets. The major mechanism involving the head-to-tail interaction of acetate units leads to both forms of sapogenins. Furthermore, after the formation of the triterpenoid hydrocarbon squalene, a branch emerges, diverges steroids in the pathway while cyclic triterpenoids another [21].

The carbohydrate portion of the molecule exhibits the water-soluble property and sapogenin exhibits the fat-soluble property, therefore, saponins have an active surface or properties of the detergent, pH affects the stability and strength of feed saponin foams, which may contribute to the development of bloat in ruminants. Saponins are extremely resistant to heat processing, and conventional cooking has no effect on their biological activity. Saponins generate a soapy lather when stirred in water.

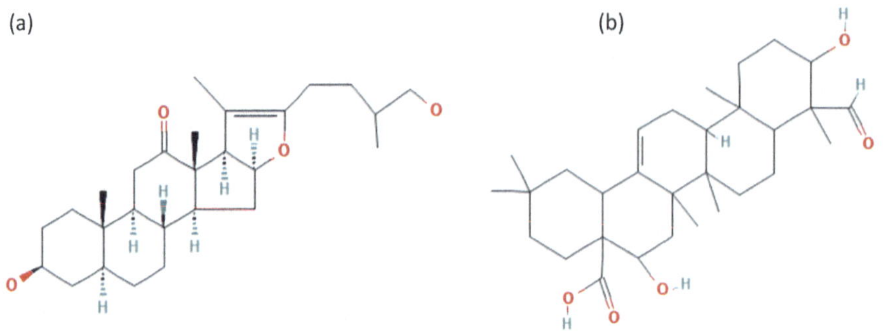

Fig. 3 Structural representation of sapogenin **a** steroid **b** triterpenoid (from *Quillaja saponaria*) (Adopted from: Pubchem)

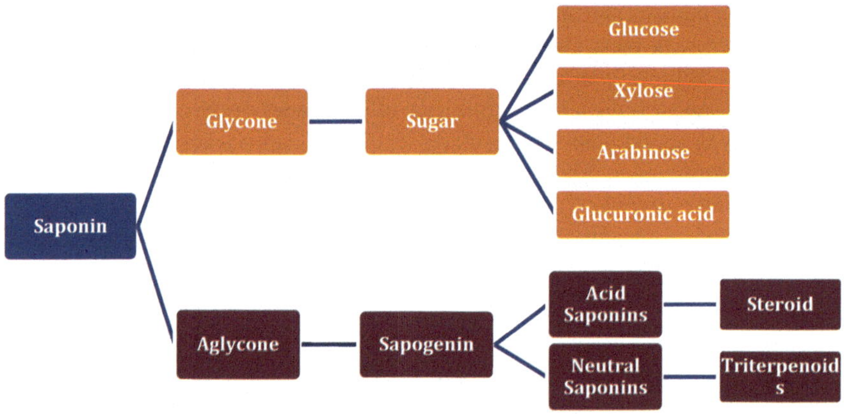

Fig. 4 Flowchart of Saponins classification

Saponins are isolated from the plant material by extraction with a polar solvent following lipid removal, such as chloroform or petroleum ether, followed by several purifying procedures. Individual saponins have been separated using a variety of chromatographic methods [57].

Saponins have been used in the pharmaceutical field as to their anti-cancerous, hypocholesterolemic and immune-stimulatory properties [23]. Other characteristics of this diverse collection of chemicals include bitter taste, hemolytic effects, and cholesterol-binding capabilities. These features are unique to saponins and are not generally exhibited by all members of the group. Some of these qualities are regarded useful from a biological standpoint, while others are deemed detrimental. The discovery that dietary saponins reduced plasma cholesterol levels in primates has the potential to minimize the incidence of coronary heart disease in people [57]. Even though saponins are nearly non-toxic to humans when consumed orally, when injected into the circulation, these are potent hemolytic agents and potential lipase activity suppressers [23, 76, 77]. Additionally, saponins are insoluble in nature, which limit their use in health care [54].

Although saponins have several impacts on microorganisms and animals, a few studies have focused on their action in plant cells. Certain saponins are antibacterial, inhibit the growth of mould, and defend the plants from insect assault. Saponins play a role in defense mechanisms of plant and have therefore been included in a wide category of defensive chemicals exist in plants as 'phyto-protectants.' Saponins compounds that are enzymes activated, under the response of pathogen infection or tissue injury, are A and B avenacosides present in *Avena sativa* (oat) [21].

3 Biological Effect of Saponins

Saponin composition in plants can vary greatly depending on age, tissue type, genetic background, physiological condition, and environmental effects. Saponins are associated with a broad array of properties, that are beneficial in some aspect and some can produce detrimental effects on human health, insecticidal, molluscicidal activity, pesticidal, antinutritional effects, allelopathic action, and as phyto-protectants that protect plants from microbes and herbivore [21]. The wide range of saponin's biological effects are discussed further below:

(a) Cell membrane permeability

Saponins possess the potential to bind to the cell membrane, and affect the permeability of the membrane by forming pores. Especially, erythrocytes are most susceptible to saponin's lytic effect. Hemolytic activity is also one property of saponin, mainly due to the binding affinity of the aglycone portion to the membrane phospholipids [9].

(b) Cholesterol metabolism

Saponins from diverse sources have been shown to lower blood cholesterol levels in a large population of animals along with humans. Large mixed micelles that are formed as a result of saponin-bile acid interaction, justify for their high level in excretion when saponin-rich foods such as ingestion soya bean, chickpea and Lucerne are taken. Blood cholesterol levels fall as a result of the faster cholesterol metabolism in the liver [49].

(c) Protein degradation

Saponins are considered to slow protein digestion by generating seldom digested saponin-protein complexes. Glycinin, an endogenous saponin, impacted the chymotrypsin hydrolysis of soya bean protein. The amalgamation of soya saponin with bovine serum albumin (BSA) improved the thermal stability of the BSA due to electrostatic and hydrophobic interactions. However, the combination of BSA and soya saponin is substantially less digestible than free BSA, depicting that the complexing with saponin has an obstructive effect [52, 64].

(d) Animal reproduction

Saponins have long been recognized to have a deleterious impact on animal reproduction, which has been attributed to their antizygotic, abortifacient and anti-implantation qualities. Saponins have been discovered to be exceptionally potent stimulators of luteinizing hormone production from in-vitro culturing of hypophysial cells [13]. However, uterine development was increased, luteinizing hormone production was decreased, and the estrous cycle was stopped when saponin-rich extracts obtained from *Combretodendron africanum* were administered into female rats.

Steroid saponins were discovered to inhibit the genes directly involved in steroidogenesis, as well as to reduce the proliferation of granulosa cells controlled by follicle-stimulating hormone in the ovarian follicle cell growth suppression which may have a mechanism similar to saponin-induced tumour cell proliferation. Saponins have been shown invitro to have both positive and negative impacts on human sperm cell survival, with particular ginseng saponins improving motility and sperm progress [16].

(e) Cold blooded animals

Saponins have lethal effect on cold-blooded animals, affecting their respiratory epithelia. Therefore, they are used as fish poison due to their active toxic effect. Quillaja saponins can potentially damage the rainbow trout's intestinal mucosa [24].

(f) Hypoglycemic activity

Fenugreek saponins are responsible for hypoglycemic activity, either by activating cells or by suppressing glucose transfer from the stomach to the small intestine and blocking glucose transport over the brush border of the small intestine [51].

(g) Hypolipidaemic activity

Saponins and high fibre content in certain plant extracts are implicated in the hypolipidaemic action. The fibre binds to cholesterol strongly, assisting in its excretion. Saponins have also demonstrated a significant level of hypolipidaemic activity. The combined action of saponins and the fibre content of the plant extract results in a decreased plasma concentration of cholesterol and lipids. As a result, the possibility of coronary heart disease, such as atherosclerosis, is reduced [53].

(h) Sarcoplasmic reticulum membrane and transverse-tubular system

Saponin enhances Ca^{+2} release from the mammalian cardiac sarcoplasmic reticulum, as well as crustacean and mammalian skeletal muscle, at low doses. The enhanced Ca^{+2} losses from mammalian skeletal sarcoplasmic reticulum following saponin treatment are concentration independent and act selectively through the ryanodine receptor [15].

(i) Anti-inflammatory effect

The considerable anti-inflammatory effect of saponins may be related to the suppression of inflammatory mediators including serotonin, prostaglandin and histamine, as well as its antioxidant capacity that limits the generation of Reactive oxygen species (ROS), essential compounds that are responsible for inflammation [58, 71].

(j) Antimicrobial activity

Saponins show antibacterial properties as they inhibit the growth of gram positive and only few inhibit the gram negative microorganisms, as to the penetrability of the cell membrane [35, 66].

(k) Antifungal activity

Antifungal property is performed by steroidal saponins due to their aglycone portion and owing to the amount and structure of monosaccharide units [75].

(l) Virucidal activity

Purified saponin combination from *Maesa lanceolata* is capable of destroying viruses. The triterpenoid sapogenin oleanolic acids inhibit HIV-1 viral replication, most likely through decreasing HIV-1 protease activity [43, 65].

Other biological activities of saponin used for biomedical purposes are, they can also potentially decreasing cancer rates by reducing blood cholesterol levels. A high saponin diet can be used to prevent dental cavities and platelet aggregation, as well as to treat hypercalciuria and as an antidote to acute lead poisoning. Saponins have been shown in an epidemiological study to have an inverse relationship with the incidence of renal stones [16]. They are also vital for their other activities such as, anticancer, antiulcerogenic, anti-malarial, anti-nociceptive, antioxidant, anthelmintic, eczema analgesic, immuno-modulatory, hepatoprotective and molluscidal [63, 70].

4 Characterization of Nanomaterials

The available techniques used frequently for the characterization of nanomaterials includes atomic force microscopy (AFM), dynamic light scattering (DLS), energy-dispersive X-ray spectroscopy (EDS), Fourier transform infrared spectroscopy (FTIR), high-resolution transmission electron microscopy (HRTEM), matrix-assisted laser desorption/ionization time-of-flight mass spectrometry (MALDI-TOF), nuclear magnetic resonance spectroscopy (NMR), scanning electron microscopy (SEM), surface-enhanced Raman spectroscopy (SERS), transmission electron microscopy (TEM), thermogravimetric analysis (TGA), X-ray photo-electron spectroscopy (XPS), X-ray diffraction (XRD), UV–visible spectroscopy (UV–vis), and zeta potential.

In some circumstances, these techniques are used exclusively to research a certain characteristic, whereas in others, they are combined. All of these approaches are evaluated in terms of accuracy, availability, affordability, selectivity, simplicity, non-destructive nature, and affinity to certain compounds. Despite their large number, the strategies are thoroughly examined. There are microscopy-based techniques that offer information on the size, shape, and crystal structure of nanomaterials (e.g., TEM,

HRTEM, and AFM). Other approaches, such as magnetic procedures, are tailored for certain categories of materials. SQUID (superconducting quantum interference device), VSM (vibrating sample magnetometer), FMR (ferromagnetic resonance), and XMCD (X-ray magnetic circular dichroism) are some of these approaches. Many different approaches are available to offer further information on the structure, elemental content, optical characteristics, and other common and more specific physical aspects of nanoparticle samples [44].

Other techniques include 3D-tomography, Brunauer–Emmett–Teller (BET), differential scanning calorimetry (DCS), electron backscatter diffraction (EBSD), electron energy loss spectroscopy (EELS), electric potential measurement (EPM), inductively coupled plasma mass spectrometry (ICP-MS), inductively coupled plasma optical emission spectroscopy (ICP-OES), low energy ion scattering (LEIS), magnetic force microscopy (MFM), nanoparticle tracking analysis (NTA), resonant mass measurement-micro-electromechanical systems (RMM-MEMS), small angle X-ray (SAX), tunable resistive pulse sensing (TRPS). While, X-ray absorption spectroscopy (XAS) is an extended technique of X-ray absorption fine structure (EXAFS) and X-ray absorption near edge structure (XANES) [44].

5 Synthesis of Nanomaterials and Their Applications

The manufactured nanomaterials are classified as carbon nanoparticles, metal nanoparticles, ceramic nanoparticles, lipid-based nanoparticles, semi-conductor nanoparticles, and polymeric nanoparticles based on their size, shape and chemical characteristics [33]. Saponins extracted from plants have been scrutinized for their application in synthesis of such nanomaterials.

Saponins extracted from plants *Phyllanthus urinaria* and *Ocimum tenuiflorum* leaf extracts were used to synthesize silver nanoparticles. Proceeding after the extraction process, Wagner's test was used to identify alkaloids, foam test for the presence of saponins, whereas, $FeCl_3$ test for the presence of tannins, phenolics and Lieberman-Burchard test is used for the identification of steroids and triterpenes. The 1 mM silver nitrate solution was mixed with tenfold higher volume of the extract. The mixture was stirred at room temperature for 8 h and then moved into a centrifuge to collect the prepared nanoparticles. After the characterization of the nanoparticles, the size distribution was calculated to be 5–61 nm, showing the ability of antifungal capability against *Fusarium oxysporum, Aspergillus niger*, and *Aspergillus flavus.* Even after having the similar phytochemical composition, it was also elucidated that extract from *Ocimum tenuiflorum* contained rich saponins, while *Phyllanthus urinaria* contained poor saponins, owing to the size of the nanoparticles formed, rich saponins help form smaller nanoparticles [47].

Silver nanoparticle synthesis was performed using *Simarouba glauca* saponins. The seeds of the plant were used to extract the saponins using [3] method, which is performed at 60°C drop-wise addition of the extract to 50 ml of 0.003 M silver nitrate solution. The coloration changes from yellow to brown depicting the production of silver nanoparticles. These newly synthesized silver nanoparticles were tested for their catalytic activity, showing the potential dye reducing capacity of dyes such as 2-chloro 4-nitro phenol, congo red, methylene blue, and methyl orange, in the presence of $NaBH_4$ (Sodium borohydride) [50].

The study involved silver nanoparticles generated from the leaf extract of *Chenopodium album L.* The sterol-based extract was obtained by a two-phase separation process involving ethyl acetate and water. Butanol and diethyl ether were used to separate the aqueous phase containing the abundant saponins. The resulting butanolic phase was rich in sterol-based compounds. The mixture of 5 ml extract with 95 ml of 1 mM silver nitrate solution was prepared and incubated at room temperature in a rotary shaker at 150 rpm in dark, and conformation of synthesis was observed by the colour change in the solution. Such silver nanoparticles synthesized using the sterol compound showed anti-acne activity and can be a potential component of the anti-acne cream [68].

Saponins derived from fenugreek were examined for their use in optimizing the synthesis of silver nanoparticles. The saponin extract from fenugreek was mixed with 1 mM silver nitrate solution, and was incubated at 37°C on a magnetic stirrer at 400 rpm for 24 h. Silver nanoparticles synthesized by this method have shown to have antibacterial properties against gram negative bacteria (*Proteus mirabilis, Pseudomonas aeruginosa, Klebsiella pneumonia, Escherichia coli, Enterobacter kobei, Ecloacae, Easburiae* and *Acinetobacter baumannii*) and gram positive bacteria (*Enterococcus faecalis* and *Staphylococcus aureus*), which are amongst the multi-drug resistant bacteria causing burn wound infection [45].

Silver and gold nanoparticles were synthesized using Platycodin D, a triterpenoid saponin extracted from *Platycodon grandiflorum* [17]. The Platycodon D was extracted from the roots of *Platycodon grandiflorum* using enzymatic transformation of the platycosides [27]. The synthesis process of gold nanoparticle is using 0.05% extract and mix it with 0.2 mM of $HAuCl_4{\cdot}3H_2O$, and a 5 min of incubation at room temperature. For the silver nanoparticles 0.01% of extract was mixed with 0.8 mM $AgNO_3$, and incubated at 80°C for 3 h with further incubation at an ambient temperature for 21 h. The size of the nanoparticles formed was elucidated using Atomic force microscopy (AFM) and HR-TEM images to be 14.94 nm for gold nanoparticles and 18.40 nm for silver nanoparticles. Gold nanoparticles synthesized by such methods showed catalytic properties to reduce 4-nitrophenol to 4-aminophenol [17].

Chitosan-Saponin nanoparticles are synthesized as the ability of such nanoparticles is to protect DNA from enzymatic degradation and providing thermostability. To prepare the chitosan-saponin nanoparticle, 0.03% of chitosan was mixed with 0.001% saponin, and stirred at 500 rpm for 5 min. After the synthesis of the nanoparticle, the pH was adjusted to 5.5 and syringe filtered, and used for encapsulating the DNA plasmid that is to be delivered [8].

Nanoemulsions are colloidal emulsions at nanoscale size ranging from 10 to 1,000 nm. Ideally, they are spherical solids with negatively charged amorphous and lipophilic surface. Components of nanoemulsion are mainly oil, aqueous phases and emulsifying agents. Castor oil, coconut oil, evening primrose oil, linseed oil, maize oil, mineral oil, olive oil, peanut oil, and other oils can be used. A crude temporary emulsion made from oil and water will split into two different phases after standing due to the coalescence of the scattered globules. Emulgents or emulsifying agents that can help these systems stay stable. Due to such properties of nanoemulsions, they are highly desirable for the drug delivery system [36]. Nanoemulsions synthesized using quillaja saponins as emulsifier assist the fabrication of nanoemulsions. Nanoemulsions prepared through high-pressure microfluidization using quillaja saponin are resistant to capsaicin degradation and stable to environmental stresses [1]. The preparation of nanoemulsions by using a high intensity ultrasonication method through the quillaja saponins extracted from cannabis, was optimized for carrier oil, surfactant, and for the hydrophilic-lipophilic balance of the surfactant solution [40]. (Table 1)

Nanowires were synthesized from the fruit extract of *Sapindus mukorossi*, and the extract is rich in saponin and is used to synthesize CuO nanowires. The 0.1 M $CuSO_4 \cdot 6H_2O$ was mixed with 2 ml of the extract and stirred for 10 min, then 0.1 M NaOH solution was mixed slowly till the pH reached 11 and stirred for 2 h. A bluish green precipitate formed in the solution is copper (II) hydroxide, which is then centrifuged and washed with de-ionized water, decanted and dried at 80°C for 24 h, and lastly placed in preheated furnace for 1 h at 300°C. The CuO nanowire synthesized

Table 1 Determination of characteristic features and appropriate characterization techniques

Characteristics features	Appropriate characterization techniques
3D aspects	3D-tomography, AFM, SEM
Agglomeration state	Cryo-TEM, DCS, DLS, SEM, TEM, UV–Vis, Zeta potential
Chemical state	EELS, XAS, XPS
Concentration	ICP-MS, DCS, PTA, RMM-MEMS, TRPS, UV–Vis
Crystal structure	EXAFS, HRTEM, STEM, XRD
Density	DCS, RMM-MEMS
Detection of nanoparticles	EBSD, SEM, STEM, TEM
Dispersion of nanoparticle	AFM, SEM, TEM
Chemical composition	ICP-MS, ICP-OES, MFM, NMR, SEM–EDX, XPS, XRD
Growth kinetics	Cryo-TEM, liquid-TEM, NMR, SAXS, TEM
Mass composition	FMR, FTIR, NMR, XPS
Magnetic properties	FMR, Magnetic susceptibility, MFM, SQUID, XMCD,
Optical properties	EELS-STEM, UV–Vis-NIR
Shape	3D-tomography, AFM, EPLS, FMR, HRTEM, TEM
Single particle properties	HRTEM, TEM, MFM, ICP-MS
Size distribution	DCS, DLS, DTA, FMR, ICP-MS, NTA, SAXS, SEM, TRPS
Structural aspects	AFM, DCS, DLS, EPLS, EXAFS, FMR, HRTEM, ICP-MS, MALDI, NMR, NTA, SEM, TEM, UV–Vis, XRD
Structural defects	EBSD, HRTEM
Surface area	BET, Liquid NMR
Surface charge	Zeta potential

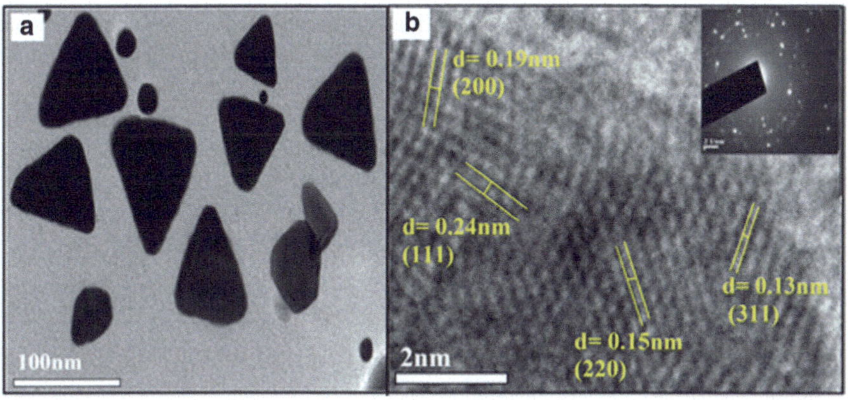

Fig. 5 Triangular silver nanocrystals shown in **a** TEM image and **b** lattice spacing with SAED pattern shown in HRTEM micrograph (adopted from [19]

showed to be highly sensitive and selective for the detection of dopamine, therefore, it can be used as a potential eco-friendly biosensor for the detection of dopamine and other biological molecules [69].

The synthesis of saponin capped triangular silver nanocrystals using the extract of fenugreek seed acts as an ecofriendly method for synthesis of nanocrystals. The nanocrystals were synthesized by mixing 2 mM silver nitrate solution with the extract at 100°C in a magnetic stirrer at 400 rpm for 1 h, later forming dark brown coloration in solution. The TEM and HR-TEM imaging of the synthesized nanoparticles are shown in Fig. 5 [19].

6 Conclusion

Saponins mediated synthesis of nanomaterials is a green synthesis approach, which is a safe, cost-effective, eco-friendly and highly stable method. Nanomaterials produced by such methods can be utilized in various fields ranging from biomedicine, bioremediation, biosensors, cosmetic industries, electronics, and pharmaceutical. The quantity of the nanoparticles formed varies with the ion reduction potential, and the plant's reducing capacity is determined by the presence of metabolites in the plant extract. The form, size, and stability of the nanoparticles are substantially determined by the concentrations of the plant extract and substrate, temperature, pH of the reaction mixture, and the exposure period. Many aspects remain unclear or partially understood, including the requisite shape, size and stability of the nanomaterials, repeatability of the synthesis process, and the precise processes involved. Surface area, form, content, optical dynamicity, chemical reactivity and mechanical strength of the nanomaterial all have a role in determining their properties for reducing the dye and their antimicrobial capacity. Some nanomaterials obtained have good affinity towards biomolecules essential for generating eco-friendly biosensors and in greater advantage for the pharmaceutical industry. The scalability of the process needs to be worked upon with the evaluation of the cost considering and the economical feasibilities. Future research can be done to examine the nanomaterial synthesized from saponins in various other sectors such are material science, food and nutrition, and also to explore its potential in more health care sectors.

References

1. Aguiar ACD, Martínez J, De PJT, Luis J, Mundo M, Mcclements DJ (2021) Influence of type of natural emulsifier and microfluidization conditions on capsicum oleoresin nanoemulsions properties and stability. J Food Process Eng 44(4):1–13
2. Alagarasi A (2011) Introduction to nanomaterials. In: Viswanathan B (ed) Nanomaterials. Narosa Publishing House, Mumbai, India, pp 2–25
3. Atarod M, Nasrollahzadeh M, Mohammad Sajadi S (2016) Euphorbia heterophylla leaf extract mediated green synthesis of Ag/TiO2 nanocomposite and investigation of its excellent catalytic activity for reduction of variety of dyes in water. J Colloid Interface Sci 462:272–279
4. Ayyub P, Chandra R, Taneja P, Sharma AK, Pinto R (2001) Synthesis of nanocrystalline material by sputtering and laser ablation at low temperatures. Appl Phys A 73:67–73
5. Azmir J, Zaidul ISM, Rahman MM, Sharif KM, Mohamed A, Sahena F, Omar AKM (2013) Techniques for extraction of bioactive compounds from plant materials: A review. J Food Eng 117(4):426–436
6. Babu B, Palanisamy S, Vinosha M, Anjali R, Kumar P, Pandi B, Tabarsa M, You S, Prabhu NM (2020) Bioengineered gold nanoparticles from marine seaweed Acanthophora spicifera for pharmaceutical uses: antioxidant, antibacterial, and anticancer activities. Bioprocess Biosyst Eng 43(12):2231–2242
7. Baig N, Kammakakam I, Falath W (2021) Nanomaterials: a review of synthesis methods, properties, recent progress, and challenges. Mater. Adv. 2:1821–1871
8. Bande F, Arshad SS, Bejo MH, Kamba AS, Omar AR (2015) Synthesis and characterization of chitosan-saponin nanoparticle for application in plasmid DNA delivery. J Nanomater:1–8
9. Bangham AD, Horne RW (1962) Action ofsaponins on biological cell membranes. Nature 196:952–953
10. Basalingappa LH, Chandravadan HP (1971) A survey of plants in Gujarat, India, for alkaloids, saponin, and tannins. In U.S.D.A. Forest Service Research Paper, pp. 1–11
11. Bayda S, Adeel M, Tuccinardi T, Cordani M, Rizzoli F (2019) The history of nanoscience and nanotechnology: from chemical-physical applications to nanomedicine. Molecules 25(1):112
12. Benelmekki M (2015) Chapter 1: an introduction to nanoparticles and nanotechnology. Designing hybrid nanoparticles. Morgan & Claypool Publishers, pp 1–14
13. Benie T, El-Izzi A, Tahiri C, Duval J, Thieulant MLT (1990) Combretodendron africanum bark extract as an antifertilityagent: estrogenic effects in vivo and LH release by cultured gonadotropecells. J Ethnopharmacol 29:13–23
14. Bernhoft A (2010) A brief review on bioactive compounds in plants. Bioactive Compds Plants-benefits Risks Man Animals 50:11–17
15. Bradley S, Launikonis D, Stephenson G (1999) Effects of ß-escin andsaponin on the transverse-tubularsystem and sarcoplasmic reticulummembranes of rat and toad skeletalmuscle. Eur J Physiol 437:955–965
16. Chen JC, Xu MX, Chen LD, ChenYN CTH (1998) Effect of Panaxnotoginseng saponins on sperm motility and progression in vitro. Phytomedicine 5:289–292
17. Choi Y, Kang S, Cha SH, Kim HS, Song K, Lee YJ, Kim K, Kim YS, Cho S, Park Y (2018) Platycodon saponins from Platycodi Radix (*Platycodon grandiflorum*) for the Green Synthesis of Gold and Silver Nanoparticles. Nanoscale Res Lett 13(1):23

18. Colomban P, Gouadec G (2005) The ideal ceramic-fibre/oxide-matrix composite: How to reconcile antagonist physical and chemical requirements ? Ann Chim Sci Mater 30(6):673–688
19. Debnath B, Sarkar S, Das R (2020) Effects of saponin capped triangular silver nanocrystals on the germination of *Pisum sativum, Cicer arietinum, Vigna radiata* seeds & their subsequent growth study. IET Nanobiotechnol 14(1):25–32
20. Deng T, Chen J, Si W, Yin M, Ma W, Liu Z (2012) Fabrication of silicon nanopore arrays using a combination of dry and wet etching. J Vac Sci Technol B: Nanotechnol Microelectron Mater, Process, Meas, Phenom 30(6):061804
21. Desai S, Desai DG, Kaur H (2009) Saponins and their biological activities. Pharma Times 41:13–16
22. Fenwick DE, Oakenfull D (1983) Saponin content of food plants and some prepared foods. J Sci Food Agr 34(2):186–191
23. Francis G, Kerem Z, Makkar HP, Becker K (2002) The biological action of saponins in animal systems: a review. Br J Nutr 88(6):587–605
24. Francis G, Makkar HPS, Becker K (2001) Antinutritional factors present in plant-derived alternate fish feed ingredients and their effects in fish. Aquaculture 199:197–227
25. Freestone I, Meeks N, Sax M, Higgitt C (2007) The Lycurgus Cup—a roman nanotechnology. Gold Bull 40(4):270–277
26. Geoffrion LD, Guisbiers G (2020) Quantum confinement: size on the grill! J Phys Chem Solids 140:109320
27. Ha IJ, Ha YW, Kang M, Lee J, Park D, Kim YS (2010) Enzymatic transformation of platycosides and one-step separation of platycodin D by high-speed countercurrent chromatography. J Sep Sci 33:1916–1922
28. Husen A (2022) Engineered nanomaterials for sustainable agricultural production, soil improvement and stress management. Elsevier Inc., 50 Hampshire St., 5th Floor, Cambridge, MA 02139, USA
29. Husen A, Iqbal M (2019) Nanomaterials and plant potential. Springer International Publishing AG, Gewerbestrasse 11, 6330 Cham, Switzerland. https://doi.org/10.1007/978-3-030-05569-1
30. Husen A, Jawaid M (2020) Nanomaterials for agriculture and forestry applications. Elsevier Inc., 50 Hampshire St., 5th Floor, Cambridge, MA 02139, USA https://doi.org/10.1016/C2018-0-02349-X
31. Husen A, Siddiqi KS (2014) Phytosynthesis of nanoparticles: concept, controversy and application. Nanoscale Res Lett 9(229):1–24
32. Husen A, Siddiqi KS (2022) Advances in smart nanomaterials and their applications. Elsevier Inc., 50 Hampshire St., 5th Floor, Cambridge, MA 02139, USA
33. Ibrahim KS (2013) Carbon nanotubes-properties and applications: a review. Carbon Lett 14:131–144
34. Iqbal P, Preece JA, Mendes PM (2012) Nanotechnology: the "top-down" and "bottom-up" approaches. In: Gale PA, Steed JW (eds) Supramolecular chemistry. John Wiley & Sons, Ltd., Chichester, UK
35. Jain AS, Surana SJ, Gokhale SB, Tatiya AU, Bothara RC (2007) Antimicrobialproperties of Eranthemum roseum(Vahl) R. Br. Iran J Pharm Res 6(2):131–133
36. Jaiswal M, Dudhe R, Sharma PK (2014) Nanoemulsion: an advanced mode of drug delivery system. 3 Biotech, 5(2), 123–127
37. Krishnan SK, Singh E, Singh P, Meyyappan M, Nalwa HS (2019) A review on graphene-based nanocomposites for electrochemical and fluorescent biosensors. RSC Adv 9(16):8778–8881
38. Kumar A, Choudhary A, Kaur H, Guha S, Mehta S, Husen A (2022) Potential applications of engineered nanoparticles in plant disease management: a critical update. Chemosphere 295:133798
39. Kumar A, Choudhary A, Kaur H, Mehta S, Husen A (2021) Metal-based nanoparticles, sensors and their multifaceted application in food packaging. J Nanobiotechnol 19(256):1–25. https://doi.org/10.1186/s12951-021-00996-0
40. Leibtag S, Peshkovsky A (2020) Cannabis extract nanoemulsions produced by high-intensity ultrasound: formulation development and scale-up. J Drug Delivery Sci Technol:101953

41. Leonhardt U (2007) Invisibility cup. Nature Photon 1:207–208 (2007)
42. Liang J, Liu Q, Li T, Luo Y, Lu S, Shi X, Zhang F, Asiri AM, Sun X (2021) Magnetron sputtering enabled sustainable synthesis of nanomaterials for energy electrocatalysis. Green Chem 23(8):2834–2867
43. Mengoni F, Lichtner M, Battinelli L, Marzi M, Mastroianni CM, Vullo V, Mazzanti G (2002) In vitro anti-HIV activity ofoleanolic acid on infected humanmononuclear cells. Planta Med 68:111–114
44. Mourdikoudis S, Pallares RM, Thanh NTK (2018) Characterization techniques for nanoparticles: comparison and complementarity upon studying nanoparticle properties. Nanoscale 10(27):12871–12934
45. Muniyan A, Ravi K, Mohan U, Panchamoorthy R (2017) Characterization and in vitro antibacterial activity of saponin-conjugated silver nanoparticles against bacteria that cause burn wound infection. World J Microbiol Biotechnol 33:147
46. Murr LE (2015) Classifications and structures of nanomaterials. In: Handbook of materials structures, properties, processing and performance. Springer, Cham, pp 719–746
47. Nguyen DH, Vo TN, Le NT, Thi DP, Thi TT (2020) Evaluation of saponin-rich/poor leaf extract-mediated silver nanoparticles and their antifungal capacity. Green Process Synth 9(1):429–439
48. Oakenfull D (1981) Saponin in food—a review. Food Chem 7(1):19–40
49. Oakenfull DG (1986) Aggregation of bile acids and saponins in aqueous solution. Aust J Chem 39:1671–1683
50. Paramesh CC, Halligudra G, Gangaraju V, Sriramoju JB, Shastri M, Kachigere HB, Habbanakuppe PD, Rangappa D, Subbegowda RK, Shivaramu PD (2021) Silver nanoparticles synthesized using saponin extract of Simarouba glauca oil seed meal as effective, recoverable and reusable catalyst for reduction of organic dyes. Results Surf Interfaces 3:100005
51. Petit PR, Sauvaire Y, Ponsin G, Manteghetti M, Fave A, Ribes G (1993) Effects of a fenugreek seed extract onfeeding behaviour in the rat: metabolicendocrine correlates. Pharmacol Biochemi Behaviour 45:369–374
52. Potter SM, Jimenez-Flores R, Pollack J, Lone TA, Berber-Jimenez MD (1993) Protein saponin interaction and itsinfluence on blood lipids. J Agric Food Chem 41:1287–1291
53. Raju J, Gupta D, Rao A, Yadava P, Baquer N (2001) Trigonella foenum graecum(fenugreek) seed powder improvesglucose homeostasis in alloxan diabeticrat tissues by reversing the alteredglycolytic, gluconeogenic and lipogenicenzymes. Molecular and CellularBiochemistry 224:45–51
54. Rejinold NS, Muthunarayanan M, Muthuchelian K, Chennazhi KP, Nair SV, Jayakumar R (2011) Saponin-loaded chitosan nanoparticles and their cytotoxicity to cancer cell lines in vitro. Carbohyd Polym 84(1):407–416
55. Roduner E (2006) Size matters: why nanomaterials are different. Chem Soc Rev 35(7):583–592
56. Santamaria A (2012) Historical overview of nanotechnology and nanotoxicology. Methods Mol Biol (Clifton, N.J.) 926:1–12
57. Savage GP (2003) SAPONINS. In: Ed. Benjamin Caballero (ed) Encyclopedia of food sciences and nutrition, 2nd edn. Academic Press, pp 5095–5098
58. Sayyah M, Hadidi N, Kamalinejad M (2004) Analgesic and anti-inflammatory activity of Lactuca sativa seed extract in rats. J Ethnopharmacol 92: 325–329
59. Shafey A (2020) Green synthesis of metal and metal oxide nanoparticles from plant leaf extracts and their applications: a review. Green Process Synth 9(1):304–339
60. Sharifi S, Behzadi S, Laurent S, Laird Forrest M, Stroeve P, Mahmoudi M (2012) Toxicity of nanomaterials. Chem Soc Rev 41:2323–2343
61. Sharma G, Nim S, Alle M, Husen A, Kim JC. 2022. Nanoparticle-Mediated delivery of flavonoids for cancer therapy: prevention and treatment. In: Kim JC, Alle M, Husen A (eds) Smart nanomaterials in biomedical applications. Nanotechnology in the Life Sciences. Springer, Cham. https://doi.org/10.1007/978-3-030-84262-8_3
62. Sharma P, Pandey V, Sharma MMM, Patra A, Singh B, Mehta S, Husen A (2021) A review on biosensors and nanosensors application in agroecosystems. Nano Res Lett 16:136. https://doi.org/10.1186/s11671-021-03593-0

63. Shi J, Arunasalam K, Yeung D, Kakuda Y, Mittal G, Jiang Y (2004) Guelph.Saponins from edible legumes:chemistry, processing, and healthbenefits. J Med Food 7(1):67–78
64. Shimoyamada M, Ikedo S, Ootsubo R, Watanabe K (1998) Effects of soybeansaponins on chymotryptic hydrolyses of soybean proteins. J Agric Food Chem 46:4793–4797
65. Sindambiwe JB, Calomme M, Geerts S, Pieters L, Vlietinck AJ, VandenBerghe DA (1998) Evaluation of biological activities of triterpenoid saponins from Maesa lanceolata. J Nat Prod 61:585–590
66. Soetan KO, Oyekunle MA, Aiyelaagbe OO, Fafunso MA (2006) Evaluation of the antimicrobial activityof saponins extract of Sorghum Bicolor L. Moench. Afr J Biotechnol 5(23):2405–2407
67. Song JY, Kim BS (2008) Rapid biological synthesis of silver nanoparticles using plant leaf extracts. Bioprocess Biosyst Eng 32(1):79–84
68. Srivastava N, Choudhary M, Singhal G, Bhagyawant SS (2020) SEM studies of saponin silver nanoparticles isolated from leaves of chenopodium album L. for in vitro anti-acne activity. Proc Natl Acad Sci India Sect B Biol Sci 90:333–341
69. Sundar S, Venkatachalam G, Kwon S (2018) Biosynthesis of copper oxide (CuO) nanowires and their use for the electrochemical sensing of dopamine. Nanomaterials 8(10):823
70. Tadros MM, Ghaly NS, Moharib MN (2008) Molluscicidal and schistosomicidalactivities of a steroidal saponincontaining fraction from Dracaenafragrans (L.). J Egypt Soc Parasitol 38(2):585–98
71. Tatiya AU, Desai DG, Surana SJ, Patil PH (2007) Anti-inflammatory and nitricoxide scavenging activity of roots of Eranthemum roseum. Indian Drugs 44(11):815–818
72. Trotta F, Mele A (2019) Nanomaterials: classification and properties. In Nanosponges, John Wiley & Sons, Ltd.: Hoboken, NJ, USA, pp 142–149
73. Walter P, Welcomme E, Hallégot P, Zaluzec NJ, Deeb C, Castaing J, Veyssière P, Bréniaux R, Lévêque JL, Tsoucaris G (2006) Early use of PbS nanotechnology for an ancient hair dyeing formula. Nano Lett 6(10):2215–2219
74. Wu Q, Miao W, Zhang Y, Gao H, Hui D (2020) Mechanical properties of nanomaterials: a review. Nanotechnol Rev 9(1):259–273
75. Yang CR, Zhang Y, Jacob MR, Khan SI, Zhang YJ, Li XC (2006) Antifungal activity of C-27 Steroidal Saponins. Antimicrob Agents Chemother 50(5):1710–1714
76. Yoshizumi K, Hirano K, Ando H, Hirai Y, Ida Y, Tsuji T, Tanaka T, Satouchi K, Terao J (2006) Lupane-type saponins from leaves of Acanthopanax sessiliflorus and their inhibitory activity on pancreatic lipase. J Agric Food Chem 54(2):335–341
77. Zhao HL, Sim JS, Shim SH, Ha YW, Kang SS, Kim YS (2005) Antiobese and hypolipidemic effects of platycodin saponins in diet-induced obese rats: evidences for lipase inhibition and calorie intake restriction. Int J Obes (Lond) 29(8):983–990
78. Zhu W, Guo Y, Ma B, Yang X, Li Y, Li P, Zhou Y, Shuai M (2020) Fabrication of highly dispersed Pd nanoparticles supported on reduced graphene oxide for solid phase catalytic hydrogenation of 1,4-bis(phenylethynyl) benzene. Int J Hydrogen Energy 45(5):8385–8395

Preparation of Nanomaterials Using Coumarin and Their Various Applications

Vinayak Adimule, Sheetal Batakurki, and Rangappa Keri

Abstract In the present study, carbon quantum dots (CDQs) modified coumarine-3-carboxylic acid (CCA) and 7-diethyl amino coumarine-3-carboxylate (DACC) were synthesized by microwave assisted method and the synthesized hybrid nanomaterials possess greater biocompatibility, water solubility, luminescent and fluorescent properties. The hybrid CDQs: CCA/DACC nanomaterials exhibit superior fluorescent properties and greater quantum yield when subjected for excitation. Also, they possess superior bio-imaging properties, fluorescent quenching and contribute for the modern diagnosis and cell identification techniques. When tested for cytotoxicity, the hybrid nanomaterial does not showed cytotoxicity against L929 mouse fibroblast. The results clearly indicated that, the hybrid nanomaterials with surface modified CQDs dispersed over CCA/DACC may become a powerful tool in the field of nano medicine and pharmacy.

Keywords Carbon quantum dots · Coumarin derivatives · Fluorescent · Cytotoxicity · Bio imaging · Synthesis · Hybrid nanomaterials

1 Introduction

Efficiency of the devices in photonics depends on the fluorescence emission of the organic compounds. Due to the interaction of the analyte and functional groups

V. Adimule (✉)
Department of Chemistry Angadi Institute of Technology and Management (AITM), Savagaon Road, Belagavi 590009, Karnataka, India
e-mail: adimulevinayak@yahoo.in; researcher@vinayakadimule.in; chemistry@aitmbgm.ac.in; research@aitmbgm.ac.in

S. Batakurki
Department of Chemistry, M. S. Ramaiah University of Applied Sciences, MSR Nagar, New BEL Road, Bangalore 560054, Karnataka, India

R. Keri
Centre for Nano and Material Sciences, Jain University (Deemed to be University), Jain Global Campus, Kanakapura, Bangalore 562112, Karnataka, India

A. Husen (ed.), *Secondary Metabolites Based Green Synthesis of Nanomaterials and Their Applications*, Smart Nanomaterials Technology,
https://doi.org/10.1007/978-981-99-0927-8_8

present in the organic compounds they show greater sensitivity, selectivity, high quantum yield [7]. The performance of the fluorescent enabled biosensors in the effective detection of biomolecules is a promising approach since it has advantages such as easy interpretation of the obtained results, grater response in short time and high reversibility. The undesirable features also exist in some of the fluorescent organic compounds such as photo bleaching phenomenon [9, 11]. The emission intensities overlap with each other therefore they cannot be employed for the multicolor detection [17]. Quantum dots (QDs) are the interesting nanomaterials due to their size less than 10 nm and they exhibit often high fluorescent emission intensities when excited with a suitable wavelength [8, 21]. In recent years many of the oxide semiconductors with nano quantum dots such as zinc selenide, cadmium selenium were robustly studied for their applications in photovoltaic [19], bio-imaging [15], photo catalysis [4], photo sensing [14]. However, the major obstacle in their application is poor water solubility before they can be used for medicine and pharmacy, which needs to be well addressed [18].

The attempts to modify the structure of the nanomaterials is by doping polymers into the surface, and such changes increase the covalence, adsorption, chelation, etc. and one of the interesting and most useful advantages is tuning the band gap of the semiconductor [3]. The conventional QDs obtained from the carbon core are water soluble and non-toxic [24]. However, carbon nano quantum dots (CNQDs) can be easily synthesized from vegetables, fruit peels, coffee grounds and even from the milk [26]. Polyphenolic compounds occurring naturally in organic compounds and dyes containing coumarin as essential constituents possess diversified biological properties such as antioxidant, antimicrobial, antitumor etc. [1, 2]. Coumarin is also characterized by fluorescent properties and is widely used in fluorescent biosensors [10, 12]. Coumarin is the precursor material for applications in the field of biomedical and industrial applications [6, 22].

In the present study, carbon quantum dots functionalized coumarin derivatives were successfully synthesized and characterized by ^{1}H-NMR (Proton nuclear magnetic resonance), FT-IR (Fourier Transform Infrared spectroscopy), SEM (Scanning Electron Microscopy), UV–visible and their biomedical and sensor applications studied in detail.

2 Synthesis of Nanomaterials Using Coumarin

The process of synthesis of nanomaterials using coumarin derivatives under the present investigation was carried out using two different steps. CCA, DACC were used for stabilizing the carbon quantum dots obtained from the lignin. Under microwave irradiation, the common synthetic process involves three different steps, namely, dehydration, carbonization and passivation as shown in Fig. 1. The rich –OH group present in the lignin yields a low quantum yield which is insufficient for the biomedical sensor applications [20]. The CCA, DACC organic molecules form coordination bonds and thus improvize the quantum yield of the NPs. Figure 2 displays

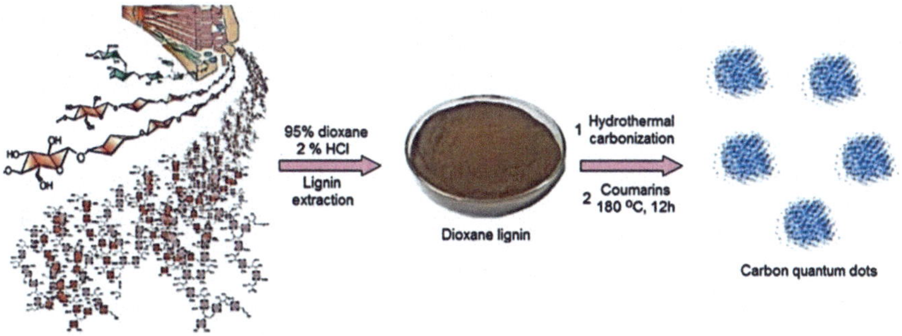

Fig. 1 General strategy for carbon quantum dots (CQDs) preparation

the ^{1}H-NMR spectral data for CCA, DACC for the products corresponding to other researcher's data [16].

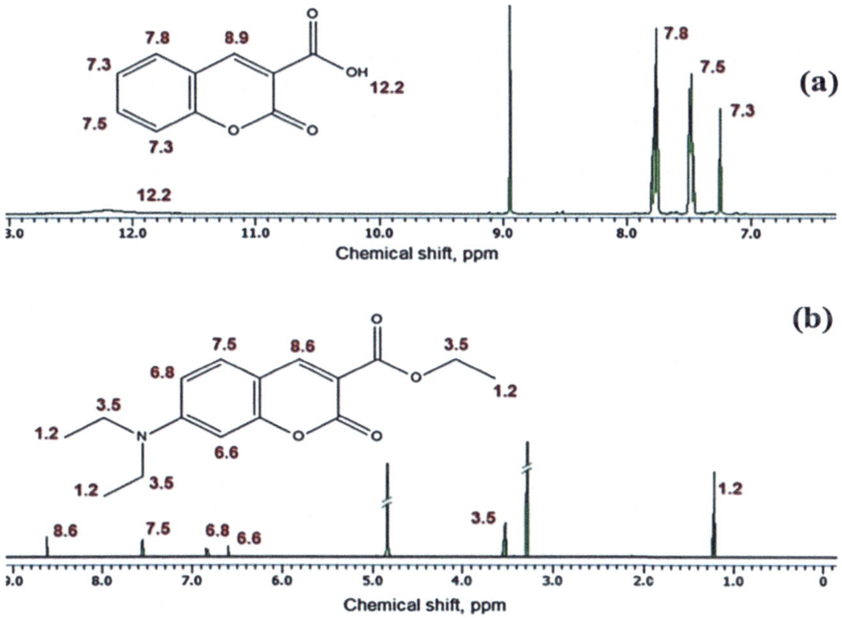

Fig. 2 NMR spectra of the obtained coumarin derivatives

Table 1 Fluorescent quantum yield, photo stability of the as-synthesized CDQs with coumarin derivatives

Samples	Fluorescence QY	Photo stability after 7 days (%)	Photo stability after 30 days (%)
CDQs-1	3.4	3.4	3.3
CDQs-1	6.7	6.6	6.4
CDQs-1	11.2	11.1	11.0
CDQs-1	11.1	11.0	10.6
CDQs-1	3.5	3.3	3.3
CDQs-1	9.8	9.6	9.5
CDQs-1	14.7	14.5	14.2
CDQs-1	18.4	18.1	17.8

2.1 CDQs synthesis

CDQs were synthesized by boiling with propylene carbonate at 180 °C in an autoclave using lignin as the raw material. H_2SO_4 is used as a carbonization agent. CCA synthesis is followed by adding 12.2 g of salicylic aldehyde, 17.68 g diethylmalonate and 0.2 mL of piperidine. The entire reaction mixture under constant stirring is irradiated with microwave for 25 min at 100C. The crude product was cooled and isolated by column chromatography. To obtain DACC, 5 g of N, N-diethyl salicylic aldehyde was mixed with 4.58 g diethylmalonate and 0.5 mL piperidine. The reaction mixture was irradiated with microwave for 6 min at 98 °C. The product was recrystallized from methanol and dried. Table 1 summarizes the various CDQs, temperature of the reaction and their yield.

3 Applications of Nanomaterials Synthesized by Coumarin Modifications

3.1 Photoluminescence (PL) and Fluorimetric Applications

In the recent investigation, PL and Fluorimetric properties of 7-acryloxycoumarin and cellulose fiber modified nanocomposites has been studied. The fluorescent emission spectrum of NPs showed deep intense emissions as compared with the 7-acryloxy coumarine derivative at the same concentration and at the same excitation in the same solvent acetone. The excitation wavelength used was 280 nm and 315 nm. The strong emission intensities at 395 nm and 465 nm appeared when excited at 315 nm as shown in Fig. 3a and b. This intense emission was observed with a higher quantum yield. This results in the applications of the coumarin modified NPs to be used for the water based anti-counter feinting ink on the cellulose fiber. Fluoroimetric and UV–visible absorption studies of the as modified coumarin NPs showed hypochromic effect (shift

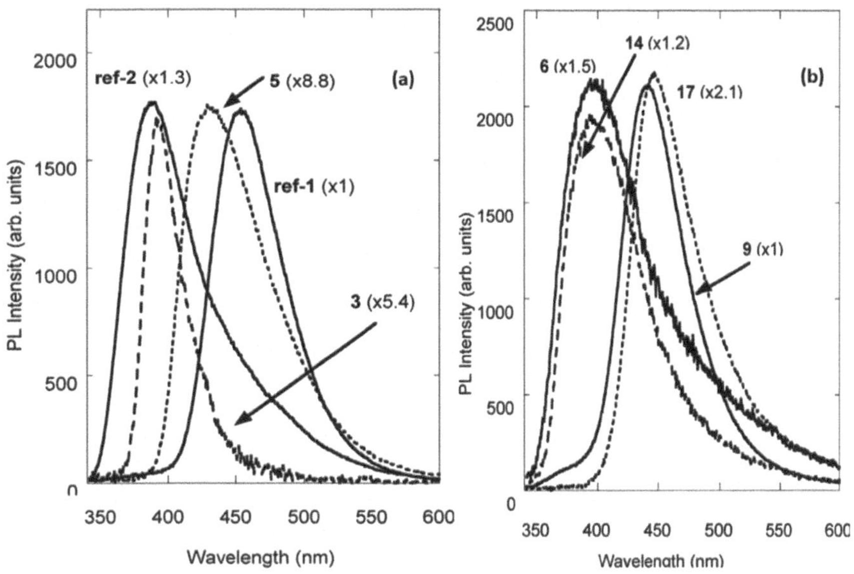

Fig. 3 **a** PL spectra of coumarin substituted derivatives when excited at 280 nm and **b** when excited at 315 nm

in the wavelength of the absorption of light towards lower wavelength region). The lower shift in the wavelength of absorption of radiation is due to delocalization of π electrons in the ring system and change in the hydroxyl group to ester group during the modification of the NPs.

3.2 Photo Crosslinking and Photo Dimerization Properties

Nanoparticles dispersed in a single chain and at the end of the chain coumarin derivatives have been modified and were studied for the photo crosslinking and photo dimerization properties. The UV–visible absorption characteristics of single chain modified NPs with coumarin showed an an absorption peak at 320 nm with decrease in the intensity of absorption with increase in the irradiation time exhibiting a growing degree of photo dimerization. The optical absorptivity spectra clearly indicates that the dimerization will reach at ~ 75% for 1 h of irradiation (Fig. 4). The photo cross linking appears to occur slowly as the reaction is also due to the presence of coumarin groups at the end of the chain makes it difficult to close distance for the dimerization to form. The change of wavelength to less than 260 nm, and dimerization reduces to 38% after 2 h of irradiation. The molecular weight of the compounds increases if the photo dimerization occurs in intermolecular and decreases if the dimerization occurs intramolecular. The formation of inter molecular dimerization can be observed for the compounds with continuous increase in the retention time.

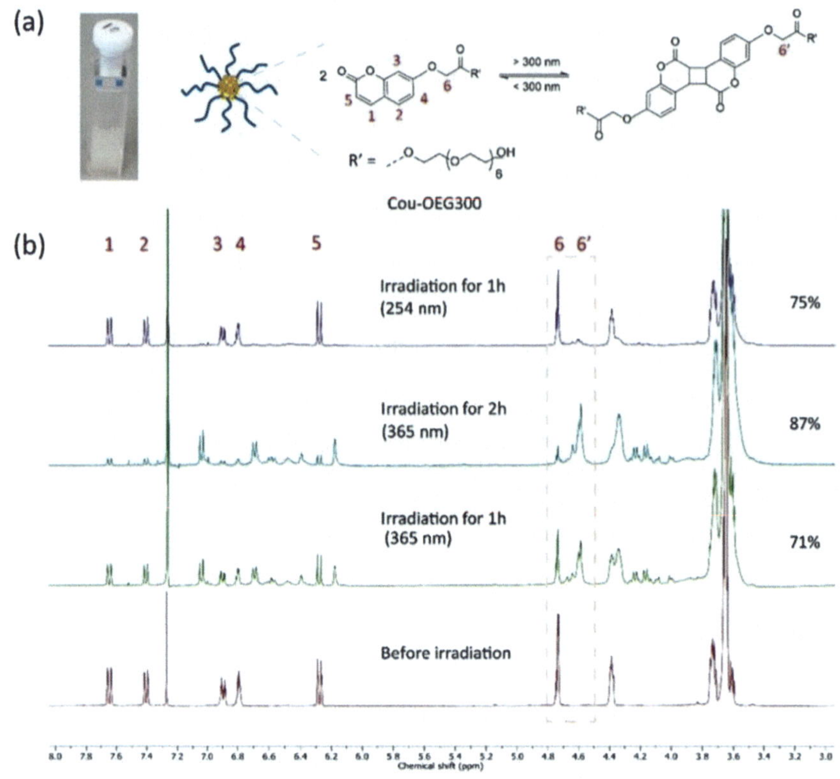

Fig. 4 **a** Photo isomerization of Coumarin substituted derivatives and **b** change in the chemical shift before and after the irradiation

3.3 Cell Imaging Properties of Modified Coumarin Derivatives

Cell imaging studies of coumarin modified Schiff base derivatives containing NPs have been investigated. Anticancer cell lines (*HeLa*) taken for the study, incubated for the time period of 5 h at 25C, are dispersed with NPs contained in DMSO (dimethyl sulfoxide). The formed cells were fixed by 4% paraformaldehyde and imaged by using the confocal laser scanni8ng microscope. The coumarin modified NPs show large stoke shifts at 270 nm and 240 nm with high quantum yield which were detected during the cell imaging studies. The cell imaging can be interpreted using fluorescence studies. The fluorescence spectra of the coumarin modified NPs can be interpreted when compounds were excited with radiation of frequency 323 nm and 416 nm. Week emission bands at 392 nm, and 529 nm appeared. Fluorescence patterns increased gradually indicating the aggregation of the NPs in the presence of Coumarin derivatives (Fig. 6). The enhanced fluorescence can be explained with the help of intra and intermolecular effects of coumarin modified NPs as well as the presence of water molecules.

3.4 *Optoelectronic Properties of Coumarin Modified Hybrid Nanocomposite Thin Films with $CoOFe_2O_4$*

Optoelectronics properties of different coumarine derivatives modified with $CoOF_2O_4$ nanocomposites were studied. As modified NPs showed optical and dielectric function. As the NPs exhibited high dielectric constant at low frequency and decreases with increase in the frequency. This behavior is quite common in case of ferrite materials. The imaginary value of the dielectric constant increases with increase in the applied frequency and decreases upon further increase in the frequency. Real and imaginary values of the optical conductivity increases proportionately with increase in the frequency and thus synthesized, modified NPs showed abrupt increase in the optical conductivity at the near UV region (Fig. 5a–c). Also the different modified coumarin NPs exhibited PL emission intensities at 485.43 nm and 477 nm (with a shift of 8 nm) when excited at 477.4 nm. The coumarin containing lone pair of electrons on the N, O elements is involved in the conjugation and form stable lower energy structures (Fig. 5d, e). The formation of lower energy structures agrees well with the energy overlap of the orbitals (HOMO) (highest occupied molecular orbital) and LUMO (lowest unoccupied molecular orbital) theory.

3.5 *Fluorescent and Photoluminescence (PL) Properties*

Fluorescent spectra of the as-modified CDQs displayed in Figs. 6a, b and 7. Depending upon the size of the quantum dots, CDQs show excellent luminescent behavior when excited with light of wavelength. The PL intensity depends upon the type of the coumarin used for the modifications and the reaction time and conditions. In each case of the modification of the CDQs, increase in the reaction time shifts the wavelength of the PL emissions to a longer side. Interestingly, the CDQs modified with DACC showed an enhanced PL spectrum which is due to the presence of the triethylene group in the structure of the derivatives [16]. One of the most important properties in biomedical application is to shift the absorbance maximum to the higher wavelength, coumarin (DACC) modified CDQs exhibit peak shift to the longer wavelength [25]. The samples with CDQs-8 and CDQs-7 exhibited greater PL emission intensity which appeared at 511 nm, 507 nm respectively as compared to the samples of CDQs-1 and CDQs-5. The complicated structure of the CDQs than the metal oxides show an interesting PL spectrum depending upon the wavelength of the excitation. Cellular level detection, visualization, in the field of bio imaging, and biosensor techniques require CDQs to be modified with coumarin derivatives.

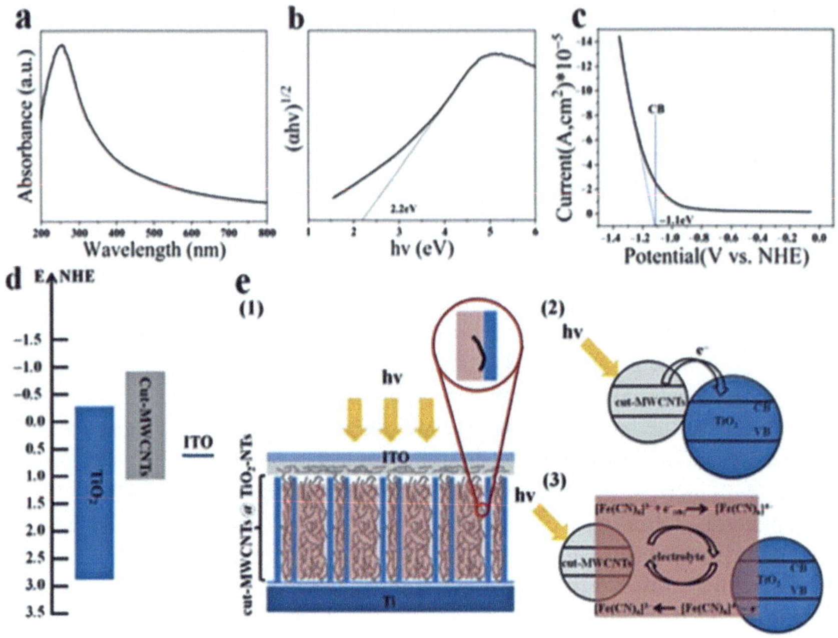

Fig. 5 **a–c** optical absorptivity of hybrid derivatives of coumarine incorporated for optoelectronic studies **d–e** typical fabrication and opto electronic property studies

The quantum yield (QY) of the CDQs synthesized with CCA/DACC modifications is higher as compared without modifications. Table 1 depicts the different CDQs modified with the CCA/DACC quantum yield with photo stability after 30 days. When irradiated with radiations of suitable intensity, considerable difference in the illumination of albumin was observed. This difference in the fluorescence phenomenon can be understood with the structure of the albumin molecule. Due to the amine group present in the DACC moiety, its interaction with CDQs leads to the phenomenon of fluorescence quenching [5]. Fluorescence quenching with different P^H values of the albumin are shown in Fig. 8. The chelating ability of the CDQs in DACC and P^H of the solvents used for the study. CDQs-6, 7 and 8 show strong interaction with the photons and CDQs-3 and 4 displayed moderate sensitivity. Difference between the synthesis procedures of the CDQs and the presence of hydroxyl, carboxyl groups are responsible for the fluorescence quenching phenomenon. PL intensity increases in an acidic environment and can be helpful for cancer cell detection and labeling [22]. Table 2 represents the parameters used during the modification of CDQs by CCA/DACC.

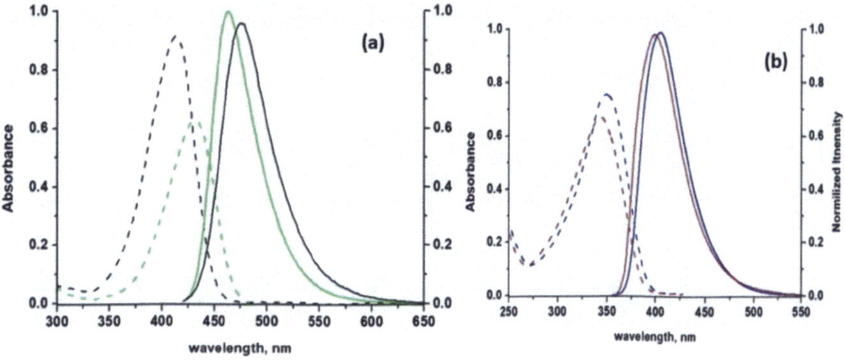

Fig. 6 **a** and **b** Fluorescence and PL spectra of coumarine substituted CDQs at different PH of the solvents

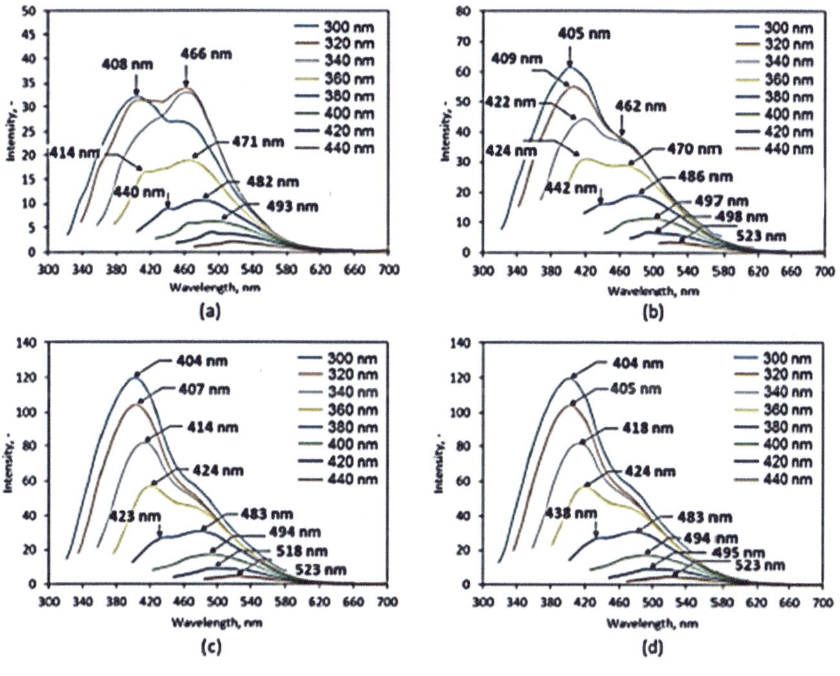

Fig. 7 Fluorescence spectra of the prepared CQDs: (**a**) CQDs-1; (**b**) CQDs-2; (**c**) CQDs-3; (**d**) CQDs-4; € CQDs-5; (**f**) CQDs-6; (**g**) CQDs-7; (**h**) CQDs-8

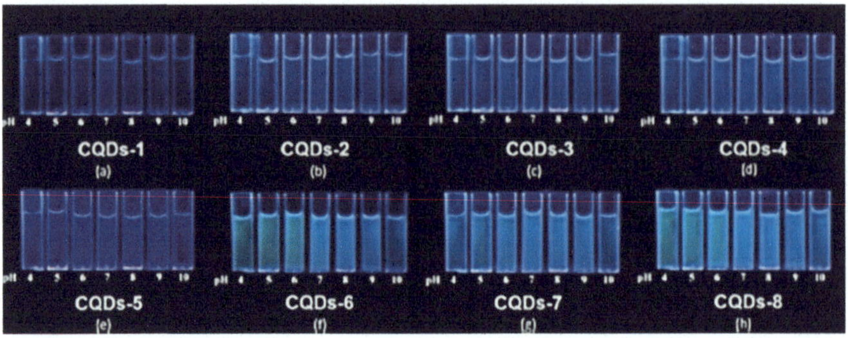

Fig. 8 Images of the prepared CQDs' fluorescence dependence on solvent pH value (excitation wavelength = 365 nm): (**a**) CQDs-1; (**b**) CQDs-2; (**c**) CQDs-3; (**d**) CQDs-4; € CQDs-5; (**f**) CQDs-6; (**g**) CQDs-7

Table 2 Parameters used during the modification of CDQs by CCA/DACC

Samples	Lignin (g), Propylene Carbonate (mL), H_2SO_4	Carbonization time (h)	Coumarin (g)	Reaction time (min)
CDQs-1	0.05;3;0.05	12	CCA/DACC	2
CDQs-1				5
CDQs-1				10
CDQs-1				20
CDQs-1				2
CDQs-1				5
CDQs-1				10
CDQs-1				20

3.6 *Biosensor Properties*

CDQs possess bio sensing applications for various ions and the extent of sensitivity and selectivity depends upon the synthesis method and modifications [23]. Figure 9 displays bio detection of glucose, fructose with the albumin-dichromate model. The regression equations were summarized in Table 3. Depending upon the type of the analyte being tested CDQs differs in their selectivity and sensitivity of the ions. CDQs-4 and CDQs-8 showed greater sensitivity as compared with other CDQs.

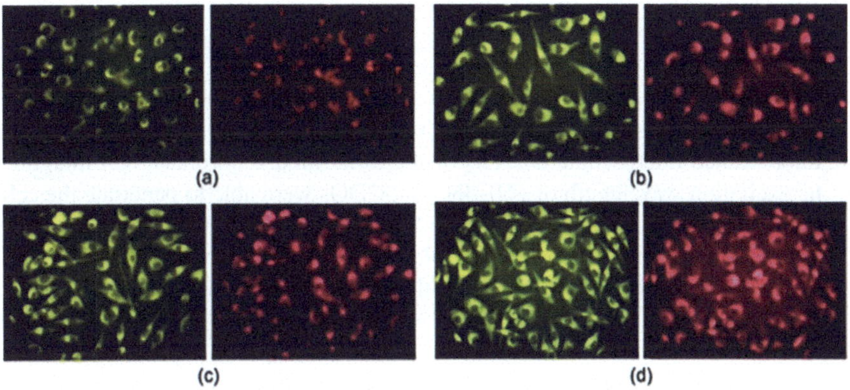

Fig. 9 Images of the prepared CQDs' fluorescence dependence on solvent pH value (excitation wavelength = 365 nm): (**a**) CQDs-1; (**b**) CQDs-2; (**c**) CQDs-3; (**d**) CQDs-4

Table 3 Coumarin-modified CQD sensing sensitivity

Sample	Analytes	Regression equation
CDQs-4	Potassium Dichromate	$y = -1874.4x + 153.16$
	Glucose	$y = -48.505x + 156.82$
	Fructose	$y = -58.180x + 156.62$
	Egg Albumin	$y = 47.451x + 155.75$
CDQs-8	Potassium Dichromate	$y = -3190.7x + 328.27$
	Glucose	$y = -279.9x + 334.29$
	Fructose	$y = -257.02x + 336.07$
	Egg Albumin	$y = 21.161x + 336.14$

3.7 *Cytotoxic Investigations*

Core C structures present in the CDQs generally show low cytotoxicity. Figure 9 displays XTT spectra of L929 mouse fibroblast commonly used for in vitro anticancer property studies. The study was commonly used for bio sensing, bio-imaging and cell labeling, visualization. Four different concentrations were used, that is, 0.05–2 mg/ml. No significant decrease in the bio availability was observed even at a high concentration of CDQs modified with CCA/DACC which proves that CDQs modified with CCA/DACC do not affect any biological properties. Thus, CDQs possesses good biocompatibility.

3.8 Cell Visualization Studies

CDQs are widely used for the bio imaging and labeling experimentation due to their fluorescence which is dependent on the excitation of the CDQs. Figure 6 displays the fluorescence microscopic analysis of the L929 mouse fibroblast cell imaging. At the excitation wavelength of 460–480 nm, CDQs were able to penetrate the cell membrane and the emission intensities of 520 nm-550 nm and 590 nm were observed. The fluorescence visualization can be explained by the functional group interaction of CDQs, and the cytoskeleton present in the protein molecules [13].

4 Conclusion

In the present investigation, CDQs modified with CCA and DACC were synthesized using the microwave synthetic method. The modified CDQs were obtained in high quantum yield and showed diversified properties. The fluorescence patterns were used for the bio-imaging and cell labeling process and on the other hand enhanced photoluminescence characteristics were employed for the optoelectronics and photonic device fabrication. Biosensor properties of the synthesized CDQs showed a linear response for the analyte and the interaction between the organic functional group is directly dependent on the structure of the analyte under investigation and the CDQs. All the synthesized CDQs exhibited no cytotoxicity against L292 mouse fibroblast and when studied for the cell visualization experimentation showed excitation dependent fluorescence emission. CDQs modified with CCA, DACC can be used widely for the biosensor device fabrication, cell labeling, and visualization, fluorescent and photon reacted applications.

References

1. Alshibl HM, Al-Abdullah ES, Haiba ME, Alkahtani HM, Awad GE, Mahmoud AH, Ibrahim BM, Bari A, Villinger, (2020) A synthesis and evaluation of new coumarin derivatives as antioxidant antimicrobialand anti-inflammatory agents. Molecules 25:3251
2. Akkol EK, Genç Y, Karpuz B, Sobarzo-Sánchez E, Capasso R (2020) Coumarins and Coumarin-related compounds. Pharmacother Cancer Cancers 12:1959
3. Atchudan R, Edison TN, Shanmugam M, Perumal S, Somanathan T, Lee YR (2021) Sustainable synthesis of carbon quantum dots from banana peel waste using hydrothermal process for in vivo bioimaging. Phys E Low-Dimens Syst Nanostruct 126:114417
4. Bajorowicz B, Kobylański MP, Gołąbiewska A, Nadolna J, Zaleska-Medynska A, Malankowska A (2018) Quantum dot-decorated semiconductor micro-and nanoparticles: A review of their synthesis, characterization and application in photocatalysis. Adv Colloid Interface Sci 256:352–372
5. Boeriu CG, Bravo D, Gosselink RJ, Van Dam JE (2004) Characterisation of structure-dependent functional properties of lignin with infrared spectroscopy. Ind Crops Prod 20:205–218

6. Cuevas JM, Seoane-Rivero R, Navarro R, Marcos-Fernández R (2020) A coumarins into polyurethanes for smart and functional materials. Polymers 12:630
7. Duong HD, Sohn O-J, Rhee JI (2020) Development of a ratiometric fluorescent glucose sensor using an oxygen-sensing membrane immobilized with glucose oxidase for the detection of glucose in tears. Biosensors 10:86
8. Huang C, Dong H, Su Y, Wu Y, Narron R, Yong Q (2019) Synthesis of carbon quantum dot nanoparticles derived from byproducts in bio-refinery process for cell imaging and in vivo bioimaging. Nanomaterials 9(3):387
9. Janus Ł, Radwan-Pragłowska J, Piątkowski M, Bogdał D (2020) Facile synthesis of surface-modified carbon quantum dots (CQDs). Biosensing Bioimaging Mater 13(15):3313.
10. Jung Y, Jung J, Huh Y, Kim D Benzo[g] (2018) coumarin-Based Fluorescent Probes for Bioimaging Applications. J Anal Methods Chem 5249765
11. Koutsogiannis P, Thomou E, Stamatis H, Gournis D, Rudolf P (2020) Advances in fluorescent carbon dots for biomedical applications. Adv Phys X 5:1758592
12. Li X, Huo F, Yue Y, Zhang Y, Yin C (2017) A coumarin-based "o_-on" sensor for fluorescence selectively discriminating GSH from Cys/Hcy and its bioimaging in living cells. Sens Actuators B Chem 253:42–49
13. Lin F, Bao YW, Wu FG (2019) Carbon dots for sensing and killing microorganisms. C J Carbon Res 5:33
14. Lou Y, Zhao Y, Chen J, Zhu J-J (2014) Metal ions optical sensing by semiconductor quantum dots. J Mater Chem C 2:595–613
15. Ma Z, Ma Y, Gu M, Huo X, Ma S, Lu Y, Ning Y, Zhang X, Tian B, Feng Z (2020) Carbon dots derived from the maillard reaction for pH sensors and Cr (VI) detection. Nanomaterials 10:1924
16. Martínez J, Sánchez L, Pérez FJ, Carranza V, Delgado F, Reyes L, Miranda R (2016) Uncatalysed production of coumarin-3-carboxylic acids: a green approach. J Chem 4678107
17. Nikazar S, Sivasankarapillai VS, Rahdar A, Gasmi S, Anumol PS, Shanavas MS (2020) Revisiting the cytotoxicity of quantum dots: an in-depth overview. Biophys Rev 12:703–718
18. Reshma V, Mohanan P (2019) Quantum dots: applications and safety consequences. J Lumin 205:287–298
19. Semonin OE, Luther JM, Beard MC (2012) Quantum dots for next-generation photovoltaics. Mater Today 15:508–515
20. Shi Y, Liu X, Wang M, Huang J, Jiang X, Pang J, Xu F, Zhang X (2019) Synthesis of N-doped carbon quantum dots from bio-waste lignin for selective irons detection and cellular imaging. Int J Biol Macromol 128:537–545
21. Suner SS, Sahiner M, Ayyala RS, Bhethanabotla VR, Sahiner N (2020) Nitrogen-doped arginine carbon dots and its metal nanoparticle composites as antibacterial agent. C J Carbon Res 6:58
22. Tsao KK, Lee AC, Racine KÉ, Keillor JW (2020) Site-specific fluorogenic protein labelling agent. Bioconjug Biomolecules 10:369
23. Xue B, Yang Y, Sun Y, Fan J, Li X, Zhang Z (2019) Photoluminescent lignin hybridized carbon quantum dots composites for bioimaging applications. Int J Biol Macromol 122:954–961
24. Yu J, Shendre S, Koh W, Liu B, Li M, Hou S, Hettiarachchi C, Delikanli S, Hernández-Martínez P, Birowosuto MD et al (2019) Electrically control amplified spontaneous emission. colloidal quantum dots. Sci Adv 5:3140
25. Zhang B, Liu Y, Ren M, Li X, Zhang W, Vajtai X, Ajayan PM, Tour JM, Wang L (2019) Sustainable synthesis of bright green fluorescent nitrogen-doped carbon quantum dots from alkali lignin. Chem SusChem 12:4202–4210
26. Zhang Y, Wu G, Ding C, Liu F, Liu D, Masuda T, Yoshino K, Hayase S, Wang R, Shen Q (2020) Surface-modified graphene oxide/lead sulfide hybrid film-forming ink for high-efficiency bulk nano-heterojunction colloidal quantum dot solar cells. Nano-Micro Lett 12:1–14

Aromatic Oil from Plants, and Their Role in Nanoparticle Synthesis, Characterization and Applications

Arundhati Singh, Vedanshi Pal, Shreyshi Aggarwal, and Manu Pant

Abstract Plant based aromatic oils are widely used for their valuable odour. They are highly valued in both domestic and foreign markets for their aroma and value-added goods like perfumes and cosmetics. In addition to having a pleasant smell, many aromatic oils offer medical properties that have raised their market value and expanded their uses in the pharmaceutical sector. Some of the aromatic oils, which are frequently utilized in product development, are extracts of Eucalyptus leaves, citrus fruit, cinnamon bark, basil plant, etc. The industry for aromatic oils is booming as they are increasingly used in the development of herbal medicines. The development of drugs using nanoparticles has become much more effective since the introduction of nanotechnology. In this method, the drug is coupled with nanoparticles and successfully delivered inside the body to treat the ailment in a target-specific manner. Aromatic oils are expected to be used in the development of nanoparticles that can then be applied to a variety of ailments. Nanoparticles might lessen the stress on plants that are mass-harvested for medicine manufacturing because they would only consume a little amount of oil. However, prior to using these formulations in the human body, it is crucial to comprehend the underlying principles of aromatic oil extraction, their chemical structure, nanochemistry, the effectiveness of nanoparticles, and their toxicology.

Keywords Aromatic oil · Nanomaterials · Synthesis · Uses

A. Singh
School of Agriculture and Environment, Bentley Perth Campus, Curtin university, Perth, Australia

V. Pal · S. Aggarwal · M. Pant (✉)
Department of Biotechnology, Graphic Era Deemed to be University, Bell Road, Clement Town, Dehradun, Uttarakhand, India
e-mail: manupant@geu.ac.in

A. Husen (ed.), *Secondary Metabolites Based Green Synthesis of Nanomaterials and Their Applications*, Smart Nanomaterials Technology,
https://doi.org/10.1007/978-981-99-0927-8_9

1 Introduction

According to WHO (World Health Organisation), more than half of the world's population depends on plant-based medicines that are widely used as antimicrobial, anticancer, immunosuppressors and cardioprotective drugs [110]. For instance, oregano oil has been shown to be effective against disease-causing *Pseudomonas* spp. and *Lactobacillus* bacteria [138]. The aromatic oils extracted from aromatic plants have played a significant role in plant-based drug development since ancient times. Aromatic oils are volatile and odoriferous, varying in odour type and range of flavours depending on the constituents present in the oil. The plant part from where the oil is extracted determines its price, and almost every part of the plant can be the source for extracting these aromatic oils [123] (Table 1).

However, mostly the essential oils (EO) are biosynthesized in leaves, as most odorous substances are found there until the flowering takes place. Once the plant starts to flower, the odorous substances travel to the flowers and assist in fertilization. Once fertilized, they accumulate in the fruits and seeds [12]. Considering the wide range of medicinal and therapeutic activities exhibited by EO, it is envisaged that they can be efficiently used for nanoparticle-based applications. In general, the metallic nanoparticle has unique characteristics and is thus being used in different sectors, and shown various applications in medicine, photothermal therapy, agriculture, forestry etc. [44–50, 53, 59, 60, 112, 113, 115] (Fig. 1)

The EOs have different functional groups in their structures like alcohol, aldehyde, ketone, lactone of terpenoids, etc. These have active groups which are reduced and further stabilized, making it a green method of nanoparticle synthesis with potential application in industrial, agricultural, biotherapeutic and other domestic areas [56, 103]. The present chapter covers different aspects of aromatic oils, their chemical structure, methods of extraction, and therapeutic properties that can be utilized for nanoparticle synthesis for effective utilization in different fields.

Table 1 Plants parts containing essential oils

Part of plant	Plant
Flower	Jasmine, rose, clove
Seeds	Almond, carrot, cumin
Leaves	Basil, sandalwood, citronella, lemon, mint, tea
Bark	Cinnamon

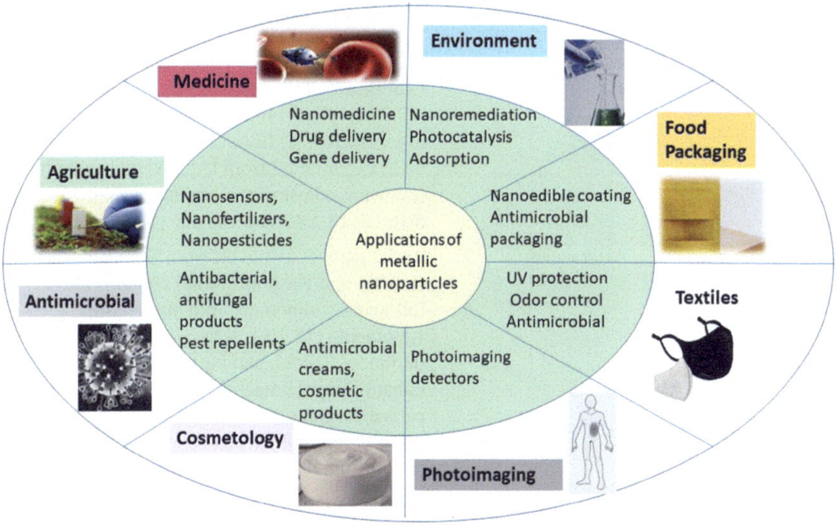

Fig. 1 Applications of metallic nanoparticles

2 Essential Oil (EO) Chemical Composition and Applications

The compounds can be broadly grouped as terpene hydrocarbons and oxygenated compounds. The most common essential oils are terpene hydrocarbons, monoterpene (C10H16), and sesquiterpenes ($C_{15}H_{24}$). The oxygenated compounds are the ones that can be derived from the terpenes like terpenoids. Other compounds include phenol, monoterpenealcohol, aldehyde, ketones, esters, oxides, lactones, and ethers [21]. The aroma depends on the arrangement of the molecules. This spatial arrangement includes simple groups like methyl, carbonyl, etc. The essential oils have spatial features, which include the components being hydrophobic, which further helps them partition the lipids from the cell membranes and mitochondria, further affecting the cellular structure and making it more permeable [22]. The active ingredients of EOs also account for their wide range of therapeutic properties (Table 2).

EOs exploited from aromatic plants are used in perfumery, flavouring, cosmetics, and spices and mainly in therapeutics. Aromatic plants and the EO extracted from them are added to the diets of broilers in the form of phytogenic feed to enhance the taste of feed for good performance [70]. Aromatic plants have also been reported to provide radioprotection due to the presence of antioxidants [102]. A study proposed that components present in the oil of *Laurusnobilis* L. such as alpha and beta-pinene, beta- ocimene and 1,8-cineole have antiviral properties and can be used to develop therapeutic against covid-19 [51]. Other uses of essential oils include increasing the shelf-life of food products. This application is one of the most sought-after because synthetic preservatives have now been recognised to have adverse side effects [41].

Table 2 Components of EOs extracted from some aromatic plants

Aromatic plant	Components of essential oil
Abies pindrow (Pinaceae)	Leaf essential oil: -pinene (16.8%), camphene (19.9%), -pinene (6.5%), myrcene (6.7%), and limonene (6.7%) from Uttarakhand, India (21%). Stem essential oil from Uttarakhand contains the following constituents: limonene (24.4%), myrcene (8.3%), and -pinene (13.8%) [86]
Achillea millefolium L. (Asteraceae)	Aerial parts essential oil from Srinagar, Kashmir (Jammu and Kashmir, India) contains the following compounds: -pinene (10.6%), 1,8-cineole (15.1%), -caryophyllene (16.2%), -terpineol (0.1%), and borneol (0.2%). Sisso, Lahaul-Spiti (Himachal Pradesh, India) aerial parts essential oil: -pinene (14%), 1,8-cineole (3.2%), -caryophyllene (12.5%), -terpineol (4.4%), and borneol (8.5 percent) [2]
Acorus calamuss L (Araceae)	Rhizome oil from Biratnagar in eastern Nepal Rhizome: (E)-asarone (1.9–4.0%) and (Z)-asarone (84–86.9%). Leaf essential oil: (Z)-asarone (78.1%) and (E)-asarone from Biratnagar, Nepal (9.9%) [2]
Aegle marmelos (L.) Corrêa (Rutaceae)	Leaf essential oil from Biratnagar, eastern Nepal, contains the following compounds: limonene (64.1%), (E)-ocimene (9.7%), and germacrene B (4.7%). [107]. Several leaf oil samples from Uttarakhand, India: limonene (31–90.3%), -phellandrene (trace-43.5%), and (E)—ocimene (0.7–7.9%)
Ageratum conyzoides L. (Asteraceae)	Ageratochromene (42.5%), demothoxyageratochromene (16.7%), and -caryophyllene are the main components of aerial parts essential oil from Kumaun, Uttarakhand, India (20.7%) [87]
Ageratum houstonianum Mill. (Asteraceae)	Ageratochromene (52.6%), demothoxyageratochromene (22.5%), and -caryophyllene are the main components of an Indian aerial parts essential oil (9.7%) [62]
Cassia tora L (Fabaceae)	Elemol (26.9%), linalool (19.6%), and palmitic acid (15.3%) are present in leaf oil from Biratnagar, Nepal [106]
Kylling abrevifolia Rottb. (Cyperaceae)	Leaf oil from Biratnagar, Nepal, with high concentrations of -cadinol (40.3%), -muurolol (19.5%), and -germadrene D-4-ol (12.5 percent) ([88]
Mentha spicata L (Lamiaceae)	Aerial parts oil from Uttarakhand contains the following constituents: carvone (76.7%), limonene (9.6%)[25]

For example, benzoic acid and sulphites have the potential to cause allergies, nitrites can produce carcinogenic nitrosamines, and butylatedhydroxyanisole and butylated-hydroxyanisole can cause rodent carcinogenicity [120]. The extracts from medicinal, aromatic plants (MAPs) have also been utilized in aquaculture. Aromatic plant-derived bioactive compounds (ABIOC) can also be used as a substitute for antibiotics. It has been demonstrated that ABIOC can change the fermentation of ruminate, improve the quality of milk and limit the growth of microbes [24]. Essential oils can be used in packaging and are a natural addition to increase the shelf-life of food, diminishing the use of chemical Preservatives [123].

The life span of an essential oil varies from plant to plant. Sandalwood and patchouli have more life spans than citrus [98]. Further, the life span depends on the conservation methods and manufacturing processes. The optimization suggests that the storage of oil should be in a hermetic and resistant glass container and should be kept at a temperature between the range of 15 to 20 °C. Under these conditions, the oil can be stored for almost 3 years [81].

3 Extraction of Aromatic Oils

Extraction of aromatic oils is necessary for their use in various fields. EOs can be extracted from different plants using multiple plant components. The methods used to produce and remove crucial oils depend on the plant material. The condition and shape of the material are other factors to consider. This leads to a loss of biological activity and natural properties. Discolouration, smell/taste, and physical changes such as increased viscosity occur in severe cases. These changes in the extracted EOs should be avoided [123].

There are more than 200 constituents in EOs from different plant species, both volatile and non-volatile. Because of their potent and efficient qualities, EOs are frequently used as antibacterial, anticancer, anti-inflammatory, and antiviral agents. For extraction purpose, traditional and cutting-edge techniques are brought into use. These include: steam distillation, hydro diffusion, solvent extraction(traditional methods), supercritical fluid extraction, subcritical fluid extraction, solventless microwave extraction (advanced techniques).

Conventional Methods

Steam distillation—A process in which EOs are isolated from aromatic plants with the help of heat to degrade and rupture cellular contents of the plant [11, 91]. Steam distillation is regarded as the most commonly used method of extraction. About 93% of the essential oil can be separated by this method, and the rest, 7%, can be isolated by other methods [78, 97]. Plant materials are either heated in steam or boiling water. This process depends on the high heating temperature of aromatic plants to decay plant composition resulting in the release of EOs.

Hydrodiffusion—This is similar to steam distillation, provided they yield steam to extract oils. The only difference between these two processes is the origin of the introduction of steam into the plant body. Steam enters the top of the plant in hydro diffusion, on the other hand, from below in steam distillation [127]. Hydrodiffusion can also occur below 100 °C steam temperature and even under less vacuum pressure. High oil yields and less time consumption are the advantages of hydrodiffusion over steam distillation. This method includes dried plant materials without damaging them at 100 °C.

Solvent extraction—The fragile materials of flora that cannot withstand a high temperature by stem distillation are exposed to conventional extraction. Here, solvents like acetone, hexane, petroleum ether, ethanol, etc. [6, 58, 93]. For the experimental setup, the solvent has to be mixed with the plant material, which can then be exposed to heat to extract the essential oil. This process is followed by filtration. This filtrate is then concentrated with the help of evaporation to get a pure form of essential oil. The filtrate is mixed with a pure form of alcohol, and the extraction of essential oil along with distillation at low temperature takes place. On the evaporation of alcohol, the aroma also evaporates as the alcohol absorbs the aroma, and we finally get the purest form of essential oil [66]. It is reported by [105] that the aqueous extract has antioxidant properties as compared to other extracts, including hexane, dichloromethane, methanol and ethyl acetate. However, if the solvent residues are not removed from the final product, this causes toxicity and allergies and affects the immune system [37].

Advanced Methods

The conventional methods have disadvantages like loss of volatile compounds, low extraction efficiency, time consumption and decomposition of esters or unsaturated compounds due to the chemical reactions [32, 73, 94]. Besides, there are hazardous solvents that are seen in the extracts, and there is a low extraction efficiency with a loss of volatile chemicals leading to the breakdown of unsaturated compounds [40, 52]. This has led to switching from the conventional method to the advanced method like using superficial fluids or microwaves to extract essential oils. Modern extraction techniques are the most promising due to their quick extraction times, minimal energy and solvent usage, and low carbon dioxide emissions [10].

Supercritical Carbon dioxidemethod—This method uses carbon dioxide for extraction, where CO_2 has a calm state and is safe and harmless in its liquid form, which is made under high pressure and is therefore used as a supercritical fluid [43, 52, 111]. Methylene chloride is used as a modifier which helps in high recovery. Supercritical fluid extraction is considered cost-effective compared to steam distillation as there is more power usage in steam distillation.

Solvent-free microwave oven—This is a quick way of extracting the EO from aromatic herbs, spices and dried seeds. This process involves isolating and concentrating the volatile compounds in a single step [14, 72]. There is a reduction in yield when using 662W [37]. It has also been reported that the microwave method is advantageous over traditional alternative methods, as it involves shorter extraction times and cold pressing [CP], and there is a better yield observed that is 0.2% vs. 0.21% for HD and 0.05% for CP [36]. Currently, the equipment used shows quicker and more efficient extraction with less wastage [36].

Microwave Gravity Diffusion—This method does not require the use of solvents or water, and no residues are formed. When [19] compared hydro diffusion and microwave gravity diffusion for the extraction of EOs from Rosemary leaves; microwave method gave the following advantages over conventional hydrodiffusion:

1. Less extraction time (about 15 min in MHC and 3 h in HD).
2. Improved antioxidant and antibacterial activities.

Farhat et al. [35] reported the superiority of microwave method over vapour diffusion method while extracting EOs from orange peel, both in terms of yield and time consumption (12 min versus 40 min). The technique has now been established to be a superior form of green technology for EO extraction [72, 82, 121]. Aromatic oils have been known to have effective antiseptic, antibacterial, antiviral, antioxidant, antiparasitic, antifungal, and insecticidal properties [15, 21, 54].

4 Therapeutic Properties of EOs

Out of the many medicinal properties of EOs, some are discussed below.

Antimicrobial activity

Plant EOs effectively prevent food deterioration and prevent or stop pathogenic microbes [68, 99]. Carvacrol, one of the chemical components of numerous EOs has been demonstrated to have a strong antibacterial impact [126]. Gram-positive bacteria's lipophilic lipoteichoic acid endings may make it easier for hydrophobic molecules from EOs to enter their cells [27]. However, the resistivity of gram-negative bacteria towards EOs is attributed to the protective function of cell wall lipopolysaccharides or outer membrane proteins, which regulate the rate of diffusion of hydrophobic substances across the lipopolysaccharide layer [21]. Carvacrol (mainly extracted from thyme and oregano [7, 63] has proven to be effective against gram-negative bacteria as it breaks the outer membrane, helps to release lipopolysaccharides, and leads to the increment of permeability of the cytoplasmic membrane for the ATP; also altering the permeability of cations (H and K) for the gram-positive bacteria [21, 126]. Similarly, oregano oil has been reported to act against gram-positive bacteria [89]. The components that are extracted from the oregano oil are thymol, carvacrol, γ-terpinene and ρ-cymene [3, 21, 64]. Another potential example is rosemary oil. The monoterpenes in the rosemary oil, which include α-pinene, β-pinene, myrcene 1,8-cineole, borneol, camphor and verbenone, account for its antibacterial activity [3, 21, 85, 104]. EOs extracted from *Callistemon comboynensis* consists of: 53.03% of 1,8-cineole, eugenol, 8.3% o methyl eugenol, α-terpineol, and carveol and shown to have antibacterial properties [1]. As per a study, EOs extracted from coriander, oregano, and rosemary were together found to inhibit *Listeria innocua*growth, while thyme oil had the potential to inhibit both *L. innocua* and *Listeria monocytogenes* [120]

Antioxidant activity

Several substances in EOs resemble plant phenols, which are recognised for their antioxidant properties. Phenolics are organic substances made of a carbon atom directly connected to a hydroxyl group (-OH) as part of an aromatic ring. Free radicals can accept the hydrogen atom from the hydroxyl group, stopping them from oxidising other substances. The redox characteristics of phenolic compounds, which enable them to serve as hydrogen donors, reducing agents, single oxidising agents, and metal chelators, are primarily responsible for their antioxidant ability [61]. In addition to quenching single-oxygen synthesis and bonding by transition metal ion catalysts, EOs also operate as a chain initiation blocker, free radical scavenger, reducing agent, peroxide termination, hydrogen abstraction continuation blocker, and more antioxidants [77, 135]. The antioxidant property of aromatic oils can also be utilised to stop the oxidation of lipids in the food system.

Thymol and carvacrol are thought to have the most antioxidant activity out of all the crucial components in the oil [29]. Other primary active components include eugenol (in clove) [133], carvacrol (in oregano) [18] m-thymol (in thyme), eucalyptol (in eucalyptus) which have antioxidant activities [20, 100, 120]. According to a study, [122], ginger and turmeric had the highest levels of DPPH radical scavenging activity among root EOs. Turmeric oil has the strongest 2,2-and-bis (3-ethylbenzothiazoline-6-sulfonic acid) (ABTS) radical scavenging activity, followed by plai and ginger. Besides phenols, plant extracts' non-phenolic antioxidants may also boost antioxidant activity [42, 84]. The antioxidant activity of aromatic oils is largely influenced by factors like the plant harvesting period, extraction process, and choice of solvent [76, 134, 136].

Anticancer activity

The anticancer potential of several EOs, demonstrated in vitro and in vivo, is one of those biological effects that have received less attention from researchers. EO made from Gannanzao oranges significantly reduced the migration and proliferation of colorectal cancer cells and liver tumours [69]. When treated preventively in a rat model, the latter is sensitive to EO from *Origanumonites* and inhibits the formation of CT26 colon tumours in vivo [118]. The in vivo anticancer properties of EO from the leaves of the Brazilian medicinal plant *Croton matourensis* were studied by [31]. In a mouse model, EO tea seeds exhibit an anti-obesity tangible impact. They can also prevent obesity, lessen physical exhaustion, and increase physical performance, suggesting a potential benefit for metabolic disorders.

Wound-healing and immunoprotective activity

EOs has shown a significant wound-healing effect in vivo and in vitro models [90]. Wound healing capacity of EO of *Eugenia dysenteric* has been recently reported [79]. The EO of garlic and some organosulfur compounds are suggested to have a phagocytic function, by stimulating neutrophils' functional activity [108]. EOs have also been shown to have antiviral activities in both animals and humans [109, 119, 128].

5 Nanoparticles (Nps) and Nano Drugs

NPs are minimal materials ranging from 1 to 100 nm, sometimes surpassing 500 nm [80]. Due to their nanoscale size and large surface area, they have distinct physical and chemical properties. They also can impart various colours due to visible range absorption. All these mentioned properties depend on their size, structure or shape [57, 65, 114]. NPs have been effectively used for drug delivery [23, 26, 137] and other medical applications like imaging, biosensors and biomarkers [4, 30, 83, 101] (Table 3). These majorly include metallic NPs like gold and silver. Gold NPs rapidly transform the strongly absorbed light into localized heat that can be used for cancer photothermal treatment [95]. They have unique optical properties and are used in electronics, biological imaging and Surface Enhanced Raman Spectroscopy (SERS) [96]. The antibacterial properties favour Ag NPs being utilized more frequently in catheters and wound dressings [8]. These NPs can also be broadly exploited in developing biosensors and chemical sensors [124]. Iron oxide nanoparticles naturally occur as magnetite, used in Ultrasound, MRI and targeted drug delivery systems [80].

Table 3 Nanoparticle-based drugs and their application

Nanoparticle material	Drug	Use	References
Carbon nanotubules	Smooth walled CNT, Doxorubiun	Delivery of drug	[74]
Dendrimers and dextrans	PAMAM dendrimer, Resouist, Sinerem	Drug delivery, contrast agent in MRI imaging	[13, 39, 132]
Gold	Aurolase (gold nanoshell), CYT-21001 (TNFαPEGylated colloidal gold particles)	Delivery of drug	[132]
Liposome	Irinotecan , Mifamurtide MTP-PE, Daunorubian	Pancreatic cancer, Osteosarcoma, Kaposi's sarcoma	[5, 67, 92]
Magnetic nanoparticles	Nanotherm (polymer coated iron oxide), TNT-Anti-Ep-CAM	Tumour imaging, drug delivery	[130, 132]
Nanoparticle-bounded albumin	Paclitaxel	Non-small cell lung cancer, breast cancer, pancreatic cancer	[28]
PEG-PLA polymeric micelle	Paclitaxel	Ovarian cancer, breast cancer, lung cancer	[129]
Polymer protein conjugate	L-aspargine	Leukemia	[125]

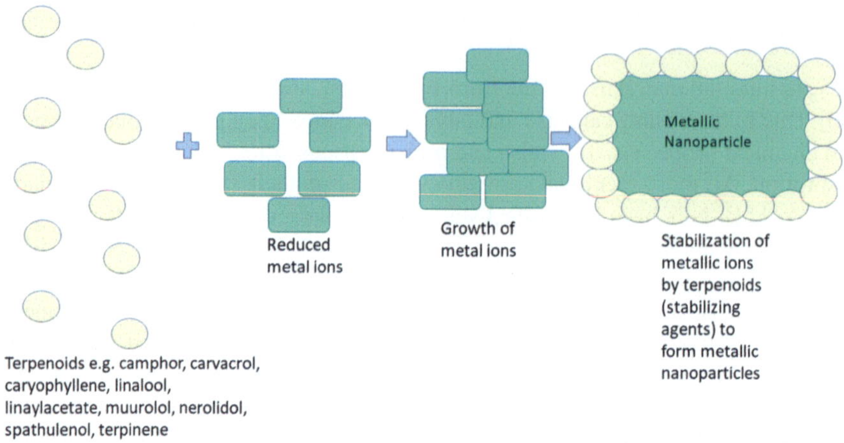

Fig. 2 Mechanism of synthesis of metallic nanoparticles from EO

6 EO Based NPs Synthesis

Use of phytomolecules in the preparation of herbal drugs has been explored by different workers. This is owing to the fact that aromatic oils have been known to have several therapeutic properties like anticancer, antinociceptive, antiphlogistic, antiviral, antibacterial, and antioxidant. The activity of these oils varies based on the source plant, chemical makeup, extraction technique, etc. EOs have different functional groups present in their structures like alcohol, aldehyde, ketone, lactone of terpenoids, etc. these have active groups which are reduced (Fig. 2) and are further stabilized. This shows their involvement in metallic nanoparticle synthesis.

Some commonly used aromatic plants for nanoparticle synthesis include hard-wood species like cinnamon, citrus, and herbs like lemon grass, basil, and mint (Table 4). Many other such aromatic oils need to be exploited for NP formulations for more effective usage in medical and other applications. The technique will not only help in efficient use of aromatic oil but also decrease the pressure on natural stands of aromatic plants which are mass-harvested for preparation of extracts for drug development.

7 Conclusion

Aromatic oils hold promise in the field of herb-based therapeutics. This has led to increased attention on these oils in the pharmaceutical industry. When combined with nanotechnology, aromatic oils can be very efficiently used in drug delivery and treatment of several diseases without any potential harmful effect on the health of the individual. The future research direction must move towards recognising the

Table 4 Nanoparticles and their activities extracted from the aromatic plants

Plant	Common name	Aromatic constituent	Nanoparticle	Activity	Key references
Anise myrtle	Ringwood	Anethole	Poly (lactic-co-glycolic acid)	Antimicrobial	[34]
Artemisia arborescens	Tree wormwood	Essential oil	Liposome	Antiviral	[17]
Cinnamomum verum	Cinnamon	Cinnamon oil	Polylactide	Antibacterial	[116]
Citrus aurantifolia	Key lime	Lime oil	Chitosan	Antibacterial	[117]
Cuminum cyminum	Cumin	Essential oil	Chitosan	Antioxidant	[55]
Curcuma longa	Turmeric	Turmeric oil	Chitosan and alginate	Antiproliferative	[55]
Cymbopogon citrates	West Indian lemon grass	Lemongrass oil	Chitosan and alginate	Antiproliferative	[55]
Eucalyptus staigeriana	lemon-scented ironbark	Eucalyptus oil	Cashew gum	Antimicrobial	[71]
Mentha spicata	Spearmint	Carvone	Poly (lactic-co-glycolic acid)	Antimicrobial	[34]
Ocimum species	Sweet basil	Eugenol	Nanoemulsion	Antifungal	[16]
Origanum dictamnus L	dittany of Crete	Thymol	Liposome	Antimicrobial	[38]
Pogostemon cablin	Patchouli	Wild patchouli oil	Nanoemulsion	Antibacterial and anti-candida	[131]
Syzygium aromaticum	Clove	Clove oil	Polylactide	Antibacterial	[116]
Thymus daenensis	Thyme	Essential oil	Nanoemulsion	Antibacterial	[75]
Zanthoxylum tingoassuiba	Fagaraarticulata	Essential oil	Liposome	Antimicrobial	[9]
Zataria multiflora	Avishan-E-Shirazi	Essential oil	Chitosan	Antifungal	[33]

phytochemicals that can help stabilize the metal nanoparticles with the help of EOs. Further understanding of the metabolic pathways can also be helpful in genetically modifying the plants to enhance the metallic nanoparticle synthesis. Besides, due importance must be given to toxicological studies on aromatic oil based nanoparticles to ensure the practical application of EO based nanotechnology.

References

1. Abdelhady MI, Aly HAH (2012) Antioxidant antimicrobial activities of *Callistemon comboynensis* essential oil. Free Radic Antioxidants 2:37–41
2. Agnihotri VK, Lattoo SK, Thappa RK, Kaul P, Qazi GN, Dhar AK, Saraf A, Kapahi BK, Saxena RK, Agarwal SG (2005) Chemical variability in the essential oil components of *Achillea millefolium* Agg. from different Himalayan habitats (India). Planta Med 71:280–283
3. Aguirre A, Borneo R, León AE (2013) Antimicrobial, mechanical and barrier properties of triticale protein films incorporated with oregano essential oil. Food Biosci 1:2–9
4. Allen TM, Cullius PR (2004) Drug delivery system: entering the mainstream. Science 303:1818–1822
5. Ando K, Mori K, Corradini N, Redini F, Heymann D (2011) Mifamurtide for the treatment of nonmetastatic osteosarcoma. Expert Opin Pharmacother 12(2):285–292
6. Areias F, Valentao P, Andrade PB, Ferreres F, Seabra RM (2000) Flavonoids and phenolic acids of sage: influence of some agricultural factors. J Agric Food Chem 48:6081–6084
7. Arrebola ML, Navarro MC, Jimenez J, Ocana FA (1994) Yield and composition of the essential oil of *Thymus serpylloides* subsp. serpylloides. Phytochemistry 36:67–72
8. AshaRani PV, Low KahMun G, Hande MP, Valiyaveettil S (2009) Cytotoxicity and genotoxicity of silver nanoparticles in human cells. ACS Nano 3:279–290
9. Asprea M, Leto I, Bergonzi MC, Bilia AR (2017) Thyme essential oil loaded in nanocochleates: encapsulation efficiency, in vitro release study and antioxidant activity. LWT 77:497–502
10. Aziz ZAA, Ahmad A, Setapar SHM, Karakucuk A, Azim MM, Lokhat D, Rafatullah M, Ganash M, Kamal MA, Ashraf GM (2018) Essential oils: extraction techniques, pharmaceutical and therapeutic potential - a review. Curr Drug Metab 19(13):1100–1110
11. Babu KGD, Kaul VK (2005) Variation in essential oil composition of rose-scented geranium (Pelargonium sp.) distilled by different distillation techniques. Flavour Frag J 20:222–231
12. Bakari M, Yusuf HO (2018) Utilization of locally available binders for densification of rice husk for biofuel production. Banats J Biotechnol 9(19):47–55
13. Barone PW, Yoon H, Ortiz-Garcia R, Zhang J, Ahn JH, Kim JH, Strano MS (2009) Modulation of single-walled carbon nanotube photoluminescence by hydrogel swelling. ACS Nano 3(12):3869–3877
14. Bayramoglu B, Sahin S, Sumnu G (2008) Solvent-free microwave extraction of essential oil from oregano. J Food Eng 88:535–540
15. Benjilali B, Ayadi A (1986) Methoded'études des propriétesantiseptiques des huilesessentielles par contact direct en milieu gelose (*Thymus capitatus, Rosmarinus officinalis, Eucalyptus globulus, Artemisia herbaalba*) Plantes Méd. Phytothér 2:155–167
16. Beyki M, Zhaveh S, Khalili ST, Rahmani-Cherati T, Abollahi A, Bayat M, Tabatabaei M, Mohsenifar A (2014) Encapsulation of menthapiperita essential oils in chitosan–cinnamic acid nanogel with enhanced antimicrobial activity against *Aspergillus flavus*. Ind Crop Prod 54:310–319
17. Bhattacharya S, Ghosh A (2009) Phytosomes: the emerging technology for enhancement of bioavailability of botanicals and nutraceuticals. Int J Health Res 2(1):141–153
18. Bounatirou S, Smiti S, Miguel MG, Faleiro L, Rejeb MN, Neffati M, Costa MM, Figueiredo AC, Barroso JG, Pedro LG (2007) Chemical composition, antioxidant and antibacterial activities of the essential oils isolated from Tunisian *Thymus capitatus* Hoff. et Link. Food Chem 105:146–155
19. Bousbia N, AbertVian M, Ferhat MA, Petitcolas E, Meklati BY, Chemat F (2009) Comparison of two isolation methods for essential oil from rosemary leaves: hydrodistillation and microwave hydrodiffusion and gravity. Food Chem 114:355–362
20. Bozin B, Mimica-Dukic N, Simin N, Anackov G (2006) Characterization of the volatile composition of essential oils of some Lamiaceae spices and the antimicrobial and antioxidant activities of the entire oils. J Agric Food Chem 54:1822–1828

21. Burt S (2004) Essential oils: their antibacterial properties and potential applications in foods– a review. Int J Food Microbiol 94(3):223–253
22. Butnariu M, Sarac I (2018) Essential oils from plants. J biotechnol Biomed 1(4):35–43
23. Calvo P, Remuoon-Lopez C, Vila-Jato JL, Alonso MJ (1997) Novel, hydropjilic chitosan-polyethylene oxide nanoparticles as protein carriers. J Appl Polym Sci 63:125–132
24. Caroprese M, Ciliberti MG, Albenzio M (2020) Application of aromatic plants and their extracts in dairy animals. In: Florou-Paneri P, Christaki E, Giannenas I (Eds) Feed additives aromatic plants and herbs in animal nutrition and health. Elsevier Inc., 50 Hampshire St., 5th Floor, Cambridge, MA 02139, USA, pp 261–277
25. Chauhan RS, Kaul MK, Shahi AK, Kumar A, Ram G, Tawa A (2009) Chemical composition of essential oils in *Mentha spicata* L. accession [IIIM(J)26] from north-west Himalayan region. Ind Crop Prod 29:654–656
26. Cooke JP, Atkins J (2016) Nanotherapeutic solutions for cardiovascular disease. Methodist Debakey Cardiovasc J 12:132–133
27. Cox SD, Mann CM, Markham JL, Bell HC, Gustafson JE, Warmington JR, Wyllie SG (2000) The mode of antimicrobial action of the essential oil of *Melaleuca alternifoli*a (tea tree oil). J Appl Microbiol 88:170–175
28. Cucinotto I, Fiorillo L, Gualtieri S, Arbitrio M, Ciliberto D, Staropoli N, Grimaldi A, Luce A, Tassone P, Caraglia M, Tagliaferri P (2013) Nanoparticle albumin bound paclitaxel in the treatment of human cancer: nanodelivery reaches prime-time? J Drug Deliv 2013:1–10
29. Dapkevicius A, Venskutonis R, Van Beek TA, Linssen JPH (1998) Antioxidant activity of extracts obtained by different isolation procedures from some aromatic herbs grown in Lithuania. J Sci Food Agric 77:140–146
30. Dash DK, Panik RK, Sahu AK, Tripathi V (2020) Role of nanobiotechnology in drug discovery, development and molecular diagnostic. In Stoytcheva M, Zlatev R (eds) Applications of nanobiotechnolgy. IntechOpen Ltd, 10 Lower Thames Street, London, United Kingdom, pp 19–32
31. De Lima EJSP, Alves RG, D'Elia GMA, Da Annunciacao TA, Silva VR, De Santos LS, Soares MBP, Cardozo NMD, Costa EV, Da Silva FMA (2019) Antitumour effect of the essential oil from the leaves of *Croton matourensis* Aubl. (Euphorbiaceae). Mol. 24:2974
32. Deng C, Yao N, Wang A, Zhang X (2005) Determination of essential oil in a traditional Chinese medicine, Fructusamomi by pressurized hot water extraction followed by liquid-phase microextraction and gas chromatography-mass spectrometry. Anal Chim Acta 536:237–244
33. Detoni CB, Cabral-Albuquerque ECM, Hohlemweger SVA, Sampaio C, Barros TF, Velozo ES (2009) Essential oil from *Zanthoxylum tingoassuiba* loaded into multilamellar liposomes useful as antimicrobial agents. J Microencapsul 26(8):684–691
34. Esfandyari-Manesh M, Ghaedi Z, Asemi M, Khanavi M, Manayi A, Jamalifar H et al (2013) Study of antimicrobial activity of anethole and carvone loaded PLGA nanoparticles. J Pharm Res 7(4):290–295
35. Farhat A, Fabiano-Tixier A-S, Visinoni F, Romdhane M, Chemat F (2010) A surprising method for green extraction of essential oil from dry spices: microwave dry-diffusion and gravity. J Chromatogr A 1217:7345–7350
36. Farhat A, Fabiano-Tixier A-S, Maataoui ME, Maingonnat J-F, Romdhane M, Chemat F (2011) Microwave steam diffusion for extraction of essential oil from orange peel: kinetic data, extract's global yield and mechanism. Food Chem 125:255–261
37. Ferhat MA, Meklati BY, Chemat F (2007) Comparison of different isolation methods of essential oil from Citrus fruits: cold pressing, hydrodistillation and microwave 'dry' distillation. Flavour Frag J 22:494–504
38. Ghosh V, Mukherjee A, Chandrasekaran N (2014) Eugenol-loaded antimicrobial nanoemulsion preserves fruit juice against, microbial spoilage. Colloids Surf B: Biointerfaces 114:392–397
39. Gindy ME, Prud'homme RK (2009) Multifunctional nanoparticles for imaging, delivery and targeting in cancer therapy. Expert Opin Drug Deliv 6(8):865–878

40. Gironi F, Maschietti M (2008) Continuous countercurrent deterpenation of lemon essential oil by means of supercritical carbon dioxide: experimental data and process modelling. Chem Eng Sci 63:651–661
41. Gómez-Estaca J, López de Lacey A, López-Caballero ME, Gómez- Guillén MC, Montero P (2010) Biodegradable gelatin-chitosan films incorporated with essential oils as antimicrobial agents for fish preservation. Food Microbiol 27(7):889–896
42. Hassimotto NMA, Genovese MI, Lajolo FM (2005) Antioxidant activity of dietary fruits, vegetables, and commercial frozen fruit pulps. J Agric Food Chem 53:2928–2935
43. Hawthorne SB, Rickkola ML, Screnius K, Holm Y, Hiltunen R, Hartonen K (1993) Comparison of hydrodistillation and supercritical fluid extraction for the determination of essential oils in aromatic plants. J Chromatogr A 634:297–308
44. Husen A (2020a) Introduction and techniques in nanomaterials formulation, *In*: Nanomaterials for Agriculture and Forestry Applications (Eds. Husen, A, Jawaid M) Elsevier Inc. 50 Hampshire St., 5th Floor, Cambridge, MA 02139, USA, pp. 1–14
45. Husen, A (2020b) Interactions of metal and metal-oxide nanomaterials with agricultural crops: an overview, In: Nanomaterials for Agriculture and Forestry Applications (Eds. Husen, A, Jawaid M) Elsevier Inc. 50 Hampshire St., 5th Floor, Cambridge, MA 02139, USA, pp. 167–197 https://doi.org/10.1016/B978-0-12-817852-2.00007-X
46. Husen A (2020c) Carbon-based nanomaterials and their interactions with agricultural crops (Eds. Husen, A, Jawaid M) Elsevier Inc. 50 Hampshire St., 5th Floor, Cambridge, MA 02139, USA, pp. 199–218 https://doi.org/10.1016/B978-0-12-817852-2.00008-1
47. Husen A (2022d) Engineered nanomaterials for sustainable agricultural production, soil improvement and stress management. Elsevier Inc., 50 Hampshire St., 5th Floor, Cambridge, MA 02139, USA
48. Husen A, Iqbal M (2019) Nanomaterials and Plant Potential. Springer International Publishing AG, Gewerbestrasse 11, 6330 Cham. https://doi.org/10.1007/978-3-030-05569-1
49. Husen A, Siddiqi KS (2022) Advances in smart nanomaterials and their applications. Elsevier Inc., 50 Hampshire St., 5th Floor, Cambridge, MA 02139, USA
50. Husen A, Siddiqi KS (2014) Phytosynthesis of nanoparticles: concept, controversy and application. Nanoscale Res Lett 9(229):1–24
51. Jahan R, Paul AK, Jannat K, Rahmatullah M (2021) Plant essential oils: possible COVID-19 therapeutics. Nat Prod Commun 16(2):1
52. Jimenez-Carmona MM, Ubera JL, Luque de Castro MD (1999) Comparison of continuous subcritical water extraction and hydrodistillation of marjoram essential oil. J Chromatogr A 855:625–632
53. Jin-Chul K, Madhusudhan A, Husen A (2021) Smart nanomaterials in biomedical applications. Springer Nature Switzerland AG, Gewerbestrasse 11, 6330 Cham. https://doi.org/10.1007/978-3-030-84262-8
54. Kaloustian J, Chevalier J, Mikail C, Martino M, Abou L, Vergnes MF (2008) Étude de six huilesessentielles: composition chimique et activitéantibactérienne. Phytothérapie 6:160–164
55. Karimirad R, Behnamian M, Dezhsetan S, Sonnenberg A (2018) Chitosan nanoparticles loaded *Citrus aurantium* essential oil: a novel delivery system for preserving the postharvest quality of *Agaricus bisporus*. J Sci Food Agric 98(13):5112–5119
56. Keller AA, McFerran S, Lazareva A, Suh S (2013) Global life cycle releases of engineered nanomaterials. J Nanopart Res 15(6):1692
57. Khan I, Saeed K, Khan I (2019) Nanoparticles: properties, applications and toxicities. Arab J Chem 12(7):908–931
58. Kosar M, Dorman HJD, Hiltunen R (2005) Effect of an acid treatment on the phytochemical and antioxidant characteristics of extracts from selected Lamiaceae species. Food Chem 91:525–533
59. Kumar RS, Sivakumar T, Sunderam RS, Gupta M, Mazumdar UK, Gomathi P, Rajeshwar Y, Saravanan S, Kumar MS, Murugesh K, Kumar KA (2005) Antioxidant and antimicrobial activities of *Bauhinia racemosa* L. stem bark. Brazil J Med Biol Res 38:1015–1024

60. Kumar A, Choudhary A, Kaur H, Mehta S, Husen A (2021) Metal-based nanoparticles, sensors and their multifaceted application in food packaging. J Nanobiotechnol 19(256):1–25. https://doi.org/10.1186/s12951-021-00996-0
61. Kumar A, Choudhary A, Kaur H, Guha S, Mehta S, Husen A (2022) Potential applications of engineered nanoparticles in plant disease management: a critical update. Chemosphere 295:133798
62. Kurade NP, Jaitak V, Kaul VK, Sharma OP (2010) Chemical composition and antibacterial activity of essential oils of *Lantana camara, Ageratum houstonianum* and *Eupatorium adenophorum*. Pharm Biol 48:539–544
63. Lagouri V, Blekas G, Tsimidou M, Kokkini S, Boskou D (1993) Composition and antioxidant activity of essential oils from oregano plants grown wild in Greece. Zeitschrift fur Lebensmittel-Untersuchung und -Forschung 197:20–23
64. Lambert RJW, Skandamis PN, Coote PJ, Nychas GJE (2001) A study of the minimum inhibitory concentration and mode of action of oregano essential oil, thymol and carvacrol. J Appl Microbiol 91:453–462
65. Laurent S, Forge D, Port M, Roch A, Robic C, Vander Elst L, Muller RN (2010) Magnetic iron oxide nanoparticles: synthesis, stabilization, vectorization, physicochemical characterizations, and biological applications. Chem Rev 108(6):2064–2110
66. Li XM, Tian SL, Pang ZC, Shi JY, Feng ZS, Zhang YM (2009) Extraction of *Cuminum cyminum* essential oil by combination technology of organic solvent with low boiling point and steam distillation. Food Chem 115:1114–1119
67. Lin HJ, Liang TL, Chang YY, Liu DZ, Fan JY, Roffler SR, Lin SY (2022) Development of irinotecan liposome armed with dual-target anti-epidermal growth factor receptor and anti-fibroblast activation protein-specific antibody for pancreatic cancer treatment. Pharm 14(1202)
68. Lis-Balchin M, Deans SG, Eaglesham E (1998) Relationship between bioactivity and chemical composition of commercial essential oils. Flavour Frag J 13:98–104
69. Liu K, Deng W, Hu W, Cao S, Zhong B, Chun J (2019) Extraction of 'Gannanzao' orange peel essential oil by response surface methodology and its effect on cancer cell proliferation and migration. Molecules 24:499
70. Li-Zhi J, Giannenas I, Dersjant-Li Y (2020) Application of aromatic plants and their extracts in diets of broiler chickens. In: Florou-Paneri P, Christaki E, Giannenas I (eds) Feed additives aromatic plants and herbs in animal nutrition and health pp 159–185. Elsevier Inc., 50 Hampshire St., 5th Floor, Cambridge, MA 02139, USA
71. Lu WC, Huang DW, Wang CR, Yeh CH, Tsai JC, Huang YT, Li PH (2018) Preparation, characterization, and antimicrobial activity of nanoemulsions incorporating citral essential oil. J Food Drug Anal 26(1):82–89
72. Lucchesi ME, Chemat F, Smadja J (2004) Solvent-free microwave extraction of essential oil from aromatic herbs: comparison with conventional hydro-distillation. J Chromatogr A 1043(2):323–327
73. Luque de Castro MD, Jiménez-Carmona MM, Fernández-Pérez V (1999) Towards more rational techniques for the isolation of valuable essential oils from plants. TrAC Trends Anal Chem 18:708–716
74. Madani SY, Naderi N, Dissanayake O, Tan A, Seifalian AM (2011) A new era of cancer treatment: carbon nanotubes as drug delivery tools. Int J Nanomed 6:2963–2979
75. Majeed H, Liu F, Hategekimana J, Sharif HR, Qi J, Ali B, Bian YY, Ma J, Yokoyama W, Zhong F (2016) Bactericidal action mechanism of negatively charged food grade clove oil nanoemulsions. Food Chem 197(Pt A):75–83
76. Malatova K, Hitimana N, Niyibizi T, Simon JE, Juliani HR (2011) Optimization of harvest regime and post-harvest handling in geranium production to maximize essential oil yield in Rwanda. Ind Crops Prod 34:1348–1352
77. Mao LC, Pan X, Que F, Fang X-H (2006) Antioxidant properties of water and ethanol extracts from hot air-dried and freeze-dried daylily flowers. Eur Food Res Technol 222:236–241

78. Masango P (2005) Cleaner production of essential oils by steam distillation. J Cleaner Prod 13:833–839
79. Mazutti da Silva SM, Rezende Costa CR, Martins Gelfuso G, Silva Guerra EN, De Medeiros YK, Gomes SM, Pic-Taylor A, Fonseca-Bazzo YM, Silveira D, Magalhães PO (2019) Wound healing effect of essential oil extracted from *Eugenia dysenterica* DC (Myrtaceae) leaves. Molecules 24:2
80. Mody VV, Siwale R, Singh A, Mody HR (2010) Introduction to metallic nanoparticles. J Pharm Bioallied Sci 2(4):282–289
81. Mohammadkhani M, Ansary SMM, Ahmadimoghadam A, Mansour S (2017) Distribution of cyanobacteria in two sirch hot springs with regards to the physicochemical traits of water. Banats J Biotechnol 8(15):83–89
82. Moradi S, Fazlali A, Hamedi H (2018) Microwave-assisted hydro-distillation of essential oil from rosemary: comparison with traditional distillation. Avicenna J Med Biotechnol 10(1):22–28
83. Nair SB, Dileep A, Rajanikant GK (2011) Nanotechnology-based diagnostic and therapeutic strategies for neuroscience with special emphasis on ischemic stroke. National Institute of Technology Calicut, India
84. Newman DJ, Gragg GM, Holbeck S, Sausville EA (2002) Natural products and derivatives as leads to cell cycle pathway targets in cancer chemotherapy. Curr Cancer Drug Targets 2:279–308
85. Okoh OO, Sadimenko AP, Afolayan AJ (2010) Comparative evaluation of the antibacterial activities of the essential oils of *Rosmarinus officinalis* L. obtained by hydrodistillation and solvent free microwave extraction methods. Food Chem 120:308–312
86. Padalia RC, Verma RS, Sundaresan V (2010) Volatile constituents of three invasive weeds of Himalayan region. Rec Nat Prod 4:109–114
87. Padalia RC, Verma RS, Chauhan A, Goswami P, Chanotiya CS (2014) Chemical analysis of volatile oils from west himalayan pindrow fir *Abies pindrow*. Nat Prod Commun 9:1181–1184
88. Paudel P, Satyal P, Khadka G, Setzer WN (2012) Leaf essential oil composition of *Kyllinga brevifolia* Rottb. from Nepal. J Essent Oil Bear Plants 15:854–857
89. Pelissari FM, Grossmann MVE, Yamashita F, Pineda EAG (2009) Antimicrobial, mechanical, and barrier properties of cassava starch-chitosan films incorporated with oregano essential oil. J Agric Food Chem 57:7499–7504
90. Perez-Recalde M, Ruiz Arias IE, Hermida EB (2018) Could essential oils enhance biopolymers performance for wound healing? A systematic review. Phytomed 38:57–65
91. Perineau F, Ganou L, Vilarem G (1992) Studying production of lovage essential oils in a hydrodistillation pilot unit equipped with a colouration system. J Chem Technol Biotechnol 53:165–171
92. Petre CE, Dittmer DP (2007) Liposomal daunorubincin as treatment for Kapsi's sarcoma. Int J Nanomed 2(3):277–288
93. Pizzale L, Bortolomeazzi R, Vichi S, Überegger E, Conte LS (2002) Antioxidant activity of sage (*Salvia officinalis* and *S. fruticosa*) and oregano (Origanumonites and *O. intercedes*) extracts related to their phenolic compound content. J Sci Food Agric 82:1645–1651
94. Pollien P, Ott A, Fay LB, Maignial L, Chaintreau A (1998) Simultaneous distillation–extraction: preparative recovery of volatiles under mild conditions in batch or continuous operations. Flavour Frag J 13:413–423
95. Prashant KJ, Ivan HS (2007) Au NPs target cancer. Nano Today 2:19–29
96. Rao RC, Kulkarni GU, Thomas JP, Edwards PP (2000) Metal nanoparticles and their assemblies. Chem Soc Rev 29:27–35
97. Reverchon E, Senatore F (1992) Isolation of rosemary oil: comparison between hydrodistillation and supercritical CO_2 extraction. Flavour Frag J 7:227–230
98. Rezaei A, Akhshabi S, Sadeghi F (2016) Evaluation of exon 17 of insulin receptor (INSR) gene and its relationship with diabetes type 2 in an Iranian population. Banats J Biotechnol 7(13):61–69

99. Rojas-Graü MA, Avena-Bustillos RJ, Olsen C, Friedman M, Henika PR, Martín-Belloso O, Pan Z, McHugh TH (2007) Effects of plant essential oils and oil compounds on mechanical, barrier and antimicrobial properties of alginate-apple puree edible films. J Food Eng 81:634–641
100. Ruberto G, Baratta MT (2000) Antioxidant activity of selected essential oil components in two lipid model systems. Food Chem 69:167–174
101. Sadrieh N, Tyner KM (2010) Nanotechnology and therapeutic delivery: a drug regulation perspective. Ther Deliv 1(1):83–89
102. Samarth RM, Samarth M, Matsumoto Y (2017) Medicinally important aromatic plants with radioprotective activity. Future Sci 3(4)
103. Sana S, Li H, Zhang Z, Sharma M, Usmani Z, Hou T, Netalaa VR, Wanga X, Gupta VK (2021) Recent advances in essential oils-based metal nanoparticles: a review on recent developments and biopharmaceutical applications. J Mol Liq 333:115951
104. Santoyo S, Cavero S, Jaime L, Ibanez E, Senorans FJ, Reglero G (2005) Chemical composition and antimicrobial activity of *Rosmarinus officinalis* L. essential oil obtained via supercritical fluid extraction. J Food Prot 68:790–795
105. Sarikurkcu C, Arisoy K, Tepe B, Cakir A, Abali G, Mete E (2009) Studies on the antioxidant activity of essential oil and different solvent extracts of Vitex agnus castus L. fruits from Turkey. Food Chem Toxicol 47:2479–2483
106. Satyal P, Woods KE, Dosoky MS, Neupane S (2012) Biological activities and volatile constituents of *Aegle marmelos* (L.). correa from Nepal. Active Passive Electr Compon 1(3):114–122
107. Satyal P, Mallik S, Gautam T, Vogler B, Setzer WN (2013) Composition and bioactivities of leaf essential oil of *Cassia tora*. Chem Nat Compd 49:553–554
108. Schepetkin IA, Kirpotina LN, Khlebnikov AI, Balasubramanian N, Quinn MT (2019) Neutrophil immunomodulatory activity of natural organosulfur compounds. Molecules 24:1809
109. Schnitzler P (2019) Essential oils for the treatment of herpes simplex virus infections. Chemotherapy 64:1–7
110. Sen S, Chakraborty R (2017) Revival, modernization and integration of Indian traditional herbal medicine in clinical practice: importance, challenges and future. J Tradit Complement Med 7(2):234–244
111. Senorans FJ, Ibanez E, Cavero S, Tabera J, Reglero G (2000) Liquid chromatographic-mass spectrometric analysis of supercritical-fluid extracts of rosemary plants. J Chromatogr A 870:491–499
112. Sharma G, Nim S, Alle M, Husen A, Kim JC. 2022. Nanoparticle-mediated delivery of flavonoids for cancer therapy: prevention and treatment. In: Kim JC, Alle M, Husen A (eds) Smart nanomaterials in biomedical applications. Nanotechnology in the life sciences. Springer, Cham. https://doi.org/10.1007/978-3-030-84262-8_3
113. Sharma P, Pandey V, Sharma MMM, Patra A, Singh B, Mehta S, Husen A (2021) A review on biosensors and nanosensors application in agroecosystems. Nano Res Lett 16:136. https://doi.org/10.1186/s11671-021-03593-0
114. Shin WK, Cho J, Kannan AG, Lee YS, Kim DW (2016) Cross-linked composite gel polymer electrolyte using mesoporous methacrylate-functionalized SiO_2 nanoparticles for lithium-ion polymer batteries. Sci Rep 6:26332
115. Siddiqi KS, Husen A (2021) Plant response to silver nanoparticles: a critical review. Crit Rev Biotechnol 42(7):773–990. https://doi.org/10.1080/07388551.2021.1975091
116. Sinico C, De Logu A, Lai F, Valenti D, Manconi M, Loy G, Bonsignore L, Fadda AM (2005) Liposomal incorporation of Artemisia arborescens L. essential oil and in vitro antiviral activity. Eur J Pharm Biopharm 59(1):161–8
117. Sotelo-Boyás M, Correa-Pacheco Z, Bautista-Baños S, Gómez y Gómez Y (2017) Release study and inhibitory activity of thyme essential oil-loaded chitosan nanoparticles and nanocapsules against foodborne bacteria. Int J Biol Macromol 103:409–414

118. Spyridopoulou K, Fitsiou E, Bouloukosta E, Tiptiri-Kourpeti A, Vamvakias M, Oreopoulou A, Papavassilopoulou E, Pappa A, Chilchia K (2019) Extraction, chemical composition, and anticancer potential of *Origanum onites* L. essential oil. Mol 24:2612
119. Tariq S, Wani S, Rasool W, Shafi K, Bhat MA, Prabhakar A, Shalla AH, Rather MA (2019) A comprehensive review of the antibacterial, antifungal and antiviral potential of essential oils and their chemical constituents against drug-resistant microbial pathogens. Microb Pathog 134:103580
120. Teixeira B, Marques A, Ramos C, Serrano C, Matos O, Neng NR, Nogueira JM, Saraiva JA, Nunes ML (2013) Chemical composition and bioactivity of different oregano (*Origanum vulgare*) extracts and essential oil. J Sci Food Agric 93(11):2707–2714
121. Tigrine-Kordjani N, Meklati BY, Chemat F (2006) Microwave 'dry' distillation as a useful tool for extraction of edible essential oils. Int J Aroma 16(3–4):141–147
122. Tongnuanchan P, Benjakul S, Prodpran T (2013) Characteristics and antioxidant activity of leaf essential oil–incorporated fish gelatin films as affected by surfactants. Int J Food Sci Technol 8(10):2143–2149
123. Tongnuanchan P, Benjakul S (2014) Essential oils: extraction, bioactivities, and their uses for food preservation. J Food Sci 79(7):R1231–R1249
124. Unser S, Bruzas I, He J, Sagle L (2015) Localized surface plasmon resonance biosensing: current challenges and approaches. Sensors 15:15684–15716
125. Varshosaz J, Anvari N (2018) Enhanced stability of L-asparaginase by its bioconjugation to poly(styrene-co-maleic acid) and ecoflex nanoparticles. IET Nanobiotechnol 12(4):466–472
126. Veldhuizen EJA, Tjeerdsma-van Bokhoven JLM, Zweijtzer C, Burt SA, Haagsman HP (2006) Structural requirements for the antimicrobial activity of carvacrol. J Agric Food Chem 54:1874–1879
127. Vian MA, Fernandez X, Visinoni F, Chemat F (2008) Microwave hydrodiffusion and gravity, a new technique for extraction of essential oils. J Chromatogr A 1190:14–17
128. Vuko E, Rusak G, Dunkic V, Kremer D, Kosalec I, Rada B, Bezic N (2019) Inhibitionof satellite RNA associated cucumber mosaic virus infection by essential oil of *Micromeria croatica* (Pers.). Schott Mol 24:1342
129. Wang YX (2011) Superparamagnetic iron oxide based MRI contrast agents: current status of clinical application. Quant Imaging Med Surg 1(1):35–40
130. Wang J, Li S, Han Y, Guan J, Chung S, Wang C, Li D (2018) Poly (ethylene glycol)- polyactide micelles for cancer therapy. Front Pharmacol 9:202
131. Wattanasatcha A, Rengpipat S, Wanichwecharungruang S (2012) Thymolnanospheres as an effective anti-bacterial agent. Int J Pharm 434(1–2):360–365
132. Wei A, Shibamoto T (2010) Antioxidant/lipoxygenase inhibitory activities and chemical compositions of selected essential oils. J Agric Food Chem 58:7218–7225
133. Wei A, Mehtala JG, Patri AK (2012) Challenges and opportunities in the advancement of nanomedicines. J Control Release 164(2):236–246
134. Wu Z, Li H, Yang Y, Zhan Y, Tu D (2013) Variation in the components and antioxidant activity of *Citrus medica* L. var. sarcodactylis essential oils at different stages of maturity. Ind Crops Prod 46:311–316
135. Yildirim A, Mavi A, Oktay M, Kara AA, Algur O, Bilaloglu V (2000) Comparison of antioxidant and antimicrobial activities of Tilia (TiliaargenteaDesf ex DC), sage (Salvia triloba l.), and black tea (*Camellia sinensis*) extracts. J Agric Food Chem 48:5030–5034
136. Zheljazkov VD, Astatkie T, Hristov AN (2012) Lavender and hyssop productivity, oil content, and bioactivity as a function of harvest time and drying. Ind Crops Prod 36:222–228
137. Zhou X, Hao Y, Yuam L, Pradhan S, Shrestha K, Pradhan O, Hongjie L, Wei L (2018) Nano-formulations for transdermal drug delivery: a review. Chin Chem Lett 29:1713–1724
138. Zinoviadou KG, Koutsoumanis KP, Biliaderis CG (2009) Physico-chemical properties of whey protein isolate films containing oregano oil and their antimicrobial action against spoilage flora of fresh beef. Meat Sci 82:338–345

Essential Oils from Plants and Their Role in Nanomaterial Synthesis Characterization and Applications

Venkata Kanaka Srivani Maddala and Sachidanand Singh

Abstract Many industries focussed on nanotechnology due to its application in various fields. Synthesis of nanomaterials by chemical and physical processes is expensive, release toxic substances into the environment, requires high energy and generates more waste. So biological sources are the best alternative for nanomaterial synthesis. There are several plants consisting of essential oils (EOs) that show antimicrobial, antiviral, anti-carcinogenic, anti-mutagenic and anti-inflammatory properties. They are used as active compounds in packaging food products and serve as natural additives in food materials, aromatherapy, cosmetic and medical industries. So synthesis of nanomaterials by plant parts is required as it is eco-friendly and sustainable. Synthesis of nanomaterials from plant EOs is more effective and eco-friendlier compared to nanomaterials made from physical and chemical processes. The present book chapter gives a detailed explanation on EOs from plant parts, its significance and synthesis of nanomaterials. It also highlights the characterization and application techniques of nanomaterial synthesized from the plant EOs.

Keywords Essential oil from plants · Significance of essential oils · Extraction of essential oils · Nanomaterial synthesis · Application of nanomaterials

1 Introduction

Essential oils (EOs) are complex liquids containing volatile and organic bioactive compounds and they are present in plants. They play an important role in cosmetics, medicine and food industries. They can be produced by glandular trichrome or other secretory tissues that are diffused onto the surface of plant organs i.e. flowers and leaves. They are the plant hormones or fluid manifestations of immune systems that

V. K. S. Maddala (✉)
Department of Chemistry, Vignans Foundation for Science Technology and Research Deemed to be University, Vadlamudi, Guntur, India
e-mail: sriani77@gmail.com

S. Singh
Department of Biotechnology, Sankalchand Patel University, Visnagar, Mehsana, Gujarat, India

A. Husen (ed.), *Secondary Metabolites Based Green Synthesis of Nanomaterials and Their Applications*, Smart Nanomaterials Technology,
https://doi.org/10.1007/978-981-99-0927-8_10

remove pests and attract insects and birds for pollination. It is the essence secreted from the cells of different plant parts and obtained by distilling or pressing. 3000 EOs are physically and chemically characterized, and out of them, 150 were manufactured in industries [1, 2]. Production of nanometer sized particles attain great interest due to its application in biomedical and other fields of science and technology [3–13]. Currently, many industries are showing special interest in nanomaterials based on transition metals due to its specific magnetic features, low melting temperature, high optical properties and mechanical resistance [14]. Even though there are many methods to synthesize nanoparticles (NPs) like physical, chemical, electrochemical, photochemical and radiation methods [15]. But they affect the environment and human health. Many research findings explained the synthesis of NPs by green plants and its parts as eco-friendly and sustainable. Synthesis of nanomaterials from plant essential oil is an eco-friendly approach and an easy process conducted at low temperature and pressure. The EOs of several plants like *Origanum vulgare, Myristica fravugrans, Thymus vulgaris, Gracilaria birdiae, Calliandra haematocephala, Rosemarinus officianalis, Cymbopogon martinii, Ocimmum gratissimum, Syzygium aromaticum, Cestrum nocturmum, lippialba, Akka ssellowiana, Psidium scatelani and Poiretia latifolia synthesize nanomaterials and show antibacterial, antiviral, antifungal, anti-inflammatory and antioxidant properties.* Nanomaterials synthesized from plants play an important role in the degradation of pollutants and possess antioxidant compounds polyphenols, sugars and amino acids. The distinctive properties of nanomaterials make them suitable for manufacturing, academic and medical domains in the areas of catalysis, food conservation, drug delivery, energy and environment, solar cells, quantum dots, photoimaging and wound repair. More preference can be given to plant essential oils that are more effective. Formation of nanomaterials from plant parts gained universal attention due to its application in various fields.

2 Essential Oils and Its Significance

EOs are the volatile compounds made from the parts of the plants like leaves, stems, flowers, roots and barks and consists of a variety of aromatic and various bioactive molecules. EOs are condensed hydrophobic liquids that consists of aromatic fragrant molecules from aromatic fragrant products formed as bioactive molecules in various plant tissues. They show antimicrobial, antiviral, anti-carcinogenic, anti-mutagenic and anti-inflammatory properties [16, 17]. EOs are used as active compounds in packaging food products due to its water vapor barrier property with hydrophobic nature. They serve as natural additives in food products and processing aid as green technology. Terpenes (monoterpenes, sesquerpenes), aromatic compounds (aldehyde, alcohol and phenols), terpenoids (isoprenoids) [18, 19] are the constituents present in EOs. They are also used for aromatherapy (alternative medicine) but there is no evidence that these EOs can treat the diseases [20]. EO were used long back due to its flavour. They are applied in food, perfume and medicinal industries. They are used in food preservation due to its antimicrobial nature against multi-resistant bacteria, an alternative to control pests and weeds. Metabolic constituents in EOs such as terpenes, alkaloids and phenolic compounds. They are also used as bio-fungicides and bio-herbicides and replace chemicals with no residual power [21]. The application of EOs in various fields is mentioned in Fig. 2.

EOs have a wide range of chemical compositions, with the primary constituents falling into the aliphatic, aromatic, and terpene families. EOs mostly comprise ternary and quaternary compounds. EOs are expensive as to obtain one drop of essential oil from rose requires 60 roses [22]. All the plants will not produce EOs and some EOs produced by the plant parts don't have therapeutic value and may be hazardous. So it is needed to take proper measures before extracting the EOs from the plant. Production of EOs from plant parts is shown in Table 1.

3 Extraction of Essential Oils from Plants

These are the liquified extractions of plants. Extraction of EOs from plant materials takes place through removal methods which are suited to the plant parts consisting of oils. Numerous methods were used to prepare EOs from plants such as hydro distillation, soxhlet extraction, microwave extraction and supercritical carbon dioxide extraction [26]. All the essential oil extraction methods are mentioned in Fig. 1. Extraction methods determine the quality of EOs and inappropriate essential oil extraction procedures may damage the chemical signature of essential oil that leads to loss of bioactivity and natural characteristics.

Steam distillation

Stream is added to the plant material containing the oil that releases plant aromatic molecules and converts into vapor and this vapor travels to the condenser and gets cooled. This cool water converts into liquid. The aromatic liquid by-products drop from the condenser and are collected in the separator. Oil and water will not mix and the essential oil floats on the water surface and is taken out.

Solvent extraction

This method uses solvents like hexane and ethanol to separate essential oil from the plant. The plant material is treated with solvent and produces a waxy like aromatic compound called concrete. This concrete combines with alcohol and oil particles are released.

The following types of solvent extraction are (i) CO_2 extraction (ii) Maceration (iii) Effleurage.

(i) CO_2 extraction: Here the pressurized CO_2 in liquid state is pumped into the chamber filled with plant material. CO_2 acts as a solvent and pulls the oil from plant material and dissolves in that solvent. $CO_{2\,is}$ brought back to natural pressure and evaporates into gas leaving the oil.

Table 1 List of plant parts containing essential oils

S. No.	Scientific name	Common name	Plant parts containing essential oils
1	*Zingiber officinale*	Ginger	Rhizome
2	*Thymus carnosus, Thymuscamphoratus*	Portuguese thyme, Camphor thyme	Flowering aerial parts [23]
3	*Citruslimon, citrus aurantifolia, citruslimonia*	Lemon, key lime, Mandarin lime	Peels of fruits [24]
4	*Boswellia ovalifoliolata*	Indian frankincense	Leaves and bark [25]
5	*Lavandula officinalis, Camellia sinensis, Cinnamomum verum, Eucalyptus golbulus, Nicotiana tobacum, Pogostimon cablin, Kunzea ericoides, Leptospermum scoparium*	Lavender, tea tree, cinnamon, blue gum, tobacco, patchouli, kanuka, manuka	Leaves
6	*Ocimum tenuiflorum, Thymus vulgaris, Salvia rosemarinus, Origanum vulgare, Mentha* x *Piperita, Rosa, Origanum majorana*	Holy basil, thyme, rosemary, peppermint, oregano, rose, marjoram	Flowers and leaves
7	*Jasminum*	Jamine	Petals, flowers and buds
8	*Boronia, Helichrysum, Syzygium aromaticum, Cananga odorata, Polianthes tubuerosa*	Boronia, curry plant, clove, Kananga, tuber rose,	Petals, flowers and buds
9	*Angelica, Zingiber officinale, Nardostachys jatamansi, Chrysopogan zizanioides*	Gardenangelica, ginger, spikenard, vetiver	Roots
10	*Elettaria cardamomum, Coffea, Daucus carota, Coriandrum sativum, Petroselinum crispum, Phaseolus vulgaris, Cuminum cyminum, Foeniculum vulgare, Pimpinella anisum, Myristica fragrans*	Cardamom, coffee, carrot, coriander, parsley, bean, cumin, fennel, anise, nutmeg	Seeds
11	*Piper nigrum, Juniperus, Pimenta dioica, Lislitsea cubeba*	Black pepper, juniper berry, allspice, may chang	Fruits/berry
12	*Cinnamomum verum, Cassia*	Cinnamon, golden shower	Bark

(continued)

Table 1 (continued)

S. No.	Scientific name	Common name	Plant parts containing essential oils
13	*Santalum album, Aniba rosaeodora, Cedrus, Bursera graveolens,Amyris*	Sandalwood, rose wood, cedar wood, Palo Santo, Sea torch wood	Wood
14	*Piceaabies, Cypressus, Abies, Pinus sylvestris*	Spruce, cypress, fir, Scotch pine	Needles
15	*Myroxylon balsanum*, *Styrax benzoin, Canarium luzonicum, Ferula gummosa, Commiphora myrrha, Boswellia sacra*	Perubalsam, benzoin, elemi, galbanum, myrrh, frankincense	Resins and gums
16	*Cymbopogon citratus, Cymbopogon nardus, Cymbopogon martinii*	Lemongrass, citronella, palm rosa	Grass

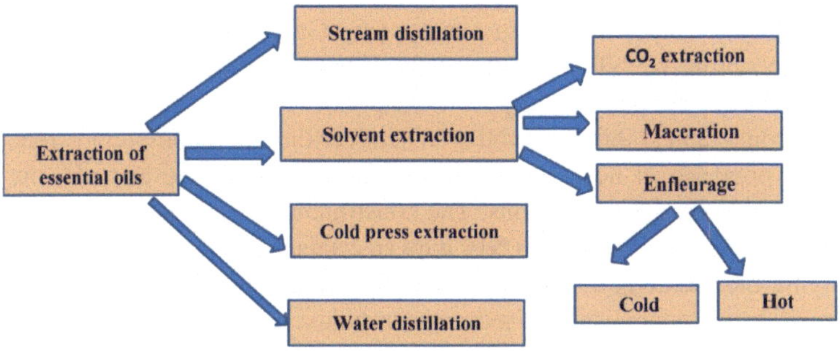

Fig. 1 Extraction of essential oils from plants and its parts

(ii) Maceration: Plant material is cut, powdered and placed in the closed vessel, the solvent is added and the mixture is shaken and leave it for one week. Liquid is strained and solid material is pressed. Strained and expressed liquids are mixed and clarified through subsidence. The base oil changes the colour after completion of macerate and the final maceration should be filtered of its plant material and kept in an air tight container for up to 12 months.

(iii) Enfleurage: Fats are used in this process. It can be done in two ways: cold and hot enfleurage. In cold enfleurage fat is spread over the glass plate and fresh flowers are added on the top of the fat layer and pressed and leave it for few weeks. Scent seeps into the fat, depleted petals are replaced and this process is repeated until the fat gets saturated. The final product is enfleurage pomade which is washed with alcohol to separate the extract. In hot enfleurage also same process as in cold enfleurage except the fats are heated.

Cold press extraction: The whole fruit is sent into the mechanical device and pierced to rupture the essential oil sacs. The EOs and pigments are run down to the device collection area. The fruits are pressed and juice and oil are squeezed. These are centrifuged to remove the solids. Oil separates from the juice layer and removed.

Water distillation

Here the plants are soaked in a container consisting of water. Water helps to prevent overheating of plants. Plant along with water is heated till the steam comes out. Oil comes out and enters into the condenser where both oil and water are collected in separate flasks and the collection of oil in the top layer of hydrosol can be separated. The extraction temperature is less than 100 °C to prevent evaporation of water and oil [27, 28].

4 Synthesis of Nano Materials from Plant Essential Oils

Nanomaterials are materials with particle sizes less than 100 nm in at least one of its dimensions and measured in nanometers. Green chemistry approaches are developed to synthesize NPs that are eco-friendly and sustainable [29]. Plants have active constituents like phenols, terpenoids, saponins, tannins, polyphenols, and vitamins that reduce and stabilize the NPs. Whenever the plant compounds react with salts in the presence of heat, temperature, rotation per minute and pH, the plant moieties reduce salts into metal ions. The constituents present in the plants synthesize metallic NPs. The synthesis of NPs with size, shape and distribution can be done by changing the methods of synthesis, stabilizers and reducing agents. Secondary constituents from plants can be used as adsorbents, reductors and capping agents of metal precursor that results in the formation of nanomaterials. This process is conducted at room temperature and are eco-friendly so don't produce any harmful derivatives. Such nanomaterial formed from the EOs can be used as antioxidant, antimicrobial, antiviral and antifungal properties [30] (Fig. 2).

Synthesis of nanomaterials from plant EOs are effective and useful in terms of their antibacterial, antifungal, antioxidant, antibiofilm and cytotoxic activities compared to the NPs synthesized from the physical and chemical processes. Silver, gold and zinc show great attention in biomedical applications, and platinum in energy storage. Silver nanoparticles are used in antibiotic agents in textile and wound dressings, medical devices, refrigerators and washing machine appliances [31]. Carbon based nanomaterials are used in drug delivery, biosensors, enzyme immobilization, bioimaging and pollutant removal [32]. Metal NPs exhibit catalytic activity and have strong adsorption capacity to remove the heavy metals [33], organic pollutants [34], inorganic anions [35] and bacteria [36]. Nanomaterials are used in the purification of air and water by adsorption, filtration and oxidation techniques. Nanosilver is an antifungal agent and can inhibit *Candida albicans, Candida glabrata, Candida parapsilosis, Candida krusei.* Gold NPs used in biosensing to detect bacteria [37], viruses, pollutant degradation [38] and monitoring pathogens [39]. Nanomaterial

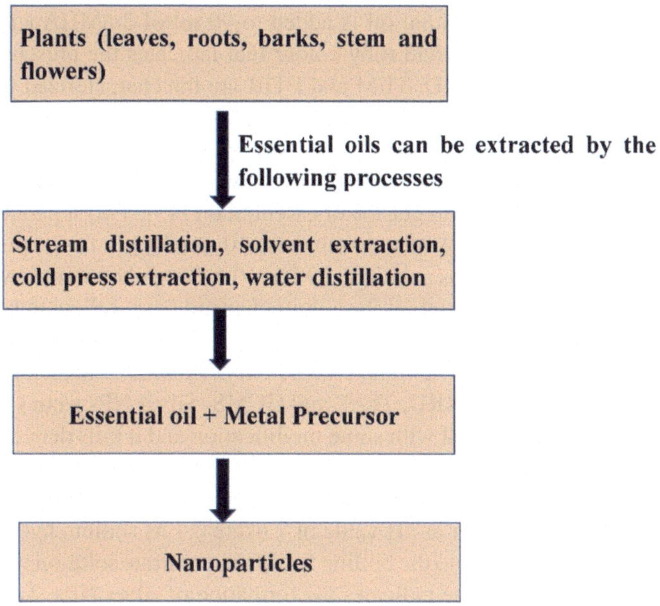

Fig. 2 Synthesis of nanoparticles from essential oils

formation can be analysed by UV–Visible spectroscopy and characterized by SEM, TEM, XRD, FTIR, DLS, and EXD.

Silver NPs can be synthesized from *Syzygium aromaticum* essential oil and it is used as a reducing agent for nanoparticle synthesis. First, the clove essential oil is diluted with acetone solution with a ratio of 1:170. 0.31 mmol.L^{-1}silver nitrate solution was prepared. To evaluate the solution at different condition, four aliquots of solution is taken and pH is adjusted to 7, 6, 9 and 10 with 0.1 mol.L^{-1}sodium hydroxide solution. 30 ML aliquot of each solution is heated with continuous stirring and 2 ml of clove essential oil is added to the boiling solution drop wise, it is heated and stirred for 30 min. The colour of the solution turns into yellowish brown indicating the formation of silver NPs. Here essential oil acts as a reducing and stabilizing agent and is characterized by UV–Vis spectroscopy, Transmission Electron Microscopy, and DLS (Dynamic light scattering [40]). Synthesis of gold nanoparticle using EOs of *Eucalyptus globulus* and *Rosemaria* occurs when 100μl of a liquor EOs of *Eucalyptus globulus* and *Rosemaria* is mixed with redistilled water: ethanol solution in 4:1 ratio for the synthesis of silver NPs. To produce gold NPs from EOs of *Eucalyptus globulus* and *R. officinalis,* the synthesis should be done at 95 °C [41]. When 1% of heated water–ethanol solution of these EOs is mixed with $HAuCl_4 \times 4\ H_2O$ to get 50.0 mg L^{-1} of gold in the reaction. UV–vis absorption revealed that gold NPs were synthesized with *E. globulus* leaf extract and its essential oil. Average production of NPs can be shown with TEM. Functional groups of organic compounds in EOs can be identified by ATR-FTIR. Key constituents can be determined by GS-MS. When

10 ml of *Mentha piperita* essential oil is added to 90 ml of 2mMHAucl$_4$ then the colour of HAucl$_4$ changes to vivid ruby colour that indicates the presence of gold NPs. UV–Vis, SEM, EDX, XRD, TEM and FTIR are the characterized techniques used in the synthesis [42].

Synthesis of gold NPs by *Diplotaxis acris* occurs when 100 μL of essential oil obtained from the flowers of *Diplotaxis acris* is dissolved in 100 mL of mixture ethanol/water (50/50) solution to get 1% of essential oil [43]. Two separate preparations were made with 10 and 10 ml and both these 10 and 20 ml of essential solution were added to HAuCL$_4$aqueous solution (1 Mm, 10 ml) and the mixture was stirred for 30 min. For 20 Ml of essential oil the colour changed from yellow to blue and in 10 Ml of essential oil gold NPs are formed and colour changed from yellow to wine red [44]. The gold NPs showed antibacterial activity. These are characterized using UV–Vis spectroscopy, FTIR, XRD, TEM, and GCMS. Silver NPs were synthesized using *C. Zeodoaria* essential oil with some modification and it kills dengue and zika virus vector aedes albopictus and the synthesis of silver NPs confirmed by UV–vis, SEM, EDX, XRD, and FTIR [45]. Essential oil of *C. Zeodoaria* in polysorbate 20 dissolves with deionized water at pH value of 7 using 0.1 M sodium hydroxide and the solution drops is dripped into the boiling 5 mM silver nitrate solution with continuous stirring and colour change indicates the formation of silver NPs. According to Sheny et al. [46], gold NPs were synthesized from EOs of *Anacardium* and characterized with the help of UV–VIS, FTIR, and TEM. Here 2 ml of essential oil was added to the 30 ml of aqueous 2.5×10^{-4} MHAu Cl$_4$ solution at 100 °C, stirred and boiled for one minute (sampleA). Colour of the solution becomes pink and AU^{3+} ions are reduced to AU^{O}, again the experiment is repeated with 5 and 8 ml oil and obtain purple and violet colour. Reduction is successful at room temperature slightly and slowly (sample D). Stock solutions of 5×10^{-3} M 4-nitrophenol and 0.25 M NaBH$_4$ were prepared and 1 ML of 4-nitrophenol mixed with 2 ml of deionized water followed by adding 1 ml of NaBH$_4$ solution [47] with stirring for 10 min and obtained bright yellow solution. To this bright yellow solution, 1 ml of AU collide was added and stirred for one minute and the solution becomes colourless.

Gold NPs were synthesized using *C. pseudomontana* [43]. Here 0.1 ml of essential oil is blended with ethanol: water solution (2.8 v/v ml). 3 ml of this mixture added to the gold solution (1 mM, 20 ml) and stirred for 30 min then the colour of the solution changes from yellow to ruby red. Gold NPs were characterized by FTIR, SEM, UV–Vis, and HR-TEM. U.V-Vis absorption measurements. Silver NPs can be synthesized from EOs of *C. aromaticus*. 3 ml of diluted essential oil is added to 30 ml of boiling 2.14X10-3 M AgNO$_3$ solution at pH value of 7 using NaOH. Colour of the solution changed to pale yellow confirming the presence of silver NPs [48]. Iron NPs were synthesized by *Satureja hortensis*. Ferrous sulphate and ferric chloride are added to the glass jar consisting of 200 ml of *Satureja hortensis* essential oil, this solution is stirred at 80 ^{0}C for half an hour and 30 ml of aqueous ammonia is pipetted slowly to the obtained suspension and continuously stirred at 80^{0}C for one hour and the FeNPs were separated and dried for one hour at 100 °C.FeNp were characterized by UV–Vis, FTIR, XRD, FE-SEM, ZP (Zete Potential) PSA (Particle size analyser) and EDX [49]. *Eucalyptus globulus* essential oil is mixed with zinc acetate dehydrate to prepare zinc NPs. XRD, DLS, FTIR, SEM, TEM, EDX and UV–Vis were used in the biosynthesis of zinc NPs [50].

Bioproduction of silver NPs are carried out from leaves of *O. Campechianum* rich in EOs and aromatic compounds. Ten grams of fresh leaf extract boiled in 50 ml of distilled water for 5 min. 3 ml of plant extract added to 50 ml of 1 mM boiled silver nitrate solution results in the formation of silver NPs and the filtrate acts as a reducing and stabilizing agent [51]. Here bioproduction of silver NPs and essential oil extract reduced the bacterial growth and essential oil extract in combination with silver NPs showed more efficiency in reducing the bacterial growth thanbioproduction of silver NPs alone [52]. Silver NPs are synthesized by using EOs of *Pogostemonis cablin* and *Aquilaria sinensis*. In this 0.01 g of each of EOs of *Pogostemonis Scablin* and *Aquilaria sinensis* in 1 ml DMSO were diluted with 10 ml of dechlorinated water and the pH of two solutions adjusted to 7 using 0.1 M of NAOH solution and this solution is dropped slowly into the boiling 100 ml of 2 mm silver nitrate solution and colourless solution changes to light yellow and finally red indicates the formation of silver NPs. Synthesis was characterized by UV–Vis, SEM, TEM, EDS, FTIR, and XRD [53].

Polymeric nanocapsules were prepared using oil in water emulsion and solvent evaporation method. Neem oil (1:1, 1.5:0.5) and Oleic acid (2:0, w/w) were dissolved in 100 ml of acetone and this solution was added to 20 ml of chloroform with 400 mg of PCL polymer (80,000 gm/mol) and the mixture was sonicated for 1 min at 100 W. Later pre-emulsion was added to the aqueous solution of polyvinyl alcohol surfactant and sonicated for eight minutes to get the emulsion. Here the rotary evaporator and the solvent gets evaporated and prepare 10 ml of emulsion characterization techniques DLS and TEM [54]. EOs from *Ocimum gratissimum* and *Pimenta racemose* were analyzed by CG-MS Zein NPs loaded in it were synthesized and their physiochemical characteristics, encapsulation efficiency, antioxidant activity and stability were evaluated. Synthesis was characterized by GCMS, FTIR and TEM. Nanoprecipitation techniques were used and Zein NPs were obtained [55]. Synthesis of NPs from EOs of plants parts, their characterization and applications are mentioned in Table 2. *Pelargonium graveolens* (6 ml) essential oil is added to 80 ml of 2 Mm of silver nitrate solution in a conical flask provided with a magnetic stirrer in it and started to run on the hot plate and within 15 min biosynthesis reaction begins and the colour of silver nitrate solution changed to dark brown in colour results in the formation of silver NPs. Characterized by UV–Vis, SEM, TEM, FTIR, XRD and Fluorescence spectroscopy [56].

Ginger essential oil act as reaction media for zinc oxide formation and also reduces Ag^+ to Ag^0. Zinc oxide silver core shell nanocomposite were synthesized using Ginger oil and the biosynthesis was characterized by UV–VIS, TEM, EDX, XRD, and FTIR [57]. Silver NPs can be synthesized from the essential oil of orange peel when 1 ml of orange peel essential oil is added to 10 ml of 0.003 M aqueous solution of silver nitrate solution with constant stirring at 70 °C for 48 h. Light brown colour obtained due to excitation of surface plasmon resonance indicates the formation of silver NPs [59]. According to Sattarahmady, [58], gold NPs are synthesized from *Citrus aurantiumL* blossoms and Rose damascene and characterized by UV–Vis,

Table 2 Synthesis of nanoparticles from plants, characterization and its applications

Name of the plant	Nanoparticals	Characterization	Applications	References
Syzygium aromaticum	Silver	UV–Vis spectroscopy, Transmission Electron Microscopy, DLS (Dynamic light scattering)	Antimicrobial	[40]
Rosmarinus officinalis and *Eucalyptus globulus*	Gold	UV–Vis, TEM, ATR-FTIR, GCMS	Antibacterial, antiviral, anti-inflammatory and antioxidant activity	[41]
Diplotaxis acris	Gold	UV–Vis spectroscopy, FTIR, XRD, TEM, GCMS	Antibacterial	
Menthapiperita	Gold	UV–Vis, SEM, EDX, XRD, TEM and FTIR	Antifungal	[42]
Curcuma zedoaria	Silver	UV–Vis, SEM, EDX, XRD, FTIR	Kills mosquitos, antiseptic, antiperiodic and anti-carcinogenic (11) Kills dengue and Zika virus vector	[45]
Anacardium occidentale	Gold	Investigated with the help of UV–VIS, FTIR, TEM	Catalytic activity	[46]
Curcuma pseudomontana	Gold	FTIR, SEM, UV–Vis, and HR-TEM. UV–Vis absorption measurements	Antibacterial, antioxidant and catalytic activity, anti-inflammatory	[43]
Essential oils of *Coleus aromatics* leaves	Silver	UV–VIS, FTIR, XRD, TEM	Antibacterial, antioxidant and catalytic activity	[48]
Satureja hortensis	Iron	UV–Vis, FTIR, XRD, FESEMED	Anticancer, antibacterial, antimicrobial	[49]

(continued)

Table 2 (continued)

Name of the plant	Nanoparticals	Characterization	Applications	References
Eucalyptus globulus	Zinc oxide	XRD. DLS, FTIR,SEM,TEM,EDX,UV–Vis	Antimicrobial	[50]
Ocimum campechianum	Silver	GCMS, UV–Vis FTIR, SEM, TEM	Antimicrobial	[52]
Pogostemonis cablin and *Aquilaria sinensis*	Silver	UV–Vis, SEM, TEM, EDS, FTIR, XRD	Antibacterial	[53]
Neem oil and oleic acid	Polymeric nanocapsules	DLS, TEM	Kills non target organisms Used as nanopesticide	
Ocimum gratissimum and *Pimenta racemosa*	Silver	FTIR, GCMS, TEM	Antioxidant	
Pelargonium Sgraveolens	Silver	UV–Vis, SEM, TEM, FTIR, XRD and Fluorescence spectroscopy	Antifungal properties	[56]
Zingiber officianalis (Ginger)	Zinc silver	Antibacterial, cytotoxicity	UV–VIS, TEM, EDX, XRD, FTIR	[57]
Citrus aurantium	Gold	UV–Vis, FE-SEM, TEM	Cytotoxicity	[58]
Orange peel essential oil	Silver	UV–Vis, FTIR, FE-SEM, XRD, GCMS, HR-TEM, EDX, DLS	Catalytic properties	[59]
Nigella sativa	Gold	Anticancer and antibacterial	UV –Vis, XRD, FTIR, TEM	[60]
Myristica fragrans seed essential oil Leaf essential oil	Silver Zinc oxide	UV–Vis, FTIR, GC–MS, XRD and TEM	Antibacterial, Cytotoxic effects in liver cells	[61]

FE-SEM, and TEM. Gold NPs from these oils diminish the toxicity of insulin fibrils in rat's pheochromocytoma PC 12 cells viability [58]. It was reported by that EOs of *Nigella sativa* synthesize gold NPs characterized by UV–Vis, FTIR, TEM, and XRD. Gold NPs shows antibacterial activity and inhibit the biofilm formation of *Staphylococcus aureus* and *vibrio harveyi.* [60]. Bioactive components present in seeds EOs of *Myristica fragrans* ie eugenol, isoelemycin, isoeugenol, myristic acid, myristicin and methoxy eugenol manufacture silver in to silver NPs and acts against *S. enterica, Ecoli.* Aromatic essential oil from its leaves admixture in acetone also synthesizes zinc NPs. Synthesis of NPs from different plant parts, their characterization and applications are mentioned in Table 2.

5 Issues in Nanoparticle Synthesis from Plant Essential Oils

Synthesis of nanomaterials level is increasing in environment and its hazards on plants, animals, birds and microbes can show indirect effects on human health. These nanomaterials can damage DNA in the cells by releasing free radicles. Lack of knowledge on nanomaterial synthesis is one of the major challenges today. Plants produce nanomaterials with different shape, size and structure, and assurance of uniformity is required. It is necessary to identify the toxicological problems of nanomaterial applications and curb the problems in future. Frequent application of these nanomaterials also affects the food chain and human health. So proper measures are required to develop to overcome these challenges. More research is needed to identify the issues during synthesis and evaluated them properly before its use [62].

6 Applications of Nanomaterials

NPs applied in various fields due to its unique characteristics of nanometer sized metal particles make it well suited for cation applications such as catalysis [63], trace substance detection [64], paint industries [65], energy storage [66], textile industries [67], sensors [68], solar cells for producing clean energy, and coating of building exterior surfaces. Another metal based NPs quantum dots are used for hyperthermia in the presence of light [69].

Today photo based nanomedicine is used in the treatment of cancer and dreadful diseases due to spatially and temporally controllable drug release localized therapy and minimally invasive nature of treatment [70]. Gold based nanomaterials are used in photothermal therapy due to unique physical and optical properties obtained from Surface Plasmon Resonance (SPR) observed in noble metal NPs due to its high surface area to volume ratio [71]. Nanomaterials are also used for bioremediation, it shows less toxic effects of microorganism and also enhances the microbial activity of specific waste and toxic material that save time and cost [72]. Nanomaterials and nanobots are used to make weapons of mass destruction and many of the armies'

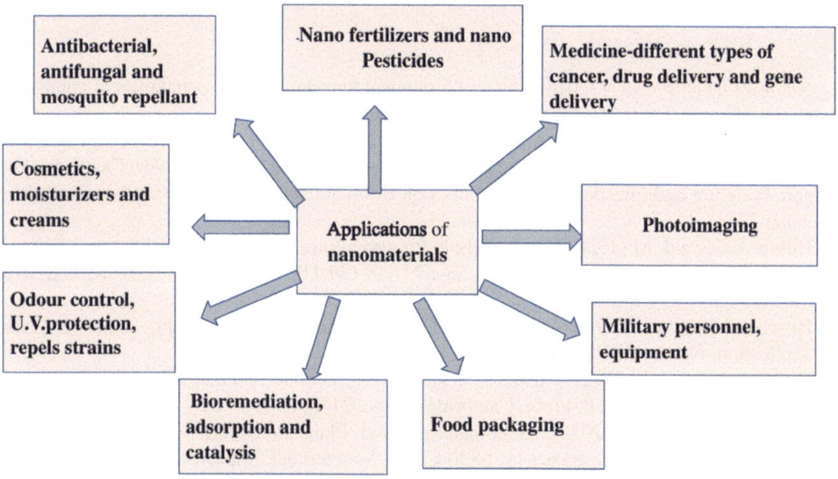

Fig. 3 Application of nanomaterials in various fields

soldiers are wiped out without fighting a single battle, this is because of unnoticed Nanopoison [73]. The applications of nanomaterials in various fields was shown in Fig. 3.

7 Conclusion

Synthesis of nanomaterials from plant EOs plays a crucial role in all the fields and is nontoxic. They help to preserve the plant, wastewater treatment and local trade development. Synthesis of NPs from green plants and its parts is a good alternative to physical and chemical methods. It is cost-effective, environmentally sustainable and doesn't require any high temperature, pressure and lethal chemicals. Nanomaterial synthesis process should be clearly understood. Changes in methods of nanomaterial synthesis may change the shape and dimensions and create a risk to the environment. It's difficult to identify the type of phytoconstituents used in the synthesis and stability of nanomaterials. The emerging issues in present and future should be solved for sustainability. All EOs are not so perfect in the synthesis of nanomaterials. They may also show an effect on plants, animals and human health if there is no proper evaluation of EOs production and nanomaterial synthesis.

References

1. Nikolova I, Georgieva N (2018) Effect of biological Sproducts on the population of aphids and chemical components in alfalfa. Banat's J Biotechnol 9(19):38–46
2. Rahimian Y, Akbari SM, Karami M, Fafghani M (2018) Effect of different levels of Fenugreek powder supplementation on performance, Influenza, Sheep red blood cell, New Castle diseases anti–bodytiter and intestinal microbial flora on Cobb 500broiler chicks. Banat's J Biotechnol 9(19):29–37
3. Husen A, Jawaid, M (2020) Nanomaterials for agriculture and forestry applications. Elsevier Inc., 50 Hampshire St., 5th Floor, Cambridge, MA 02139, USA. https://doi.org/10.1016/C2018-0-02349-X
4. Husen A, Siddiqi KS (2014) Phytosynthesis of nanoparticles: concept, controversy and application. Nanoscale Res Lett 9(229):1–24
5. Husen A, Siddiqi KS (2022) Advances in smart nanomaterials and their applications. Elsevier Inc., 50 Hampshire St., 5th Floor, Cambridge, MA 02139, USA
6. Husen, A, Iqbal, M (2019). Nanomaterials and Plant Potential. Springer International Publishing AG, Gewerbestrasse 11, 6330 Cham, Switzerland. https://doi.org/10.1007/978-3-030-05569-1
7. Tiwari PM, Vig K, Dennis VA, Singh SR (2011) Functionalized gold nanoparticles and their biomedical applications. Nanomaterials 1(1):31–63
8. Husen A, Pandey SBS (2020) Nanofertilizers for smart agricultural practices. National Seminar (Under ICAR, New Delhi, India) on Smart-Horticulture. Agricultural University, Jhalawar, Kota, Rajasthan, India, pp 203–215
9. Jin-Chul K, Madhusudhan A, Husen A (2021) Smart nanomaterials in biomedical applications. Springer Nature Switzerland AG, Gewerbestrasse 11, 6330 Cham, Switzerland
10. Kumar A, Choudhary A, Kaur H, Guha S, Mehta S, Husen A (2022) Potential applications of engineered nanoparticles in plant disease management: a critical update. Chemosphere 295:133798
11. Kumar A, Choudhary A, Kaur H, Mehta S, Husen A (2021) Metal-based nanoparticles, sensors and their multifaceted application in food packaging. J Nanobiotechnology 19(256):1–25. https://doi.org/10.1186/s12951-021-00996-0
12. Sharma G, Nim S, Alle M, Husen A, Kim JC (2022) Nanoparticle-mediated delivery of flavonoids for cancer therapy: prevention and treatment. In: Kim JC, Alle M, Husen A (eds) Smart nanomaterials in biomedical applications. Nanotechnology in the life sciences. Springer, Cham. https://doi.org/10.1007/978-3-030-84262-8_3
13. Sharma P, Pandey V, Sharma MMM, Patra A, Singh B, Mehta S, Husen A (2021) A review on biosensors and nanosensors application in agroecosystems. Nano Res Lett 16:136. https://doi.org/10.1186/s11671-021-03593-0
14. Verma C, Ebenso EE, Quraishi MA (2019) Transition metal nanoparticles in ionic liquids: synthesis and stabilization. J Mol Liq 276:826–849
15. Mittal AK, Chisti Y, Banerjee UC (2013) Synthesis of metallic nanoparticles using plant extracts. Biotechnol Adv 31(2):346–356
16. Sharmeen JB, Mahomoodally FM, Zengin G, Maggi F (2021) Essential oils as natural sources of fragrance compounds for cosmetics and cosmeceuticals. Molecules 26(3). https://doi.org/10.3390/molecules26030666
17. Souto EB, Fernandes AR, Martins-Gomes C, Tiago E, Coutinho AD, Lucarini M, Selma B, Souto AM, Silva AS (2020) Nanomaterials for skin delivery of cosmeceuticals and pharmaceuticals. Appl Sci 10(5):1594
18. Bakkali F, Averbeck S, Averbeck D, Idaomar M (2008) Biological effects of essential oils—a review. Food Chem Toxicol 46:446–475
19. Mohamed AA, El-Emary GA, Ali HF (2010) Influence of some citrus essential oils on cell viability, glutathione-s-transferase and lipid peroxidation in Ehrlich ascites Carcinoma cells. J Am Sci 6(10):820–826

20. Lee MS, Choi J, Posadzki P, Ernst E (2012) Aromatherapy for health care: an overview of systematic reviews. Maturitas 71(3):257–260
21. Ootani1 MA, Aguiar1 RW, Ramos1 ACC, Brito1 DR, da Silva2 JB, Cajazeira JP (2013) Use of essential oils in agriculture. J Biotechnol Biodivers 4(2):162–174
22. Ouis N, Hariri A (2018) Antioxidant and antibacterial activities of the essential oils of *Ceratonia siliqua*. Banat's J Biotechnol 9(17):13–23
23. Zuzarte M, Alves-Silva JM, Alves M, Cavaleiro C, SalgueiroL CruzMT (2018) New insights on the anti-inflammatory potential and safety profile of *Thymus carnosus* and *Thymus camphoratus* essential oils and their main compounds. J Ethnopharmacol 225:10–17
24. Amorim JL, SimasDLR PMMG, Moreno DSA, Alviano CS, da Silva AJR, Dias Fernandes P (2016) Anti-inflammatory properties and chemical characterization of the essential oils of four citrus species. PLoS ONE 11(4):e0153643
25. Geetha V, Chakravarthula SN (2018) Chemical composition and anti-inflammatory activity of *Boswellia ovalifoliolata* essential oils from leaf and bark. J For Res 29(2):373–381
26. Tongnuanchan P, Benjakul S (2014) Essential oils: extraction, bioactivities, and their uses for food preservation. J Food Sci 79:1231–1249
27. Öztekin S, Martinov M (2007) Medicinal and aromatic crops, harvesting, drying and processing, Haworth food and agricultural products press TM. An Imprint of the Haworth Press, Inc., 10 Alice Street, Binhamton, New York 13904–1580 USA
28. Ranjitha J, Vijiyalakshmi S (2014) Facile methods for the extraction of essential oil from the plant species A. Int J Pharm Sci Res 5(4):1107–1115
29. Shumail H, Khalid S, Ahmad I, Khan H, Amin S, Ullah B (2021) Review on green synthesis of silver nanoparticles through plants. Endocrine, Metabolic and Immune Disorders-Drug Targets (Formerly Current Drug Targets-Immune, Endocrine and Metabolic Disorders 21(6):994–1007
30. Pathania D, Sharma M, Kumar S, Thakur P, TorinoE JD, Thakur S (2021) Essential oil derived biosynthesis of metallic nano-particles: implementations above essence. Sustain Mater Technol 30:e00352
31. Ndeh NT, Maensiri S, Maensiri D (2017) The effect of green synthesized gold nanoparticles on rice germination and roots. Adv Nat Sci Nanosci Nanotechnol 8(3):035008
32. Chen M, QinX ZG (2017) Biodegradation of carbon nanotubes, graphene and their derivatives. Trends Biotechnol 35(9):836–846
33. Tang WW, Zeng GM, Gong JL, Liang J, Xu P, Zhang C, Huang BB (2014) Impact of humic/fulvic acid on the removal of heavy metals from aqueous solutions using nanomaterials: a review. Sci Total Environ 468:1014–1027
34. Yan J, Han L, Gao W, Xue S, Chen M (2015) Biochar supported nanoscale zerovalent iron composite used as persulfate activator for removing trichloroethylene. Bioresour Technol 175:269–274
35. Liu Fn, Yang JH, Zuo J, Ma D, Gan L, Xie B, Wang P, Yang B (2014) Graphene-supported nanoscale zero-valent iron: removal of phosphorus from aqueous solution and mechanistic study. J Environ Sci 26(8):1751–1762
36. Kalhapure RS, Sonawane SJ, Sikwal DR, Jadhav M, Rambharose S, ChunderikaMocktar TG (2015) Solid lipid nanoparticles of clotrimazole silver complex: an efficient nano antibacterial against *Staphylococcus aureus* and MRSA. Colloids Surf B 136:651–658
37. Phillips RL, Miranda OR, You CC, Rotello VM, Bunz UH (2008) Rapid and efficient identification of bacteria using gold-nanoparticle–poly (para-phenyleneethynylene) constructs. Angew Chem Int Ed Engl 47:2590–2594
38. Alvarez-Puebla RA, dos Santos Jr DS, Aroca RF (2007) SERS detection of environmental pollutants in humic acid–gold nanoparticle composite materials. Analyst 132(12):1210–1214
39. Li F, Zhao Q, Wang C, Lu X, Li XF, Le XC (2010) Detection of Escherichia coli O157: H7 using gold nanoparticle labeling and inductively coupled plasma mass spectrometry. Anal Chem 82:3399–3403
40. Maciel MVOB, Almeida AR, Machado MH, de Melo APZ, da Rosa CG, de Freitas DZ, NoronhaCM TeixeiraGL, deArmas RD, Barreto PLM (2019) *Syzygium aromaticum* L. (Clove) essential oil as a reducing agent for the green synthesis of silver nanoparticles. Open J Appl Sci 9:45–54

41. Shukla R, Nune SK, Chanda N, Katti K, Mekapothula S, Kulkarni RR, Welshons WV, Kannan R, Katti KV (2008) Soybeans as a phytochemical reservoir for the production and stabilization of biocompatible gold nanoparticles. Small 4(9):1425–1436
42. Thanighaiarassu RRP, Sivamai R, Devika BN (2014) Green synthesis of gold nanoparticles characterization by using plant essential oil *Mentha piperita* and their antifungal activity against human pathogenic fungi. J Nanomedicine Nanotechnol 5(5):1–6
43. Muniyappan N, Nagarajan NS (2014) Green synthesis of gold nanoparticles using *Curcuma pseudomontana* essential oil, its biological activity and cytotoxicity against human ductal breast carcinoma cells T47D. J Environ Chem Eng 2(4):2037–2044
44. Oueslati MH, Tahar LB, Harrath AH (2020) Catalytic, antioxidant and anticancer activities of gold nanoparticles synthesized by kaempferol glucoside from *Lotus leguminosae*. Arab J Chem 13(1):3112–3122
45. Sutthanont N, Attrapadung S, Nuchprayoon S (2019) Larvicidal activity of synthesized silver nanoparticles from *Curcuma zedoaria* essential oil against *Culex quinquefasciatus*. Insects 10(1):27
46. Sheny DS, Mathew J, Philip D (2012) Synthesis characterization and catalytic action of hexagonal gold nanoparticles using essential oils extracted from *Anacardium occidentale*. Spectrochim Acta Part A Mol Biomol Spectrosc 97:306–310
47. Wu F, Yang Q (2011) Ammonium bicarbonate reduction route to uniform gold nanoparticles and their applications in catalysis and surface-enhanced Raman scattering. Nano Res 4(9):861–869
48. Vilas V, Daizy SPhilip, Joseph Mathew, (2016) Essential oil mediated synthesis of silver nanocrystals for environmental, anti-microbial and antioxidant applications. Mater Sci Eng C 61:429–436
49. Ahmadi S, Fazilati M, Nazem H, Mousavi SM (2021) Green synthesis of magnetic nanoparticles using *Satureja hortensis* essential oil toward superior antibacterial/fungal and anticancer performance. Biomed Res Int 8822645:1–14
50. Obeizi Z, Benbouzid H, Ouchenane S, Yılmaz D, Culha M, Bououdina M (2020) Bio synthesis of Zinc oxide nanoparticles from essential oil of Eucalyptus globulus with antimicrobial and anti-biofilmactivities. MaterialsToday Commun 25:101553
51. Anuradha G, Syama Sundar B, Kumar JS, Ramana MV (2014) Synthesis and Characterization of Silver Nanoparticles from *Ocimum basilicum L. var. thyrsiflorum*. Eur J Acad Essays 1.5:5–9
52. Bomma M, Okafor F, Mentreddy SR, Nyochembeng L, Rangari VK, Khan S (2020) The chemical composition, characterization, and combination effect of *Ocimum campechianum* leaf essential oils and bio-produced silver nanoparticles against *Listeria monocytogenes* and *Escherichia coli*. Int J Curr Eng Technol 10.4:518–526
53. Ga 'al H, Fouad H, Mao G, Tian J, Jianchu M (2018) Larvicidal and pupicidal evaluation of silver nanoparticles synthesized using *Aquilaria sinensis* and *Pogostemon SScablin* essential oils against dengue and zika viruses vector Aedes albopictus mosquito and its histopathological analysis. Artif Cells Nanomedicine Biotechnol 46(6):1171–1179
54. Pasquoto-Stigliani T, Estefânia VR, Campos Jhones L, Oliveira CMG, Silva N-J, Guilger M, Troost J et al (2020) Nanocapsules containing neem (*Azadirachta indica*) oil: development, characterization, and toxicity evaluation. Sci Rep 7(1):1–12
55. De Melo APZ, Da Rosa CG, SganzerlaWG NMR, Noronha CM, Maciel MVDOB, Barreto PLM (2019) Synthesis and characterization of zein nanoparticles loaded with essential oil of *Ocimum gratissimum* and *Pimenta racemosa*. Mater Res Express 6(9):095084
56. Arassu RR, Thanighai BN (2018) *Pelargonium graveolens* plant leaf essential oil mediated green synthesis of Silver nanoparticles and its antifungal activity against human pathogenic fungi. J Pharmacogn Phytochem 7:1778–1784
57. Azizi S, Mohamad R, Rahim RA, Moghaddam AB, Moniri M, Ariff A, Saad WZ, Namvab F (2016) ZnO-Ag core shell nanocomposite formed by green method using essential oil of wild ginger and their bactericidal and cytotoxic effects. Appl Surf Sci 384:517–524
58. Sattarahmady NV, Firozabad R, Nazari-Vanani AN (2018) Investigation of amyloid formation inhibition of chemically and biogenically from *Citrus aurantium L.* blossoms and *Rosedamascena* oils of gold nanoparticles: Toxicity evaluation in rat pheochromocytoma PC12 cells. Int J Biol Macromol 112:703–711

59. Veisi H, Dadres N, Mohammadi P, Hemmati S (2019) Green synthesis of silver nanoparticles based on oil-water interface method with essential oil of orange peel and its application as nanocatalyst for A3 coupling. Mater Sci Eng C 105:110031
60. Manju S, Malaikozhundan B, Vijayakumar S, Shanthi S, Jaishabanu A, Ekambaram P, Vaseeharan B (2016) Antibacterial, antibiofilm and cytotoxic effects of *Nigellasativa* essential oil coated gold nanoparticles. Microb Pathog 91:129–135
61. Ashokan AP, Paulpandi M, Dinesh D, Murugan K, Vadivalagan C, Benelli G (2017) Toxicity on dengue mosquito vectors through *Myristica fragrans*-synthesized zinc oxide nanorods, and their cytotoxic effects on liver cancer cells (HepG2). J Cluster Sci 28(1):205–226
62. Dekkers S, Oomen AG, Bleeker EAJ, Vandebriel RJ, Micheletti C, Cabellos J, Janer G, Fuentes N, Vázquez-Campos S, Borges T, Silva MJ, Prina-Mello A, Movia D, Nesslany F, Ribeiro AR, Leite PE, Groenewold M, Cassee FR, Wijnhoven SWP (2014) Towards a nanospecific approach for risk assessment. Regul Toxicol Pharmacol 80:46–59
63. Edison TNJI, Sethuraman MG (2013) Electrocatalytic reduction of benzyl chloride by green synthesized silver nanoparticles using pod extract of *Acacia nilotica*. ACS Sustain Chem Eng 1(10):1326–1332
64. Nirala NR, Pinker N, Desitti C, Shtenberg G (2019) Milk haptoglobin detection based on enhanced chemiluminescence of gold nanoparticles. Talan 197:257–263
65. Tornero AC, Fernández M, Blasco G, Chiquirrín Azqueta MC, Acevedo F, Salazar CC, Ramos López SJ (2018) Antimicrobial ecological waterborne paint based on novel hybrid nanoparticles of zinc oxide partially coated with silver. Prog Org Coat 121:130–141
66. Zhang Q, Uchaker E, Candelaria SL, Cao G (2013) Nanomaterials for energy conversion and storage. Chem Soc Rev 42(7):3127–3171
67. Huang X, Bao X, Wang Z, Hu Q (2017) A novel silver-loaded chitosan composite sponge with sustained silver release as a long-lasting antimicrobial dressing. RSC Adv 7(55):34655–34663
68. Galal A, Nada F, Atta EH (2012) Probing cysteine self-assembled monolayers over gold nanoparticles–towards selective electrochemical sensors. Talanta 93:264–273
69. Chu, M, Pan X, Zhang D, Wu Q, Peng J, Hai W (2012) The therapeutic efficacy of CdTe and CdSe quantum dots for photothermal cancer therapy. Biomaterials 33(29):7071–7083
70. Rai P, Mallidi S, Zheng X, Rahmanzadeh R, Mir Y, Elrington S, Khurshid A, Hasan T (2010) Development and applications of photo-triggered theranostic agents. Adv Drug Deliv Rev 62(11):1094–1124
71. Chen J, Glaus C, Laforest R, Zhang Q, Yang M, Gidding M, Michael J, Welch YX (2010) Gold nanocages as photothermal transducers for cancer treatment. Small 6(7):811–881
72. Rizwan Md, Man Singh, Chanchal K, Mitra, Roshan K, Morve (2014) Eco friendly application of nanomaterials: nano bioremediation. J Nanoparticles 431787:1–8
73. Kharat DKH, Muthurajan PB (2006) Present and futuristic military applications of nanodevices. Synth React Inorg Met-Org Nano-Met Chem 36(2):231–235

Plant Leaf-Based Compounds and Their Role in Nanomaterials Synthesis and Applications

Lipi Pradhan, B. Mounika, and Sudip Mukherjee

Abstract Nanotechnology is the management of substances on a scale of 1–100 nm. Nanotechnology-focused approaches have encouraged researchers to manufacture several drug delivery, bioimaging and biosensing platforms. It has similarly made it feasible to build a nanoparticle-based drug delivery system used for various drugs. The environment and public health are at risk due to the typical chemical synthesis methods utilized to create the nanomaterials. Here in we have focused on nanomaterials and their green manufacturing using plant-based leaf extracts that have emerged as a key area in biomedical research. Green chemistry-based synthesized nanomaterial due to its low cost, minimal danger to people and the environment are utilized widely in health care and medicine as well as drug development. Moreover, practices of plant-based extract are an inexpensive, biocompatible and easy to scale-up method for the design and development of metal nanoparticles from their respective precursors. In this current book chapter, we review the current progresses of green chemistry facilitated synthesis of nanoparticles and their biomedical functions including cancer theranostics, antibacterial, antifungal, biosensing etc.

Keywords Nanomaterials · Plant-leaf based compounds · Green chemistry · Metallic nanoparticles · Phytoconstituents

1 Introduction

The process of altering matter using various physical or chemical methods to create chemicals with specific uses is known as nanotechnology. It may alternatively be described as a microscopic-sized particle with at least one dimension considerably less compared to 100 nm [20]. Early illness identification can improve the effectiveness of medicines, including quick diagnosis and treatment. The manipulation of materials between 1 and 100 nm is known as nanotechnology (nm). Researchers now

L. Pradhan · B. Mounika · S. Mukherjee (✉)
School of Biomedical Engineering, Indian Institute of Technology (BHU), Varanasi 221005, UP, India
e-mail: sudip.bme@iitbhu.ac.in

A. Husen (ed.), *Secondary Metabolites Based Green Synthesis of Nanomaterials and Their Applications*, Smart Nanomaterials Technology,
https://doi.org/10.1007/978-981-99-0927-8_11

have the means to create diverse medication delivery and biosensing platforms thanks to nanotechnology-driven techniques. Additionally, it has enabled the creation of several medication delivery systems at the nanoscale [59, 60]. Green nanotechnology is the ideal way towards reducing the adverse impacts of nanomaterial manufacturing and use, hence reducing the riskiness of nanotechnology. There are hundreds of such goods on the market, the vast majority of which are included in standard personal care items [11, 15, 29, 39]. Green nanotechnology has an inventive impact on nanomaterials-based products by eliminating overall pollution.

Recently biosynthesis has become a more affordable and ecologically responsible way for creating nanoparticles than chemical and physical processes [36, 37, 44, 46, 53, 58, 61]. Plants based extracts seem to be the biggest solution amongst the various biological alternatives [23-27, 29, 33, 36, 37, 44, 75]. Plants act as the "chemical workshops" of nature. They are economical and involve low upholding costs. Green biosynthesis of nanoparticles offers advancement as it is cost-efficient. Utilizing plant extracts is the foundation of plant-mediated green production of nanoparticles. The use of phyto-MNPs in the medical, pharmaceutical, and aesthetic areas is still a long way off.

The conversion of metal ions into nanoparticles (NPs) is greatly aided by terpenoids, tannins, alkaloids, steroids, saponins and polyphenols [64, 65]. Functional groups found in aqueous plant leaf extract have previously been used for possible usage as reduction substances. The green chemistry-based synthesis of NPs is a solitary-step procedure that uses lesser energy for the fabrication of eco-friendly nanoparticles. Nanoparticles can alter agricultural litter and food into energy and several other beneficial products. Chemical and physical techniques are being used to prepare NPs. In this book chapter, we will discuss recent advancements in plant leaf-based synthesis of nanomaterials and their potential applications in healthcare and medicine.

2 Chemical Synthesis of Nanoparticles (NPs)

Nanoparticles are created utilizing certain synthesis methods and environmental factors. Surface controlling agents (SCAs) are introduced before, during, or immediately after precipitate formation. To prevent agglomeration, these SCAs obstruct the nucleating and developing particle. It has been found that few nanoparticles are utilized as catalysts to create coverings for polycarbonate that are transparent, as well as an overcoating using polymerizable NPs to create coatings that are ultrahard and anti-reflective. These instances highlight the materials' potential for use in industrial and medicinal settings. Metal ions can be reduced in aqueous or nonaqueous solutions using chemical reducing agents. The manufacture of nanoparticles often involves the use of reducing agents including sodium borohydride ($NaBH_4$), polyethyleneimine (PEI), sodium citrate, peptides, ascorbate, oligopeptides and elemental hydrogen [45, 47, 50, 67, 85].

The chemically assisted nanoparticle synthesis is that it makes it possible to produce particles with precise dimensions, compositions, and architectures that may be used widely in a range of scientific domains, including catalysis, drug delivery, bioimaging, data storage, and biosensing. Furthermore, the process of producing chemical-based nanomaterials is straightforward to predict.

3 Green Synthesis of (NPs)

Steroids, terpenoids, alkaloids, phenolic acids, polyphenols, saponins, alkaloids, proteins and tannins, perform a major part in mediating the conversion of metallic ions into nanoparticles [21]. As a result, functional groups such as amine, polyphenols, and carboxylic acids that are present in aqueous plant leaf extract were previously utilized for use as reducing substrates during the synthesis of AgNPs. Plant-assisted synthesis methods for creating nanoparticles have a higher efficiency than chemical and physical methods [16, 17, 19, 22, 36]. The importance of plant leaf extracts for converting metal ions to metal oxides has been highlighted in recent investigations. Numerous techniques have been utilized to synthesize NPs utilizing the green components of different plants and algae. These methods are quite effective and preferable than chemical methods since they don't use harmful substances or chemicals that might harm a range of organisms besides the intended target. This is a green tactic that turns garbage into a raw nutritional supplement for the procedure of synthesis. The fundamental resources of this approach are regenerative. A tailored medicine delivery technique using green NPs has been recently developed. Since these NPs are below 100 nm in size, they can be utilized to administer drugs with precision. The energy industry may gain profit from green nanotechnology. Researchers are now more aware of how different plants and their byproducts may be used to synthesize NPs. NPs made from organic resources were broadly applied in the field of 3D culture models and cancer theranostics [35, 38, 48].

3.1 Green Synthesis of (NPs) by Plant-Leaf Extracts

In recent years, green synthesis using leaf extracts were wellknown for their widespread availability and therapeutic value. Iron nanoparticles are created using leaf extracts from *Mangifera indica*, *Mangifera edodes*, and other plants. *Azadiracta indica*, *Geranium Acalypha indica* leaf, *Magnolia champaca*, *Murraya koenigii*, and to evaluate its medicinal potential in domestic wastewater. A variety of plant leaf extracts, including *Camellia sinensis* leaf, *Aloe vera* leaf, *Magnolia kobus* and *Diopyros kaki* leaf, *Coriandrum sativum* leaf, *Cinnamomum camphora* leaf, *Sorbus aucuparia* leaf, *Gliricidia sepium* leaf, and Rose leaf, are used by researchers to make silver nanoparticles that are widely manufactured.

3.2 Green Synthesis of (NPs) by Different Phytochemicals

The reduction and stabilization of meta-based nanomaterials make great use of polysaccharides and physiologically effective plant-based products as scaffolds. Biocomposites with unique applications in nanotechnology and nanomedicine may be made possible by the use of phytochemicals in the green chemistry mediated production of silver (AgNPs) and gold (AuNPs) nanoparticles. The ecologically friendly manufacturing of AuNPs and AgNPs uses two kinds of biologically active substances: phytochemicals and polysaccharides. Phytochemicals and physiologically active polysaccharides may create nanoparticles with cooperative effects by blending their biological actions with the synthesized nanomaterials. Early analyses of the cooperative functions of these phytochemicals, polysaccharides and metal nanocomposites are too summarized. Various other phytochemicals including terpenoids, tannins, alkaloids, steroids, saponins and polyphenols are identified to perform an essential function for the synthesis and stabilization of the green synthesis of metal nanoparticles.

Many researchers worked together and identified that biologically derived resources play a significant role in reduction and stabilization process. Several researcher groups demonstrated that AgNPs can be produced from transitional Ag + ionic complex using different phenolic hydroxyl groups that converted it to AgNPs upon oxidation to quinine. SDS-PAGE study was conducted to study the role of protein in the biosynthesis of AgNPs using *Olax scandens* (OX) leaf extracts as a reduction and stabilization agent. Findings revealed that the presence of low molecular weight protein from OX leaf extract was missing from synthesized AgNPs [57].

4 Green Chemistry-Based Synthesis of Metal Nanoparticles

In bio nanotechnology, the green production of nanoparticles (NPs) utilizing live cells is a unique and promising approach. A bottom-up method called "green synthesis" is comparable to chemical reduction and uses a non-toxic reagent. The reducing ingredient is replaced with a natural extract like plant leaf fruit or crop leaves to produce metal or metal oxide nanomaterials. Green synthesis technique allows for the environmentally benign, economically viable, and biologically secure synthesis of nanoparticles from microbes and plants. Metallic nanoparticles may be produced by living things like plants, cyanobacteria, actinomycetes, viruses, yeasts, and algae either intra- or extracellularly.

In recent years, it came to point that use of the chemical and synthetic compounds causes many hazardous and life-threatening cases to the human life, which is not environmentally healthy. So, as a concern green synthesis is a tremendous process as ecofriendly, biological and safe for the nature. In this process, there are minimal chances of toxicity, which is better for global point. In a diversity of contemporary scientific and technological applications, metal nanoparticles (MNPs) are extensively utilized. MNPs can be made using traditional chemical processes or ecofriendly synthesis techniques. Nanosized metals including copper (CuNPs), palladium (PtNPs), aluminium oxide NPs (Al_2O_3 NPs), iron (FeNPs), zinc (ZnO NPs), gold (AuNPs), and silver (AgNPs) are used to make metal-based synthetic nanomaterials (Ag). Catalysts, sensor elements, optical gadgets, and biological applications may all be made with these materials.

AuNPs

It is well known that gold nanoparticles (AuNPs) have several medical uses. There are several ways to make AuNPs, and they can typically be allocated into two groups: chemical approaches and physical methods. There are considerable efforts to look for alternate, more cost-effective, and large-scale environmentally friendly technologies, such as green synthesis. AuNPs are one of the popular and most commonly synthesized material because of their different functionalities and novel surface plasmon properties, which can be utilized in numerous viewpoints. AuNPs can be combined with various therapeutic agents including drugs, antibodies, proteins and oligonucleotides to upgrade their capabilities. The synthesis of AuNPs is mostly carried out in acidic cases though it is stable in an acidic setting. Gold nanoparticles were synthesized employing four separate plant-based extracts for reduction and stabilization. Various researchers demonstrated green chemistry-based approaches of synthesizing AuNPs by applying plant leaf extracts that were utilized for different biomedical applications [16, 19, 28, 55, 66, 77].

AgNPs

Different types of biological sources for producing AgNPs are described in the previous decade by different research groups. AgNP's size, shape and application depend on different reaction factors under which they are being developed. Due to its exceptional protection against a variety of pathogens as well as the emergence of medication resistance, silver nanoparticles (AgNPs) have drawn a lot of attention. Because of their superior conductivity, AgNPs are used in inks, adhesives, electrical devices, pastes, and other products. Many research groups including ours have utilized plant leaf extracts for the synthesis of AgNPs that were utilized for many biomedical applications including cancer theranostics, antibacterial activities etc. [34, 43, 49, 53, 54, 57, 61, 76].

PtNPs

Production of platinum nanoparticles (PtNPs) from the plant based bioresources would be an added advantage in recycling the waste material. Attempts have been made to synthesize and characterize platinum nanoparticles using plant extracts. Platinum nanoparticle (PtNPs) creation utilizes a bioagent derived from a profoundly obtrusive earthly weed coral plant (*Antigononl eptopus*) [14]. To influence the PtNPs formation, concentrates of the plant's three essential parts—leaves, stems, and roots were explored. The created PtNPs electron micrographs showed that they contained particles with monodispersed round and polydispersed morphologies, with measurements going from 5 to 190 nm. There are several other published papers that demonstrated the biological synthesis of PtNPs using plant leaf extracts for applications in healthcare and medicine [32, 78].

CuNPs

Application of plants, fungi, algae, and microorganisms for the preparation of effective, biocompatible metallic nanomaterials has witnessed an expanding development in the past. Copper nanoparticles (CuNPs) were synthesized by a green chemistry mediated process applying cotton fabric fibers in aqueous reaction conditions. This green synthesis technique is ecofriendly, economical and can be applied on a greater scale. These nanoparticles can be characterized by various physicochemical techniques including FTIR, XRD, DSC, and TGA analysis. Biosynthesized CuNPs can be utilized for different biomedical applications including antibacterial and antifungal activities [51, 52, 72, 75].

ZnNPs

An essential element for healthy plant growth and development is zinc. Farmers must augment the Zn intake of their crops by applying fertilizer to the soil or the leaves. Decreased leaf volume with interveinal necrosis and wavy leaf margins are signs of Zn deficiency in plants. Economical and environmentally sustainable ways to improve soil fertility include using nanofertilizers with increased solubility, soil accessibility, and transmission that are given with lower treatment rates. Zinc and zinc oxide nanoparticles (NP) can be synthesized using both chemogenic and biological methods. Chemogenic synthesis of NPs is extremely noxious and causes numerous ecological concerns including water and soil contamination and greater deposition in the food chains. More lately, green chemistry mediated synthesis process applied for the preparation of ZnNPs as it is environmentally healthy and were used for several biomedical applications [2, 5, 17, 31, 79].

FeNPs

Nanoparticles are used in many different industries, including agrochemicals, cosmetics, and medicines. One of the possible plants for the production of nanoparticles is Catharanthus roseus. It was utilized to create iron nanoparticles that were antibacterial toward *E. coli* and color-degrading toward methyl orange dye. *Devath* et al. has showed green synthesis of iron NPs using various leaf extracts for the treatment of local wastewater [10]. In this work, leaf extracts are used as a reducing agent for the production of iron nanoparticles, and how well they clean household wastewater [10]. FeNPs were developed utilizing different plants, including *Moringa oleifera, Eriobotrya japonica, Peltophorum pterocarpum*, and *Mentha piperita. Catharanthus roseus* is a blossoming plant that belongs to the Apocynaceae family tree. It comprises several bioactive compounds such as vinblastine, vincristine, reserpine, and vinpocetine which possess several medicinal properties and are used in different biomedical applications. Several other reports showed the green synthesis of FeNPs for various biomedical applications [3, 80, 84].

Carbon dot

Carbon dots are materials with dimensions less than ten nanometers. They have the peculiar property of biocompatibility and hence have numerous biomedical applications. It is widely used for bioimaging, diagnostic, drug delivery, etc. CD nanoparticles of various shapes and sizes can be synthesized from green synthesis approach using plant extracts as a reducing agent which reduces the overall cost of the material and minimizes the toxicity and side effects. Various groups including ours have developed green synthesized carbon dots using plant leaf extracts and other bioresources for applications in cancer theranostics, biosensing etc. [13, 73].

Apart from these above mentioned nanoparticles other nanoparticles including chromium oxide NPs, tin oxide NPs, nickel NPs, palladium NPs etc. are also synthesized by biosynthesis approaches and were utilized for several biomedical applications [18, 38, 81, 82].

5 Applications of Green Synthesis Nanoparticles in Health Care and Medicine

5.1 Anticancer and Drug Delivery

Cancer develops after genes that control normal cell cycle and cell division are transformed. Conventional chemotherapeutical agents have decreased bioavailability, irregular circulation, and substantial unfavorable impacts. Green synthesized nanoparticles are economically cheap, safe, and non-toxic in nature. Metallic nanomaterials are extremely useful due to their selective electrical, optical, and reactant properties. 60% of presently utilized anticancer treatments were obtained from various natural resources. Live green plants can similarly be employed to synthesize various types of nanomaterials using plant-based extracts. Recently many research groups including ours utilized plant leaf-based resources to develop nanoparticles and demonstrated their applications as an anticancer agent [56–58, 61, 71]. Muthukumaran and coworkers made ZnO nanoparticles with the help of *Tecoma castanifolia* leaf extract using Green synthesis that were used for inhibition of A549 (lung) cell lines because of its anticancer activity [74]. Vallinayagan et al. synthesized silver nanoparticles (AgNPs) using *Naringi crenulata* leaves extract as reducing agents that had a profound influence on the treatment of HER2 + breast cancer cells by inhibiting cell proliferation [83]. Tian and team green synthesized carbon dots using leaves extract of *Morus alba* (mulberry leaves) for delivery of anticancer drugs to the target site [73].

5.2 Antibacterial

The improvement of innovative and powerful bactericidal medications is of key medical significance. Metal-based nanomaterials have been demonstrated to be encouraging and complementary to the existing antibiotics. Nanoparticles can interact with different key cellular organelles and biomolecules including DNA, cell membrane, ribosomes, proteins, enzymes, and lysosomes. Several research groups synthesized biosynthesized AgNPs using plant leaf extracts and demonstrated their antibacterial applications [4, 5, 7, 8]. Paliwal and team found leaf extract of *Parkia speciosa* can be used for green chemistry-based synthesis of AgNPs with antimicrobial properties which can be applied for the treatment of various diseases [70]. Punniyakotti et al. synthesized copper nanoparticles (CuNPs) using leaf extracts of *Cardiospermum halicacabum*. They found CuNPs have antibacterial properties, and it disrupts the cell wall of bacteria disturbing their growth inside the host body [69]. Mukherjee et al. exhibited the green chemistry-based synthesis of AgNPs using *Olax scandens* leaf extract and showed that the silver nanoparticles can be used in (4-in-1) applications including antibacterial properties, anticancer, cell imaging and as a biocompatible delivery system [57] (Fig. 1).

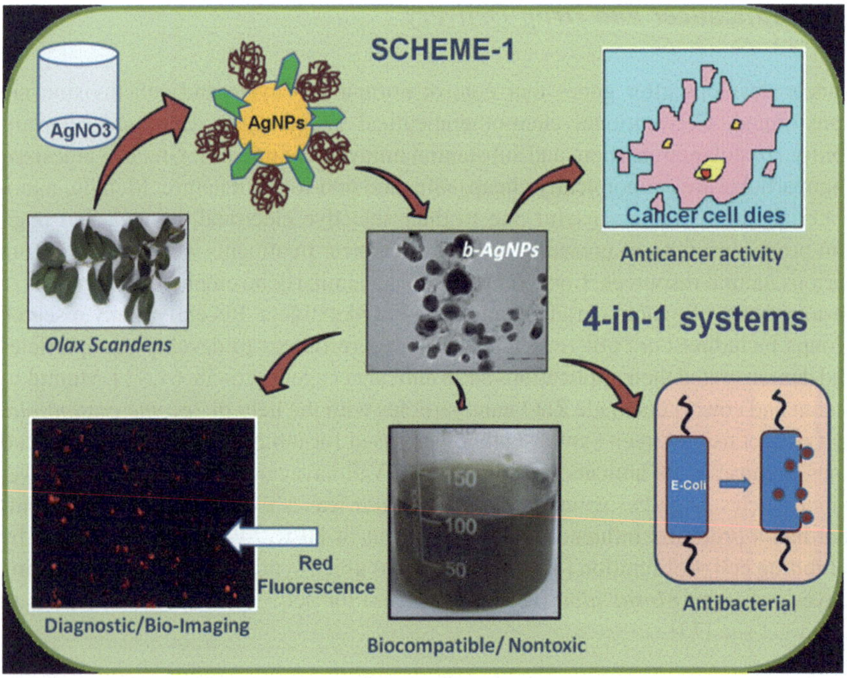

Fig. 1 Overall presentation for synthesis, characterization and biomedical applications (diagnostic, anticancer antibacterial applications) of biosynthesized silver nanoparticles (b-AgNPs) using *Olax scandens* leaf extract. Adapted and reprinted with permission ©2014, Ivy Spring [57]

5.3 Antiviral

Nanotechnology centered methodologies are being examined for their encouraging antibacterial and antiviral characteristics. Nanomaterials synthesized utilizing plant-based extracts are a competent contender of novel antiviral therapeutic agents against different types of viruses including hepatitis B virus, HIV, and Chikungunya virus. Nanoparticle synthesis methodologies has been incorporated with biosynthesis strategies to improve the outcome of the bioinspired nanomaterials for possibility as antiviral agents. Plant extract-mediated preparation of nanoparticles makes use of different natural products that can generate nanoparticles with higher purity. Qurashi and team published a nice review that summarized various green synthesis-based nanoparticle synthesis and their applications as antiviral agents [63]. Abdelkhalek et al. produced Zinc Oxide (ZnONPs) nanoparticles using green synthesis by taking leaves extract of *Mentha spicata* as a reducing agent. They demonstrated antiviral characteristics of ZnONPs against Tobacco mosaic virus (TMV) which can lead to effective management of viral diseases in plants [1]. Zeina and group synthesized zinc oxide and silver NPs using leaves extract of olive (*Olea europaea*) and honey as reducing agent and stated their antiviral property which can be used against bovine herpesvirus-1 (BoHV-1) viral infection for both in vivo and in vitro [86].

5.4 Antifungal

Synthesis of nanomaterials are performed utilizing mainly three distinct strategies such as physical, chemical, and biological processes. Plant-based nanotechnology could be economical, biocompatible, reliable, and ecofriendly in nature. Green processes have recently been considered a significant branch of nanotechnology. Green mediated generation of nanomaterials has been emerging as an alternative route. Various green sources, including microorganisms, fungi, viruses and plant extracts, have been reported. Dridi and team green synthesized silver nanoparticles (AgNPs) using leaves extract of *Melia azedarach* as a reducing agent. The synthesized AgNPs showed antifungal activity that inhibited the growth of Verticillium dahlia [34]. Trivedi and coworkers synthesized cost-effective and ecofriendly copper nanoparticles (CuNPs) using *Celastrus paniculatus* leaves extract for reducing copper. It was found that CuNPs has great antifungal property against *Fusarium oxyporum* [52]. Zhu et al. made green synthesis of zinc nanoparticles (ZnNPs) utilizing leaves extract of *Cinnamomum camphora* synthesized ZnNPs showed antifungal activity against *Alternaria alternate*. It has the ability to inhibit spore germination of *A. alternate* along with cell membrane disruption that leads to leakage of essential proteins and nucleic acid causing death of fungal pathogen [87].

5.5 Biosensing

Metal nanoparticles (MNPs) have constantly been a topic of importance as a result of their superior conductivity, large surface area, characteristics plasmonic phenomenon, multifunctional applications etc. Properties of MNPs improve to magnify and control the light at the nanoscale level. Green leaves contain several naturally active phytochemicals including proteins, polyphenols, terpenoids, carbohydrates, lipids, and enzymes that assist in the formation of silver nanoparticles (AgNPs) by the reduction of silver ions. Most studies have explored the antibacterial activity of AgNPs by applying different leaves such as guava, aloe vera, neem, and lemon. Rambabu and team developed silver nanoparticles (AgNPs) with the help of silver nitrate and leaf extracts of *Ocimum tenuiflorum*. These green synthesized AgNPs were found to have glucose sensing ability which can provide its application in the diagnosis of diabetes reducing time and treatment costs [9]. Ismail et al. showed the biosynthesis of silver nanoparticles using *Duranta erecta* extract and used for biosensing of hexavalent chromium and ammonia in an aqueous solution [30]. Green synthesis of gold nanoparticles were performed that was utilized for the selective and quantitative detection of cerium using various analytical techniques [68].

5.6 Phototherapy

Photothermal therapy (PTT) manipulated the photothermal agents (PTAs) to transform light energy into hyperthermia for eliminating tumor tissues. RGD peptides could exclusively identify and attach to integrin receptor proteins on the surface of the tumor cell. Nanoparticles can be synthesized by using various reactants such as polymers, organic molecules and metals. Silver nanoparticles due to their usefulness against different illnesses and ecological functions are well known. *Datura suaveolens* and *Verbena tenuisecta* were selected for the synthesis of silver nanoparticles that demonstrated effective photodynamic impact when treated to lung cancer cells (A549) due to DNA damage [6]. Plant comprises various types of chemical ingredients including fatty acid esters, fatty alcohols, and volatile oils. Cervic et al. published a report showing anticancer activity of b-AgNPs using *Cynara scolymus* leaf extracts mediated synthesis of b-AgNPs and showed the cytotoxic potential using PDT[12]. Findings revealed that when AgNPs were injected to the MCF-7 breast cancer cells it damaged the mitochondria and led to the generation of ROS that eventually decreased cell migration and proliferation and up-regulation of apoptotic proteins such as Bax, caspase-3 with inhibition of Bcl-2. Kiran et al. successfully developed nanoparticles of gold (AuNPs) from leaf extracts of *Moringa oleifera*. The synthesized AuNPs have contributed a lot in the photothermal therapy of cancer 100,714.

5.7 *Theranostics*

Gold nanoparticles (AuNPs) were extensively investigated and are well-established for their biomedical applications [42]. AuNPs can be produced by exploiting plants' secondary metabolites, such as phytochemicals found in plant extracts. Knowledge of knowing the mechanisms inherent in the therapeutic properties of AuNPs will assist in the development of individualized medicines and remedies for cancer. Healthy cells in the body change from their natural condition and divide uncontrolled in cancer. Engineering, biology, chemistry, and physics are all included in the interdisciplinary area of nanotechnology. Drug delivery to the target has significantly boosted thanks to the use of nanoparticles in cancer treatment. Numerous anticancer treatments based on nanotechnology have received approval from the US FDA and EMA. Gupta and coworkers developed gold nanoparticles (AuNPs) using *Pimenta dioica* leaf extracts as a reducing agent of gold. Studies revealed that AuNPs showed good photothermal activity that can be applied for cancer theranostics applications [40]. Gupta and team biosynthesized superparamagnetic Iron oxide nanoparticles (SPIONs) from leaf extract of *Pimenta dioica*. It was found that due to superparamagnetic properties SPIONs act as a theranostics agent to target cervical cancer (HeLa) cells [41]. Our group utilized *Olax scandens* leaf extract to develop biosynthesized gold and silver nanoparticles that were utilized as cancer theranostics agents [57, 62] (Fig. 2).

6 Advantages of Green Synthesis Over Chemical Synthesis

Green synthesis methods are ideal for an eco–friendly, cost–effective, and biologically safe process optimization. Nanotechnology can have a considerable influence on the development of more 'greener' and 'cleaner' technologies with substantial health and ecological advantages. The applications of nanotechnology are being investigated for their possibility to deliver remarkable resolutions to control, diminish, and clean-up air, water, and land pollution.

7 Conclusion

In this present book chapter, we have summarized various green synthesis-based nanoparticles that don't utilize unsafe substances or lessen the reliance on such components and their biomedical applications. Because of the inescapable profit capacity of plants plant-based materials, and naturally dynamic biomolecules, the biosynthesis of nanoparticles have shown to be a unique and more advantageous pathway when contrasted with its chemical counterparts. Numerous studies have demonstrated their viability against different diseases. Hence, we strongly believe

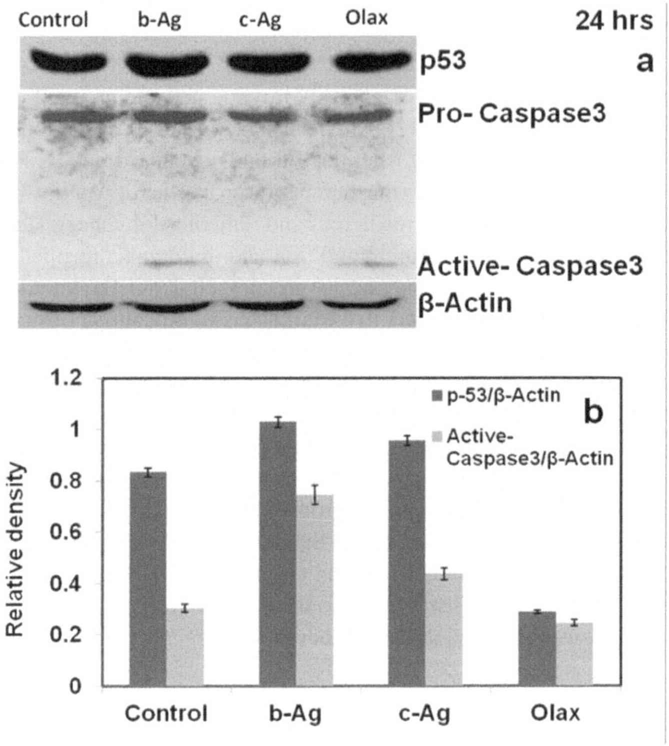

Fig. 2 Western blot analysis of b-AgNPs treated B16 cells show up-regulation of p53 and active Caspase-3; (b) Quantification of signal intensity of each of the protein was normalized with the corresponding β-Actin signal shows increase in p53 & active caspase-3. Adapted and reprinted with permission ©2014, Ivy Spring [57]

that green chemistry-based nanomaterials will pave a way for alternative therapeutic approaches in near future.

Acknowledgements Dr Sudip Mukherjee acknowledge IIT (BHU) for providing Seed Grant Support for the work (OH-35). The authors have no other relevant affiliations or financial involvement with any organization or entity with a financial interest in or financial conflict with the subject matter or materials discussed in the manuscript apart from those disclosed. Dr. Sudip Mukherjee acknowledges the director of IIT (BHU) for constant support.

References

1. Abdelkhalek A, Al-Askar AA (2020) Green synthesized ZnO nanoparticles mediated by *Mentha Spicata* extract induce plant systemic resistance against tobacco mosaic virus. Appl Sci, 10
2. Al-Shabib NA, Husain FM, Ahmed F, Khan RA, Ahmad I, Alsharaeh E, Khan MS, Hussain A, Rehman MT, Yusuf M (2016) Biogenic synthesis of Zinc oxide nanostructures from *Nigella sativa* seed: prospective role as food packaging material inhibiting broad-spectrum quorum sensing and biofilm. Sci Rep 6:36761
3. Alam T, Khan RAA, Ali A, Sher H, Ullah Z, Ali M (2019) Biogenic synthesis of iron oxide nanoparticles via *Skimmia laureola* and their antibacterial efficacy against bacterial wilt pathogen *Ralstonia solanacearum*. Mater Sci Eng C 98:101–108
4. Ali SG, Ansari MA, Alzohairy MA, Alomary MN, Alyahya S, Jalal M, Khan HM, Asiri SMM, Ahmad W, Mahdi AA (2020) Biogenic gold nanoparticles as potent antibacterial and Antibiofilm nano-antibiotics against *Pseudomonas aeruginosa*. Antibiotics 9:100
5. Anand GT, Renuka D, Ramesh R, Anandaraj L, Sundaram SJ, Ramalingam G, Magdalane CM, Bashir A, Maaza M, Kaviyarasu K (2019) Green synthesis of ZnO nanoparticle using *Prunus dulcis* (Almond Gum) for antimicrobial and supercapacitor applications. Surfaces Interfaces 17:100376
6. Ashraf A, Fatima N, Shahzadi I, Tariq H, Shahzadi A, Yameen MA, Iqbal J, Rafi M (2020) *Datura suaveolens* and *Verbena tenuisecta* mediated silver nanoparticles, their photodynamic cytotoxic and antimicrobial evaluation. World J Microbiol Biotechnol, 36
7. Ballottin D, Fulaz S, Cabrini F, Tsukamoto J, Duran N, Alves OL, Tasic L (2017) Antimicrobial textiles: biogenic silver nanoparticles against Candida and Xanthomonas. Mater Sci Eng, C 75:582–589
8. Castro L, Blázquez ML, González F, Muñoz JÁ, Ballester A (2015) Biosynthesis of silver and platinum nanoparticles using orange peel extract: characterisation and applications. IET Nanobiotechnol 9:252–258
9. Dayakar T, Rao KV, Park J, Sadasivuni KK, Rao KR, Rambabu NJ (2018) Non-enzymatic biosensing of glucose based on silver nanoparticles synthesized from *Ocimum tenuiflorum* leaf extract and silver nitrate. Mater Chem Phys 216:502–507
10. Devath CP, Thalla AK, Katte SY (2016) Green synthesis of iron nanoparticles using different leaf extracts for treatment of domestic waste water. J Clean Prod 139:1425–1435
11. Ebrahiminezhad A, Taghizadeh S, Ghasemi Y, Berenjian A (2018) Green synthesized nanoclusters of ultra-small zero valent iron nanoparticles as a novel dye removing material. Sci Total Environ 621:1527–1532
12. Erdogan O, Abbak M, Demirbolat GM, Birtekocak F, Aksel M, Pasa S, Cevik O (2019) Green synthesis of silver nanoparticles via *Cynara scolymus* leaf extracts: the characterization, anticancer potential with photodynamic therapy in MCF7 cells. Plos One, 14
13. Gaddam RR, Mukherjee S, Punugupati N, Vasudevan D, Patra CR, Narayan R, Kothapalli RVSN (2017) Facile synthesis of carbon dot and residual carbon nanobeads: implications for ion sensing, medicinal and biological applications. Mater Sci Eng C-Mater Biol Appl 73:643–652
14. Ganaie SU, Abbasi T, Abbasi SA (2018) Biomimetic synthesis of platinum nanoparticles utilizing a terrestrial weed *Antigonon leptopus*. Part Sci Technol 36:681–688
15. Ghoreishi SM, Behpour M, Khayatkashani M (2011) Green synthesis of silver and gold nanoparticles using Rosa damascena and its primary application in electrochemistry. Physica E 44:97–104
16. Gopinath K, Kumaraguru S, Bhakyaraj K, Mohan S, Venkatesh KS, Esakkirajan M, Kaleeswarran P, Alharbi NS, Kadaikunnan S, Govindarajan M (2016) Green synthesis of silver, gold and silver/gold bimetallic nanoparticles using the *Gloriosa superba* leaf extract and their antibacterial and antibiofilm activities. Microb Pathog 101:1–11

17. Hameed S, Khalil AT, Ali M, Numan M, Khamlich S, Shinwari ZK, Maaza M (2019) Greener synthesis of ZnO and Ag–ZnO nanoparticles using *Silybum marianum* for diverse biomedical applications. Nanomedicine 14:655–673
18. Hassan D, Khalil AT, Solangi AR, El-Mallul A, Shinwari ZK, Maaza M (2019) Physiochemical properties and novel biological applications of *Callistemon viminalis*-mediated α-Cr2O3 nanoparticles. Appl Organometallic Chem, e5041
19. He S, Guo Z, Zhang Y, Zhang S, Wang J, Gu N (2007) Biosynthesis of gold nanoparticles using the bacteria *Rhodopseudomonas capsulata*. Mater Lett 61:3984–3987
20. He X, Deng H, Hwang H-M (2019) The current application of nanotechnology in food and agriculture. J Food Drug Anal 27:1–21
21. Huang J, Lin L, Sun D, Chen H, Yang D, Li Q (2015) Bio-inspired synthesis of metal nanomaterials and applications. Chem Soc Rev 44:6330–6374
22. Huang J, Wang W, Lin L, Li Q, Lin W, Li M, Mann S (2009) A general strategy for the biosynthesis of gold nanoparticles by traditional chinese medicines and their potential application as catalysts. Chem Asian J 4:1050–1054
23. Husen A, Siddiqi KS (2022) Advances in smart nanomaterials and their applications. Elsevier Inc., 50 Hampshire St., 5th Floor, Cambridge, MA 02139, USA
24. Husen A, Iqbal M (2019). Nanomaterials and plant potential. Springer International Publishing AG, Gewerbestrasse 11, 6330 Cham, Switzerland. https://doi.org/10.1007/978-3-030-05569-1
25. Husen A, Jawaid, M (2020) Nanomaterials for agriculture and forestry applications. Elsevier Inc., 50 Hampshire St., 5th Floor, Cambridge, MA 02139, USA https://doi.org/10.1016/C2018-0-02349-X
26. Husen A (2019) Natural product-based fabrication of zinc oxide nanoparticles and their application. In: Husen A, Iqbal M (eds) N*anomaterials* and plant potential. Springer International Publishing AG, Gewerbestrasse 11, 6330 Cham, pp 193–291 https://doi.org/10.1007/978-3-030-05569-1_7
27. Husen A (2020) Introduction and techniques in nanomaterials formulation. In: Husen, A, Jawaid M (eds) Nanomaterials for agriculture and forestry applications. Elsevier Inc. 50 Hampshire St., 5th Floor, Cambridge, MA 02139, USA, pp 1–14
28. Husen A (2017) Gold nanoparticles from plant system: synthesis, characterization and their application. In: Ghorbanpourn M, Manika K, Varma A (eds) Nanoscience and plant–soil systems, vol 48. Springer International Publishing AG, Gewerbestrasse 11, 6330 Cham pp 455–479
29. Iqbal J, Abbasi BA, Batool R, Khalil AT, Hameed S, Kanwal S, Ullah I, Mahmood T (2019) Biogenic synthesis of green and cost effective cobalt oxide nanoparticles using *Geranium wallichianum* leaves extract and evaluation of in vitro antioxidant, antimicrobial, cytotoxic and enzyme inhibition properties. Mater Res Express
30. Ismail M, Khan MI, Akhtar K, Khan MA, Asiri AM, Khan SB (2018) Biosynthesis of silver nanoparticles: a colorimetric optical sensor for detection of hexavalent chromium and ammonia in aqueous solution. Physica E 103:367–376
31. Itroutwar PD, Govindaraju K, Tamilselvan S, Kannan M, Raja K, Subramanian KS (2019) Seaweed-based biogenic ZnO nanoparticles for improving agro-morphological characteristics of rice (*Oryza sativa* L.). J Plant Growth Regul, 1–12
32. Jan H, Gul R, Andleeb A, Ullah S, Shah M, Khanum M, Ullah I, Hano C, Abbasi BH (2021) A detailed review on biosynthesis of platinum nanoparticles (PtNPs), their potential antimicrobial and biomedical applications. J. Saudi Chem. Soc, 25
33. Javid A, Bashir A (2019) Bioinspired synthesis of silver nanoparticles for the control of dengue and malaria vectors. Digest J Nanomater Biostr (DJNB), 14
34. Jebril S, Ben Jenana RK, Dridi C (2020) Green synthesis of silver nanoparticles using Melia azedarach leaf extract and their antifungal activities: In vitro and in vivo. Mater Chem Phys, 248
35. Jha AK, Prasad K, Prasad K, Kulkarni A (2009) Plant system: nature's nanofactory. Colloids Surf B 73:219–223

36. Kanwar R, Rathee J, Salunke DB, Mehta SK (2019) Green nanotechnology-driven drug delivery assemblies. ACS Omega 4:8804–8815
37. Karuppaiya P, Satheeshkumar E, Tsay HS (2019) Biogenic synthesis of silver nanoparticles using rhizome extract of *Dysosma pleiantha* and its antiproliferative effect against breast and human gastric cancer cells. Molec Biol Reports, 1–10
38. Khan MM, Harunsani MH, Adedeji AR (2019) Plant-Assisted Fabrication of SnO_2 and SnO_2-based nanostructures for various applications. Springer, Nanomaterials and Plant Potential
39. Khandel P, Yadaw RK, Soni DK, Kanwar L, Shahi SK (2018) Biogenesis of metal nanoparticles and their pharmacological applications: present status and application prospects. J Nanostr Chem 8:217–254
40. Kharey P, Dutta SB, Gorey A, Manikandan M, Kumari A, Vasudevan S, Palani IA, Majumder SK, Gupta S (2020) Pimenta dioica mediated biosynthesis of gold nanoparticles and evaluation of its potential for theranostic applications 5:7901–7908
41. Kharey P, Indoliya A, Gupta R, Poddar R, Sharma D, Gupta S (2022) Near-infrared active superparamagnetic iron oxide nanoparticles for magnetomotive optical coherence tomography imaging and magnetic hyperthermia therapeutic applications. J Magnet Magnet Mater, 549
42. Kotcherlakota R, Nimushakavi S, Roy A, Yadavalli HC, Mukherjee S, Haque S, Patra CR (2019) Biosynthesized gold nanoparticles. in vivo study of near-infrared fluorescence (NIR)-based bio-imaging and cell labeling applications. ACS Biomater Sci Eng 5:5439–5452
43. Krishnaraj C, Jagan EG, Rajasekar S, Selvakumar P, Kalaichelvan PT, Mohan N (2010) Synthesis of silver nanoparticles using *Acalypha indica* leaf extracts and its antibacterial activity against water borne pathogens. Colloids Surf B 76:50–56
44. Kumar N, Balamurugan A, Balakrishnan P, Vishwakarma K, Shanmugam K (2020) Biogenic nanomaterials: synthesis and its applications for sustainable development. Biogenic nanoparticles and their use in agro-ecosystems. Springer
45. Kusada K, Mukoyoshi M, Wu D, Kitagawa H (2022) Chemical synthesis, characterization and properties of multi-element nanoparticles. Angew Chem Int Ed Engl
46. Li S, Shen Y, Xie A, Yu X, Qiu L, Zhang L, Zhang Q (2007) Green synthesis of silver nanoparticles using *Capsicum annuum L.* extract. Green Chem 9:852–858
47. Ling D, Lee N, Hyeon T (2015) Chemical synthesis and assembly of uniformly sized iron oxide nanoparticles for medical applications. Acc Chem Res 48:1276–1285
48. Liu H, Sun Y, Yu T, Zhang J, Zhang X, Zhang H, Zhao K, Wei J (2019) Plant-mediated biosynthesis of iron nanoparticles-calcium alginate hydrogel membrane and its eminent performance in removal of Cr(VI). Chem Eng J 378:122120
49. Lotha R, Shamprasad BR, Sundaramoorthy NS, Ganapathy R, Nagarajan S, Sivasubramanian A (2018) Zero valent silver nanoparticles capped with capsaicinoids containing *Capsicum annuum* extract, exert potent anti-biofilm effect on food borne pathogen Staphylococcus aureus and curtail planktonic growth on a zebrafish infection model. Microb Pathog 124:291–300
50. Lu X, Rycenga M, Skrabalak SE, Wiley B, Xia Y (2009) Chemical synthesis of novel plasmonic nanoparticles. Annu Rev Phys Chem 60:167–192
51. Lv Q, Zhang B, Xing X, Zhao Y, Cai R, Wang W, Gu Q (2018) Biosynthesis of copper nanoparticles using Shewanella loihica PV-4 with antibacterial activity: novel approach and mechanisms investigation. J Hazard Mater 347:141–149
52. Mali SC, Dhaka A, Githala CK, Trivedi R (2020) Green synthesis of copper nanoparticles using *Celastrus paniculatus* Willd. leaf extract and their photocatalytic and antifungal properties. Biotechnol Reports 27:e00518
53. Manjumeena R, Duraibabu D, Sudha J, Kalaichelvan PT (2014) Biogenic nanosilver incorporated reverse osmosis membrane for antibacterial and antifungal activities against selected pathogenic strains: An enhanced eco-friendly water disinfection approach. J Environ Sci Health Part A 49:1125–1133
54. Marcato PD, Nakasato G, Brocchi M, Melo PS, Huber SC, Ferreira IR, Alves OL, Durán N (2012) Biogenic silver nanoparticles: antibacterial and cytotoxicity applied to textile fabrics. J Nano Res Trans Tech Publ, 69–76

55. Mares-Briones F, Rosas G, Structure and stability of gold nanoparticles synthesized using schinus molle L. extract. J Cluster Sci, 1–9
56. Mittal AK, Kumar S, Banerjee UC (2014) Quercetin and gallic acid mediated synthesis of bimetallic (silver and selenium) nanoparticles and their antitumor and antimicrobial potential. J Colloid Interface Sci 431:194–199
57. Mukherjee S, Chowdhury D, Kotcherlakota R, Patra S et al (2014) Potential theranostics application of bio-synthesized silver nanoparticles (4-in-1 system). Theranostics 4:316
58. Mukherjee S, Dasari M, Priyamvada S, Kotcherlakota R, Bollu VS, Patra CR (2015) A green chemistry approach for the synthesis of gold nanoconjugates that induce the inhibition of cancer cell proliferation through induction of oxidative stress and their in vivo toxicity study. J Mater Chem B 3:3820–3830
59. Mukherjee S, Kotcherlakota R, Haque S, Bhattacharya D, Kumar JM, Chakravarty S, Patra CR (2020a) Improved delivery of doxorubicin using rationally designed PEGylated platinum nanoparticles for the treatment of melanoma. Mater Sci Eng C-Mater Biol Appl 108:110375
60. Mukherjee S, Kotcherlakota R, Haque S, Das S, Nuthi S, Bhattacharya D, Madhusudana K, Chakravarty S, Sistla R, Patra CR (2020b) Silver prussian blue analogue nanoparticles: rationally designed advanced nanomedicine for multifunctional biomedical applications. ACS Biomater Sci Eng 6:690–704
61. Mukherjee S, Patra CR (2017) Biologically synthesized metal nanoparticles: recent advancement and future perspectives in cancer theranostics. Future Sci
62. Mukherjee S, Vinothkumar B, Prashanthi S, Bangal PR, Sreedhar B, Patra CR (2013) Potential therapeutic and diagnostic applications of one-step in situ biosynthesized gold nanoconjugates (2-in-1 system) in cancer treatment. RSC Adv 3:2318–2329
63. Naikoo GA, Mustaqeem M, Hassan IU, Awan T, Arshad F, Salim H, Qurashi A (2021) Bioinspired and green synthesis of nanoparticles from plant extracts with antiviral and antimicrobial properties: a critical review. J Saudi Chem Soc, 25
64. Ovais M, Khalil AT, Islam NU, Ahmad I, Ayaz M, Saravanan M, Shinwari ZK, Mukherjee S (2018) Role of plant phytochemicals and microbial enzymes in biosynthesis of metallic nanoparticles. Appl Microbiol Biotechnol, 1–16
65. Ovais M, Raza A, Naz S, Islam NU, Khalil AT, Ali S, Khan MA, Shinwari ZK (2017) Current state and prospects of the phytosynthesized colloidal gold nanoparticles and their applications in cancer theranostics. Appl Microbiol Biotechnol 101:3551–3565
66. Philip D (2010) Green synthesis of gold and silver nanoparticles using *Hibiscus rosa sinensis*. Physica E: Low-Dimensional Syst Nanostr 42:1417–1424
67. Polarz S, Roy A, Merz M, Halm S, Schroder D, Schneider L, Bacher G, Kruis FE, Driess M (2005) Chemical vapor synthesis of size-selected zinc oxide nanoparticles. Small 1:540–552
68. Priyadarshini E, Pradhan N, Panda PK, Mishra BK (2015) Biogenic unmodified gold nanoparticles for selective and quantitative detection of cerium using UV–vis spectroscopy and photon correlation spectroscopy (DLS). Biosens Bioelectron 68:598–603
69. Punniyakotti P, Panneerselvam P, Perumal D, Aruliah R, Angaiah S (2020) Anti-bacterial and anti-biofilm properties of green synthesized copper nanoparticles from *Cardiospermum halicacabum* leaf extract. Bioprocess Biosyst Eng 43:1649–1657
70. Ravichandran V, Vasanthi S, Shalini S, Shah SAA, Tripathy M, Paliwal N (2019) Green synthesis, characterization, antibacterial, antioxidant and photocatalytic activity of Parkia speciosa leaves extract mediated silver nanoparticles. Results Phys, 15
71. Safaepour M, Shahverdi AR, Shahverdi HR, Khorramizadeh MR, Gohari AR (2009) Green synthesis of small silver nanoparticles using geraniol and its cytotoxicity against fibrosarcoma-wehi 164. Avicenna J Med Biotechnol 1:111–115
72. Sathiyavimal S, Vasantharaj S, Bharathi D, Saravanan M, Manikandan E, Kumar SS, Pugazhendhi A (2018) Biogenesis of copper oxide nanoparticles (CuONPs) using *Sida acuta* and their incorporation over cotton fabrics to prevent the pathogenicity of Gram negative and Gram positive bacteria. J Photochem Photobiol B 188:126–134
73. Shao YY, Zhu CY, Fu ZF, Lin K, Wang YD, Chang YX, Han LF, Yu HY, Tian F (2020) Green synthesis of multifunctional fluorescent carbon dots from mulberry leaves (*Morus alba L.*) residues for simultaneous intracellular imaging and drug delivery. J Nanoparticle Res, 22

74. Sharmila G, Thirumarimurugan M, Muthukumaran C (2019) Green synthesis of ZnO nanoparticles using *Tecoma castanifolia* leaf extract: characterization and evaluation of its antioxidant, bactericidal and anticancer activities. Microchem J 145:578–587
75. Siddiqi KS, Husen A (2020) Current status of plant metabolite-based fabrication of copper/copper oxide nanoparticles and their applications: a review. Biomater Res 24(11):1–15
76. Siddiqi KS, Husen A, Rao RAK (2018) A review on biosynthesis of silver nanoparticles and their biocidal properties. J Nanobiotechnol 16(14):1–28
77. Siddiqi KS, Husen A (2017) Recent advances in plant-mediated engineered gold nanoparticles and their application in biological system. J Trace Elem Med Biol 40:10–23
78. Siddiqi KS, Husen A (2016) Green synthesis, characterization and uses of palladium/platinum nanoparticles. Nanoscale Res Lett 11(482):1–13
79. Siddiqi KS, Rahman A, Tajuddin, Husen A (2018) Properties of zinc oxide nanoparticles and their activity against microbes. Nanoscale Res Lett 13(141):1–13
80. Siddiqi KS, Rahman A, Tajuddin, Husen A (2016) Biogenic fabrication of iron/iron oxide nanoparticles and their application. Nanoscale Res Lett 11(498):1–13
81. Thema F, Manikandan E, Gurib-Fakim A, Maaza M (2016) Single phase Bunsenite NiO nanoparticles green synthesis by *Agathosma betulina* natural extract. J Alloy Compd 657:655–661
82. Thovhogi N, Park E, Manikandan E, Maaza M, Gurib-Fakim A (2016) Physical properties of CdO nanoparticles synthesized by green chemistry via *Hibiscus Sabdariffa* flower extract. J Alloy Compd 655:314–320
83. Vallinayagam S, Rajendran K, Sekar V (2021) Green synthesis and characterization of silver nanoparticles using *Naringi crenulate* leaf extract: key challenges for anticancer activities. J Molecu Struct, 1243
84. Wang T, Lin J, Chen Z, Megharaj M, Naidu R (2014) Green synthesized iron nanoparticles by green tea and eucalyptus leaves extracts used for removal of nitrate in aqueous solution. J Clean Prod 83:413–419
85. Winnischofer H, Rocha TC, Nunes WC, Socolovsky LM, Knobel M, Zanchet D (2008) Chemical synthesis and structural characterization of highly disordered N colloidal nanoparticles. ACS Nano 2:1313–1319
86. Zeedan GSG, El-Razik KAA, Allam AM, Abdalhamed AM, Zeina HAA (2020) Evaluations of potential antiviral effects of green zinc oxide and silver nanoparticles against *Bovine Herpesvirus-1*. Adv Animal Veterinary Sci, 8
87. Zhu WJ, Hu CM, Ren YF, Lu Y, Song YP, Ji Y, Han C, He,JY (2021) Green synthesis of zinc oxide nanoparticles using *Cinnamomum camphora* (L.) Presl leaf extracts and its antifungal activity. J Environ Chem Eng, 9

Flower-Based Compounds and Their Role in Nanomaterials Synthesis and Applications

Harshita Shand, Rittick Mondal, Soumendu Patra, Paulami Dam, Suvankar Ghorai, and Amit Kumar Mandal

Abstract Flowers are an essential part of the plant which are rich in a variety of secondary compounds like terpenoids, flavonoids, coumarins, sterols and xanthones. Floral therapy was an integral part of ancient Ayurveda in curing several diseases. Presently Nanotechnology is an evolving branch of science that serves a variety of applications. The techniques and protocols involved in the NPs synthesis are costly, hazardous, and demand high energy input. Plants and herbal-based compounds can provide natural, economical, and sustainable methods for the biogenesis of NPs. Flowers are rich in various phytochemicals that can serve as a vital precursor molecule of NPs biogenesis. The NPs obtained from floral precursor shows unique characteristics and can have various applications. They can be used for therapeutic purposes in treating breast cancer, gastric adenocarcinoma, and human colorectal adenocarcinoma. Metallic NPs synthesized from floral extract have shown extinguishing antioxidant and α-amylase activity and had a promising effect on diabetes. Flower-based metallic NPs shows antimicrobial and insecticidal properties.

H. Shand · S. Patra · S. Ghorai
Department of Microbiology, Raiganj University, North Dinajpur 733134, West Bengal, India

R. Mondal · P. Dam · A. K. Mandal (✉)
Chemical Biology Laboratory, Department of Sericulture, Raiganj University, North Dinajpur 733134, West Bengal, India
e-mail: amitmandal08@gmail.com

A. K. Mandal
Centre for Nanotechnology Sciences, Raiganj University, North Dinajpur 733134, West Bengal, India

A. Husen (ed.), *Secondary Metabolites Based Green Synthesis of Nanomaterials and Their Applications*, Smart Nanomaterials Technology,
https://doi.org/10.1007/978-981-99-0927-8_12

Keywords Nanoparticles · Flower extract · Anticancer · Phytochemicals · Antimicrobial · Insecticidal

Abbreviations

nm	Nanometer
NPs	Nanoparticles
AuNPs	Gold nanoparticles
AgNPs	Silver nanoparticles
CdS NPs	Cadmium sulfide nanoparticles
Cd NPs	Cadmium nanoparticles
TiO_2 NPs	Titanium dioxide nanoparticles
GC-MS	Gas Chromatography-Mass Spectrometry
MCF-7	Michigan Cancer Foundation-7
MgO NPs	Magnesium oxide nanoparticles
FeO NPs	Ferrous oxide nanoparticles
Pd NPs	Palladium nanoparticles
UV–vis spectroscopy	Ultraviolet-visible spectroscopy
FTIR	Fourier Transform Infrared Spectroscopy
SEM	Scanning electron microscopy
TEM	Transmission electron Microscopy
EDX	Energy-dispersive X-ray spectroscopy
XRD	X-ray diffraction
DLS	Dynamic light scattering spectroscopy

1 Introduction

Nanotechnology is a developing stream of science with huge applications in biomedicine, pharmaceuticals, sensors, catalysis, cosmetics, agriculture, textile products, etc. The concept of nanotechnology was first termed by American physicist Richard Feynman [7]. Nanotechnology includes understanding and manipulating matter at the atom and molecule levels. The ultra-small size of NPs (1–100 nm) and their large surface-to-volume ratio brings ardour to the researchers in the usage of NPs [40, 54]. The techniques involved in NPs synthesis are time-consuming, overpriced, and toxic to the ecosystem due to the use of harmful compounds. Because of this researchers are trying to develop an environment-friendly way to synthesize NPs which includes using plant parts to make the process cost-effective, simple, and sustainable [20, 21, 50]. Researchers are using various plants and herbal extracts for the capping and reducing agents in the NPs biogenesis [17–19, 52]. Purified plant compounds like cellulose and glucose have also been exploited in the biogenesis of NPs. Plant parts such as leaves, flowers, gums, seeds, latex, roots, pulp, bark, and fruits have also been exploited for NPs biogenesis [22]. Plant products contain phytochemicals. Flowers have an attractive visual property that stimulates the cerebral network and sensory motors. The ancient Indian Ayurveda and Siddha system has reported the medicinal use of some flowers. Flowers are a rich source of spices, vegetables, flowering agents, and raw materials. Flowers are rich in various phytochemicals like flavonoids, tannins, coumarins, xanthones, sterol, etc. which can be used as raw-material for the synthesis of NPs [6]. *Hibiscus rosa sinesis* is a rich source of saponins, tannins, indole alkaloids, terpenoids, and reducing sugars [30, 46]. Anthocyanin is another phytochemical that imparts colors to different plant parts, mostly flowers. It has anticancer, antioxidant properties [8]. Anthocyanin is soluble in water which makes them a suitable precursor for NP synthesis [57]. Kaempferol is a kind of flavonoid present in various flowers that can be used in the generation of AuNPs [47]. Researchers in 2019 have used terpenoids from *Tussilago farfara* for AgNPs and AuNPs biogenesis [32]. *Punica granatum* (a native shrub of Afghanistan and China) poses hemostatic properties. In Ayurveda, Puccinia flowers were used in diabetes treatment, and in Chinese medicine, it has been used as a treatment for hair related issues [59]. Flowers of Pomegranate are rich in gallic acid, ellagic acid, ethyl blevifolin carboxylate, triterpens, oleanolic acid, ursolic acids, polyphenols maslinic acid, and asiatic acid. Tea flowers are a rich source of catechins [34]. The tea flowers extract possess antioxidant properties.

2 Green Synthesis of NPs

Green synthesis of NPs provides an easy, efficient, economical, and eco-friendly method for the production of metallic NPs from different extracts (cell or cell-free) from bio-resources. The idea of green synthesis of NPs highlights the use of eco-friendly reducing agents and innocuous material for NP integrity [53]. Compounds like polyphenols, sugars, water, vitamins, and peptides can also be used for nanoparticle synthesis. Prokaryotic cells are also used to synthesize various NPs like CdS, gold, silver oxide, silver, and titanium dioxide (TiO_2). Some fungi have also been used to synthesize CdS and TiO_2 NPs. Recently fungi and algae are also used for NPs synthesis. Plant-based NPs are most stable and monodispersed compared to microbial NPs. Green synthesis of NPs from flowers provides a basic advantage over other biological NPs synthesis processes as synthesizing NPs from the microbial system is a laborious and time-consuming job as it involves maintaining microorganisms in pure cultures in aseptic conditions. Also, the segregation of NPs from microbial culture is a tedious job during downstream processing.

3 Different Types of Flower-Mediated-NPs and it's Characterization

3.1 Metallic NPs

3.1.1 AgNPs

Silver NPs possess a high surface area which gives them their unique catalytic actions and atomic behavior in comparison with other bulk counterparts [61]. AgNPs synthesis is a dual-step process and involves Ag^+ ions reduction to Ag° and after the aggregation and stabilization step, oligomeric cluster formation of colloidal AgNPs takes place [39]. The reduction process happens in presence of a bio-catalyst like floral extracts. Researchers have used *Rosa santana* (aqueous extract) for AgNPs assembly. The genesis of AgNP was confirmed by UV–vis and the absorption peak at 438 nm was obtained. TEM studies have shown the NP shape to be spherical and a size of 6.52–25.24 nm with a particle size of 14.98 nm and an average zeta potential of about −26.50 mV showing high stability of NPs [23]. *Fritillaria* flower extract has been used for the reduction and stabilization of AgNPs [16]. SEM and TEM analysis indicated spherical AgNPs with an average size of 10 nm.

3.1.2 AuNPs

Gold NPs (AuNPs) have attained extensive consideration due to its size, biocompatible nature, optical properties, and good shape [14]. AuNPs possess several sizes and morphology, and due to this are employed in the field of medical science for drug carriers, tumor detectors, and radiotherapy dose enhancers. Researchers have synthesized AuNPs using *Lonicera Japonicas* floral extract. Color change from light-yellow to wine red and the absorption at 530–580 nm in UV–Vis spectroscopy indicated the AuNPs synthesis. FITR analysis confirmed the presence of phenols, alcohols, 1° amines, carboxylic acid, and alkynes in the floral extracts. TEM studies showed spherical as well as triangular and hexagonal NPs with sizes ranging from 10–40 nm. XRD confirmed the crystallization of formed AuNPs [45].

3.2 Other Metallic Nanoparticles

Metallic NPs like Copper (Cu), Cadmium (Cd), Titanium (Ti), Iron (Fe), Zinc (Zn), etc. have evolved as a new group of nanoparticles because of their unique application in the investigation. Aqueous floral extract of *Hibiscus sabdariffa* has been used to fabricate FeNPs. TEM analysis showed the shape (spherical) and size (100 nm) of NPs. The occurrence of anthocyanin in the extract was confirmed by FTIR [3]. Marigold flower extract has been used in the biogenesis of Cd NPs synthesis. 88 ml of cadmium chloride and 12 ml of petal extract were mixed to obtain a yellow-colored nanoparticle solution. Observation under the fluorescent microscope showed sphere-shaped NPs [15]. Flowers of *Mangifera oleifera* have been used for the biogenesis of Pd NPs. Determination of the chemical composition of the floral extract was done using SEM, FTIR, EDX, TEM, and GC–MS, which revealed the presence of palmitic acid, docosane, tricosane, tetracosane, pentacosane, hexacosane, phthalate. TEM studies revealed the size of NPs to be ranging from 10–50 nm [4]. Flower, stem, and leaf extracts of *Gnidia glauca* have been used for Cu NPs synthesis. The change in color (pale blue to brown) confirms Cu-NP generation. HR-TEM studies have revealed the synthesis of 5 nm spherical nanoparticles from the floral extract of *Gnidia glauca.* A characteristic peak was obtained at ~3400–3420 cm^{-1} in the FTIR spectrum [43].

3.3 Metallic Oxide Nanoparticles

Magnesium oxide NPs (MgO NPs) have been synthesized using aqueous floral extract of *Rosmarinus officinalis.* MgO NPs were obtained by continuous stirring at 70 °C for 4 h. EDS confirmed the presence of Mg (35.55%) and O (64.45%) [1] *Matricaria chamomilla* commonly known as Chamomile flowers along with olive leaves, and tomato fruits have been used for the biogenesis of ZnO NPs. Properties of ZnO NPs were determined by UV–Vis spectroscopy, XRD, FTIR, SEM, and TEM techniques. The synthesized NPs size ranged from 51.2 ± 3.2 nm [9]. Floral extract of *Avicennia marina* was used for the synthesis of FeO NPs. The NPs ranged from 30–100 nm as revealed by the SEM study and SPR peak at 295-301 nm in UV–Vis spectroscopy was revealed [27]. *Hibiscus rosa Sinensis* floral extract was utilized for the production of TiO_2 NPs. XRD studies confirmed the size of the fabricated NPs was 7 nm. SEM analysis revealed the spherical NPs and no agglomeration was there. The floral phytochemicals were used in the capping and stabilization of NPs [31]. Aqueous extracts of *Calotropis gigantean* have also been used for the synthesis of TiO_2 NP synthesis. SEM exposed the aggregated spherical shape of NPs with a mean size of 160–220 nm [38]. ZnO NPs have been produced by zinc nitrate and *Aspalathus linearis* floral extracts of [12]. Floral extracts of *Jacaranda mimosifolia* were employed forthe fabrication of of ZnO-NPs. The size of the NPs ranged from 2–4 nm [48]. Flower of *Trifolium pratense* was used for the fabrication of ZnO NP and the biosynthesized NPs were spherical in shape with sizes ranging from 60–70 nm as revealed by XRD studies. SEM studies showed particles ranging from 100–190 nm [13]. ZnO NPs produced from Bougainvillea flower extracts were characterized by UV–Vis and SEM which exhibited the size of NPs was 40 nm [2].

4 Applications of Nanoparticles Synthesized from Floral Extracts

4.1 Antimicrobial Activity

The NPs can enter the bacterial cell membrane and affect their metabolic pathways, influencing cell activity. NPS can associate with bacterial cell components like DNA, proteins, enzymes, lysozymes, and ribosomes and lead to oxidative stress, and variation in the permeability of cell membrane. It can also denature or deactivate bacterial proteins [58]. To adsorb the NPs, the gram-positive and gram-negative bacterial cell wall acts in a diversified way [33]. The lipopolysaccharide in gram-negative bacteria provides a negative charge that can play an important role in attracting NPs. Gram-positive bacterial cell wall contains teichoic acid that allows circulation of NPs over the phosphate molecular chain and avoids its aggregation. Metal and metal oxide NPs derived from floral extract have shown enhanced antimicrobial activity. AgNPs synthesized from *Ipomoea digitata* flower extract exhibited broad-spectrum antimicrobial properties [56] (Table 1). MgO NPs synthesized by floral extract of *Rosmarinus officinalis* showed promising results against the pathogen causing blight disease in rice. It has reduced *Xanthomonas oryzae* pv, *Oryzae* growth, biofilm forming activity, and motility by damaging the cellular integrity of the bacterial cell and leaks its cellular content [1]. AuNPs synthesized from *Plumeria alba* flower extract displayed enhanced antibacterial activity, showing a synergistic interaction with antibiotics. AuNPs collaborative to vancomycin and norfloxacin displayed higher activity against *Aspergillus flavus* [41]. *Catharanthus roseus* flower extract was explored for the biogenesis of AgNPs that showed activity against *Bacillus subtilis, Staphylococcus aureus, E. coli, Klebsiella pneumoniae* and *Pseudomonas putida* [36].

4.2 Antioxidant Activity

Flowers are rich in phytochemicals that possess antioxidant properties. *Cassia angustifolia* floral extract contains phenols, carbonyls, alkyl halides, nitro compounds, alkane, and other aromatic phytochemicals that can act as reducing, stabilizing and capping agents for AgNP synthesis [11]. AgNPs synthesized from *Bauhinia variegata* floral extract exhibited excellent antioxidant and inhibitory effects against a-amylase enzyme activity and can be used as a nano-drug for the treatment of diabetic conditions [25]. AgNPs fabricated with *Allamanda cathartica* and *Couroupita guianensis* floral extract have shown potential antioxidant activities [28]. CuNPs synthesized from *Gnidia glauca* and *Plumbago zeylanica* flowers have shown remarkable effects as an anti-diabetic agent [24].

Table 1 Showing flowering plant names, NPs obtained and their functions in various activities

Plant name	NPs synthesised	Function of NPs	Key references
Aerva lanata	AgNPs	Exhibits broad-spectrum antibacterial, antioxidant and catalytic property	[44]
Rosa floribunda charisma	MgO NPs	Antibacterial activity against skin disease related pathogens (*Pseudomonas aeruginosa, Staphylococcus epidermidis, Streptococcus pyogenes*)	[62]
Plumeria pudica Jacq	AgNPs	Anticancer and larvicidal activity	[55]
Abelmoschus esculentus	AgNPs	Anticancer and antimicrobial activity	[10]
Zephyranthes candida	AgNPs	Potent anti-diabetic, antioxidant, antiinflammatory and anticancer activities	[26]
Hylotelephium telephium	CuO-NPs & ZnO NPs	Antioxidant and antibacterial activity	[29]
Ipomoea digitata	AgNPs	Effective antimicrobial activity against both pathogenic gram-positive and gram-negative bacteria	[56]
Rosmarinus officinalis	MgO NPs	Reduced bacterial growth of *Xanthomonas oryzae* pv. *Oryzae* causative agent of blight disease in rice	[1]
Gnidia glauca and Plumbago zeylanica	CuNPs	Can be used as a potential candidate in anti-diabetic nanomedicine development	[24]
Scrophularia striata	AgNPs	As a potential therapeutic in breast cancer treatment	[35]
Lonicera japonica	AuNPs	Cytotoxicity against cancer cell lines	[45]
Bougainvillea	ZnO NPs	Anticancer activity against breast cancer cell line (MCF-7)	[2]
Bauhinia variegata	AgNPs	Inhibitory effects against a-amylase enzyme activity and can be used as diabetes treatment	[25]
Mimosa pudica	AuNPs	Catalytic activity against reduction of 4-nitrophenol to 4-aminophenol	[37]

(continued)

Table 1 (continued)

Plant name	NPs synthesised	Function of NPs	Key references
Catharanthus roseus	AgNP	Antibacterial activity against *Bacillus subtilis, E. coli, Klebsiella pneumoniae, Pseudomonas putida,* and *Staphylococcus aureus*	[36]
Chryasanthemum indicum L	TiO_2 NPs	Effective against the larvae of *Haemaphysalis bispinosa* and *Rhipicephalus microplus*	[38]
Plumeria alba	AuNP	Enhanced antibacterial activity against *E. Coli*	[41]

4.3 Anticancer Activity

Researchers have found that AgNPs fabricated from *Scrophularia striata* show toxicity against the human breast cancer cell line (MCF-7) and can be used as a promising agent in breast cancer treatment [35]. AuNPs synthesized using *Lonicera japonica* showed cytotoxicity against cancer cell lines [45]. *Tussilago farfara* flower mediated AgNPs and AuNPs showed anticancer properties against human pancreas ductal adenocarcinoma, gastric adenocarcinoma cells, human colorectal adenocarcinoma cells (HT-29) [32]. *Bougainvillea* flower mediated ZnO-NPs showed anticancer activity against MCF-7 [2].

4.4 Insecticidal Activity

Nanoparticles that are derived from flowers have also shown insecticidal activity. Marigold petal extract mediated cadmium nanoparticles (Cd NPs) (10 ppm) have shown a 68.9% death rate against *Aedes albopictus* [15]. AgNPs synthesized from *Chryasanthemum indicum* L. floral extract have shown promising results against *Anopheles stephensi* mosquito larvae and pupae [5]. TiO_2 NPs synthesized from *Calotropis gigantean* flower extract were also shown promising results against the larvae of *Haemaphysalis bispinosa* and *Rhipicephalus microplus* [38].

4.5 Catalytic Activity

Herbicides, insecticides, and synthetic dyestuffs are commonly produced by using 4-nitrophenol and its derivatives which is toxic to the environment and act as water pollutant [51]. The reduction product of 4-nitrophenol acts as a mediator for sulfur dyes, precursors for antipyretic and analgesic drugs, rubber antioxidants, paracetamol, etc. [60]. $NaBH_4$ has been used as a metal catalyst for AuNPs, AgNPs, CuO-NPs, and Pd NPs [49]. AuNPs synthesized from *Mimosa pudica* and *Mangifera indica* flowers extract exhibit high catalytic activity reduction of 4-nitrophenol to 4-aminophenol [37]. AgNPs synthesized from *Ipomoea digitata* flower exhibited a remarkable reduction of methylene blue dye [56]. NPs can be used for waste-waters treatment [42] (Scheme 1).

5 Conclusion

Flower-based NPs synthesis is cost-effective, nontoxic, environment-friendly, and has distinctive properties. Flowers are augmented with various bioactive molecules and have the great potential ability to reduce the metal ions and stabilize the NPs. The spherical NPs have several applications and showed greater stability over a long time. Furthermore, flower-mediated-NPs have shown potential application as an antibacterial, antioxidant, anticancer, agent, etc. As a result, it is expected that flower-based biogenic synthesis of nanomaterials would improve our prospects and understanding for effective formulation and applications in various disciplines of science and technology.

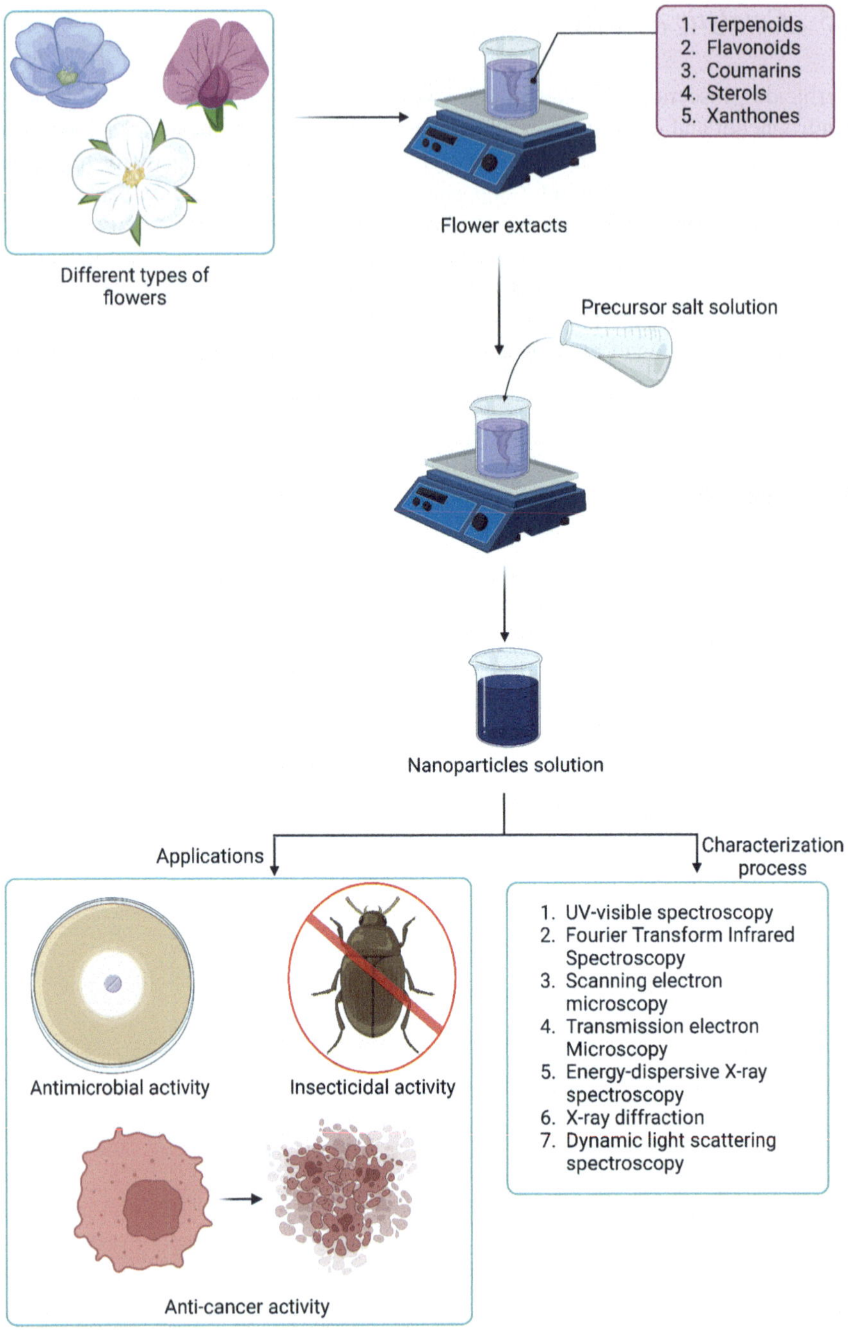

Scheme 1 Flower-based NPs synthesis and it's application and characterization process

References

1. Abdallah Y, Ogunyemi SO, Abdelazez A, Zhang M, Hong X, Ibrahim E, Hossain A, Fouad H, Li B, Chen J (2019) The green synthesis of MgO nano-flowers using *Rosmarinus* officinalis L. (Rosemary) and the antibacterial activities against Xanthomonas oryzae pv. oryzae. BioMed Res Int, 5620989
2. Ahmar Rauf M, Oves M, Ur Rehman F, Rauf Khan A, Husain N (2019) Bougainvillea flower extract mediated zinc oxide's nanomaterials for antimicrobial and anticancer activity. Biomed Pharmacother, 116–108983
3. Alshehri A, Malik MA, Khan Z, Al-Thabaiti SA, Hasan N (2017) Biofabrication of Fe nanoparticles in aqueous extract of *Hibiscus sabdariffa* with enhanced photocatalytic activities. RSC Adv 7:25149–25159
4. Anand K, Tiloke C, Phulukdaree A, Ranjan B, Chuturgoon A, Singh S Gengan, RM (2016) Biosynthesis of palladium nanoparticles by using *Moringa oleifera* flower extract and their catalytic and biological properties. J Photochem Photobiol B Bio 165:87–95
5. Arokiyaraj S, Dinesh Kumar V, Elakya V, Kamala T, Park SK, Ragam M, Saravanan M, Bououdina M, Arasu MV, Kovendan K, Vincent S (2015) Biosynthesized silver nanoparticles using floral extract of *Chrysanthemum indicum* L.—potential for malaria vector control. Environ Sci Pollut Res Int 22:9759–9765
6. Bachheti RK, Fikadu A, Bachheti A, Husen A (2020) Biogenic fabrication of nanomaterials from flower-based chemical compounds, characterization and their various applications: a review. Saudi J Biol Sci 27(10):2551–2562
7. Bhattacharyya D, Singh S, Satnalika N (2009) Nanotechnology, big things from a tiny world: A review. Int J u-e- Serv Sci Technol 2:29–38
8. Bowen-Forbes CS, Zhang Y, Nair MG (2010) Anthocyanin content, antioxidant, anti-inflammatory and anticancer properties of blackberry and raspberry fruits. J Food Compos Anal 23:554–560
9. Chung IM, Park, Seung-Hyun K, Thiruvengadam M, Rajakumar G (2016) Plant-mediated synthesis of silver nanoparticles: their characteristic properties and therapeutic applications. Nanoscale Res Lett, 11–40
10. Devanesan S, AlSalhi MS (2021) Green synthesis of silver nanoparticles using the flower extract of Abelmoschus esculentus for cytotoxicity and antimicrobial studies. Int J Nanomed 16:3343–3356
11. Devaraj B, Bhuvaneshwari V (2019) evaluation of the cytotoxic and antioxidant activity of phyto-synthesized silver nanoparticles using *Cassia angustifolia* flowers. BioNanoScience 9
12. Diallo A, Ngom BD, Park E, Maaza, M (2015) Green synthesis of ZnO nanoparticles by *Aspalathus linearis*: structural & optical properties. J Alloys Compd 646:425-430
13. Dobrucka R, Długaszewska J (2016) Biosynthesis and antibacterial activity of ZnO nanoparticles using Trifolium pratense flower extract. Saudi J Biol Sci 23:517–523
14. Elia P, Zach R, Hazan S, Kolusheva S, Porat Z, Zeiri Y (2014) Green synthesis of gold nanoparticles using plant extracts as reducing agents. Int J Nanomed 9:4007–4021
15. Hajra A, Dutta S, Mondal NK (2016) Mosquito larvicidal activity of cadmium nanoparticles synthesized from petal extracts of marigold (*Tagetes* sp.) and rose (*Rosa* sp.) flower. J Parasit Dis 40:1519–1527
16. Hemmati S, Rashtiani A, Zangeneh MM, Mohammadi P, Zangeneh A, Veisi H (2019) Green synthesis and characterization of silver nanoparticles using fritillaria flower extract and their antibacterial activity against some human pathogens. Polyhedron 158:8–14
17. Husen A (2017) Gold nanoparticles from plant system: synthesis, characterization and their application. In: Ghorbanpourn M, Manika K, Varma A (eds) Nanoscience and plant–soil systems, vol 48. Springer International Publishing AG, Gewerbestrasse 11, 6330 Cham pp 455–479
18. Husen A (2019) Natural product-based fabrication of zinc oxide nanoparticles and their application. In: Husen A, Iqbal M (eds) N*anomaterials* and plant potential. Springer International

Publishing AG, Gewerbestrasse 11, 6330 Cham, pp 193–291. https://doi.org/10.1007/978-3-030-05569-1_7
19. Husen A (2020) Introduction and techniques in nanomaterials formulation. In: Husen, A, Jawaid M (eds) Nanomaterials for agriculture and forestry applications. Elsevier Inc. 50 Hampshire St., 5th Floor, Cambridge, MA 02139, USA, pp 1–14
20. Husen A, Siddiqi KS (2022) Advances in smart nanomaterials and their applications. Elsevier Inc., 50 Hampshire St., 5th Floor, Cambridge, MA 02139, USA
21. Husen, A, Iqbal, M (2019). Nanomaterials and plant potential. Springer International Publishing AG, Gewerbestrasse 11, 6330 Cham, Switzerland. https://doi.org/10.1007/978-3-030-05569-1
22. Husen, A, Jawaid, M (2020) Nanomaterials for agriculture and forestry applications. Elsevier Inc., 50 Hampshire St., 5th Floor, Cambridge, MA 02139, USA. https://doi.org/10.1016/C2018-0-02349-X
23. Jahan I, Erci F, Isildak I (2019) Microwave-assisted green synthesis of non-cytotoxic silver nanoparticles using the aqueous extract of *Rosa santana* (rose) Petals and their antimicrobial activity. Anal Lett 52:1860–1873
24. Jamdade DA, Rajpali D, Joshi KA, Kitture R, Kulkarni AS, Shinde VS, Bellare J, Babiya KR, Ghosh S (2019) *Gnidia glauca* and *Plumbago zeylanica*-mediated synthesis of novel copper nanoparticles as promising anti-diabetic agents. Adv Pharmacol Sci 9080279
25. Johnson P, Krishnan V, Loganathan C, Govindhan K, Raji V, Sakayanathan P, Vijayan S, Sathishkumar P, Palvannan T (2018) Rapid biosynthesis of *Bauhinia variegata* flower extract-mediated silver nanoparticles: an effective antioxidant scavenger and α-amylase inhibitor. Artif Cells Nanomed Biotechnol 46:1488–1494
26. Kaliammal R, Parvathy G, Maheshwaran G, Velsankar K, Devi VK, Krishnakumar M, Sudhahar S (2021) *Zephyranthes candida* flower extract mediated green synthesis of silver nanoparticles for biological applications. Adv Powder Technol 32:4408–4419
27. Karpagavinayagam P, Vedhi C (2019) Green synthesis of iron oxide nanoparticles using *Avicennia marina* flower extract. Vacuum 160:286–292
28. Karunakaran G, Jagathambal M, Gusev A, Minh NV, Kolesnikov E, Mandal AR, Kuznetsov D (2016) Nitrobacter sp. extract mediated biosynthesis of Ag_2O NPs with excellent antioxidant and antibacterial potential for biomedical application. IET nanobiotechnol 10:425–430
29. Karunakaran G, Jagathambal M, Kumar GS, Kolesnikov E (2020) *Hylotelephium telephium* flower extract-mediated biosynthesis of CuO and ZnO nanoparticles with promising antioxidant and antibacterial properties for healthcare applications. Jom 72:1264–1272
30. Khan ZS, Shinde VN, Bhosle NO, Nasreen S (2010) Chemical composition and antimicrobial activity of angiospermic plants. Middle-East J Sci Res 6:56–61
31. Kumar PSM, Francis AP, Devasena T (2014) Biosynthesized and chemically synthesized titania nanoparticles: comparative analysis of antibacterial activity. J Environ Nanotechnol 3:73–81
32. Lee YJ, Song K, Cha SH, Cho S, Kim YS, Park Y (2019) Sesquiterpenoids from *Tussilago farfara* flower bud extract for the eco-friendly synthesis of silver and gold nanoparticles possessing antibacterial and anticancer activities. Nanomaterials, 9–819
33. Lesniak A, Salvati A, Santos-Martinez MJ, Radomski MW, Dawson KA, Aberg C (2013) Nanoparticle adhesion to the cell membrane and its effect on nanoparticle uptake efficiency. J Am Chem Soc 135:1438–1444
34. Lin YS, Wu SS, Lin JK (2003) Determination of tea polyphenols and caffeine in tea flowers (*Camellia sinensis*) and their hydroxyl radical scavenging and nitric oxide suppressing effects. J Agric Food Chem 51:975–980
35. Mameneh R, Shafiei M, Aidy A, Karimi E, Badakhsh B, Abbasi N (2019) Toxicity study of silver nanoparticles synthesized from aqueous flower extract of Scrophularia striata on MCF-7 human breast cancer cell line. Pharmacogn Mag 14:66–72
36. Manisha DR, Alwala J, Kudle KR, Rudra MPP (2014) Biosynthesis of silver nanoparticles using flower extracts of *Catharanthus roseus* and evaluation of its antibacterial efficacy. World J Pharm Pharm Sci. 3:877–885
37. Mapala K, Pattabi M (2017) *Mimosa pudica* flower extract mediated green synthesis of Au-NPs. NanoWorld J 3:44–50

38. Marimuthu S, Rahuman AA, Jayaseelan C, Kirthi AV, Santhoshkumar T, Velayutham K, Bagavan A, Kamaraj C, Elango G, Iyappan M, Siva C, Karthik L, Rao KV (2013) Acaricidal activity of synthesized titanium dioxide nanoparticles using *Calotropis gigantea* against *Rhipicephalus microplus* and *Haemaphysalis bispinosa*. Asian Pac J Trop Med 6:682–688
39. Mashwani ZU, Khan MA, Khan T, Nadhman A (2016) Applications of plant terpenoids in the synthesis of colloidal silver nanoparticles. Adv Colloid Interface Sci 234:132–141
40. Mondal R, Yilmaz MD, Mandal AK (2021) Green synthesis of carbon nanoparticles: characterization and their biocidal properties. In: Handbook of greener synthesis of nanomaterials and compounds Elsevier, pp 277–306
41. Nagaraj B, Malakar B, Divya TK, Krishnamurthy NB, Liny P, Dinesh R (2012) Environmental benign synthesis of gold nanoparticles from the flower extracts of *Plumeria alba* Linn. (Frangipani) and evaluation of their biological activities. Int J Drug Dev Res 4:144–150
42. Nayan V, Onteru SK, Singh D (2018) *Mangifera Indica* flower extract mediated biogenic green gold nanoparticles: efficient nanocatalyst for reduction of 4- nitrophenol. Environ Prog Sustain Energy 37:283–294
43. Ogunyemi SO, Abdallah Y, Zhang M, Fouad H, Hong X, Ibrahim E, Masum M, Hossain A, Mo J, Li B (2019) Green synthesis of zinc oxide nanoparticles using different plant extracts and their antibacterial activity against *Xanthomonas oryzae* pv. *oryzae*. Artif Cells Nanomed Biotechnol 47:341–352
44. Palithya S, Gaddam SA, Kotakadi VS, Penchalaneni J, Golla N, Krishna SB, Naidu CV (2022) Green synthesis of silver nanoparticles using flower extracts of *Aerva lanata* and their biomedical applications. Part Sci Technol 40:84–96
45. Patil MP, Bayaraa E, Subedi P, Piad LL, Tarte NH, Kim G (2019) Biogenic synthesis, characterization of gold nanoparticles using *Lonicera japonica* and their anticancer activity on HeLa cells. J Drug Deliv Sci Technol 51:83–90
46. Puckhaber LS, Stipanovic RD, Bost GA (2002). Analyses for flavonoid aglycones in fresh and preserved Hibiscus flowers. In: Jules J, Anna W (eds) Trends in new crops and new uses, vol 56. ASHS Press, pp 563–21
47. Raghavan BS, Kondath S, Anantanarayanan R, Rajaram R (2015) Kaempferol mediated synthesis of gold nanoparticles and their cytotoxic effects on MCF-7 cancer cell line. Process Biochem 50:1966–1976
48. Sharma D, Sabela MI, Kanchi S, Mdluli PS, Singh G, Stenstrom TA, Bisetty K (2016) Biosynthesis of ZnO nanoparticles using *Jacaranda mimosifolia* flowers extract: synergistic antibacterial activity and molecular simulated facet specific adsorption studies. J Photochem Photobiol B Bio 162:199–207
49. Sharma JK, Akhtar MS, Ameen S, Srivastava P, Singh G (2015) Green synthesis of CuO nanoparticles with leaf extract of *Calotropis gigantea* and its dye-sensitized solar cells applications. J Alloys Compd 632:321–325
50. Siddiqi KS, Husen A (2020) Current status of plant metabolite-based fabrication of copper/copper oxide nanoparticles and their applications: a review. Biomater Res 24(11):1–15
51. Singh J, Dutta T, Kim KH, Rawat M, Samddar P, Kumar P (2018) "Green" synthesis of metals and their oxide nanoparticles: applications for environmental remediation. J Nanobiotechnol 16:84
52. Some S, Das S, Mondal R, Gangopadhyay M, Basak GK (2021) Medicinal plant extract mediated green synthesis of metallic nanoparticles: a review. Int J Plant Env 7:119–132
53. Some S, Mondal R, Dam P, Mandal AK (2022) Synthesis of biogenic silver nanoparticles using medicinal plant extract: A new age in nanomedicine to combat multidrug-resistant pathogens. In: Abd-Elsalam KA (eds) Green synthesis of silver nanomaterials. Elsevier, pp 359–387
54. Some S, Sarkar B, Biswas K, Jana TK, Bhattacharjya D, Dam P, Mondal R, Kumar A, Deb AK, Sadat A, Saha S, Kati A, Ocsoy I, Franco OL, Mandal A, Mandal S, Mandal AK, Ince IA (2020) Bio-molecule functionalized rapid one-pot green synthesis of silver nanoparticles and their efficacy toward the multidrug-resistant (MDR) gut bacteria of silkworms (*Bombyx mori*). RSC Adv 10:22742–22757

55. Suriyakala G, Sathiyaraj S, Gandhi AD, Vadakkan K, Rao UM, Babujanarthanam R (2021) *Plumeria pudica* Jacq. flower extract-mediated silver nanoparticles: characterization and evaluation of biomedical applications. Inorg Chem Commun 126:108470
56. Varadavenkatesan T, Selvaraj R, Vinayagam R (2019) Dye degradation and antibacterial activity of green synthesized silver nanoparticles using *Ipomoea digitata* Linn. flower extract. Int J Environ Sci Technol 16:2395–2404
57. Wang BC, He R, Li ZM (2010) The stability and antioxidant activity of anthocyanins from blueberry. Food Technol Biotech, 48–42
58. Wang L, Hu C, Shao L (2017) The antimicrobial activity of nanoparticles: present situation and prospects for the future. Int J Nanomed 12:1227–1249
59. Wang R, Wang W, Wang L, Liu R, Ding Y, Du L (2006) Constituents of the flowers of *Punica granatum*. Fitoterapia 77:534–537
60. Woo Y, Lai DY (2012) Aromatic amino and nitro-amino compounds and their halogenated derivatives. In: Bingham E, Cohrssen B, Powell CH (eds) Patty's toxicology. Wiley Hoboken, pp 1–96
61. Xu ZP, Zeng QH, Lu GQ, Yu AB (2006) Inorganic nanoparticles as carriers for efficient cellular delivery. Chem Eng Sci 61:1027–1040
62. Younis IY, El-Hawary SS, Eldahshan OA, Abdel-Aziz MM, Ali ZY (2021) Green synthesis of magnesium nanoparticles mediated from Rosa floribunda charisma extract and its antioxidant, antiaging and antibiofilm activities. Sci Rep 11:1–5

Seed-Based Oil in Nanomaterials Synthesis and Their Role in Drug Delivery and Other Applications

Vijayalakshmi Selvakumar, Ramachandran Chelliah, Kaliyan Barathikannan, Fazle Elahi, Momna Rubab, Simpy Sanyal, Su-Jung Yeon, and Deog-Hwan Oh

Abstract Due to their natural and secure status, widespread consumer acceptability, and diverse functional capabilities, essential and vegetable oils have seen a rise in popularity over the past ten years in food and medical industrial applications. The development of novel therapeutics or efficient agents based on plant oils has, however, been hampered by issues with limited stability and/or diminished efficacy. Because of the need for plant oil and limitations of the stability, researchers search for alternate solutions like encapsulation. In encapsulation techniques nanocarriers act as a barrier between oils and environment, it will be helpful for the control of oil release, improving stability, reducing toxicity and increasing convenience. In this chapter, we focused on vegetable oil and essential oil with polymeric, liposomes and solid lipid nanoparticles. At the same time, vegetable and essential oil manufacturing, and research regarding the plant oils and nanocarriers are highlighted.

Keywords Essential oil · Vegetable oil · Encapsulation · Polymeric nanoparticle · Liposome · Solid lipid nanoparticle

---

V. Selvakumar · R. Chelliah (✉) · K. Barathikannan · F. Elahi · M. Rubab · S.-J. Yeon · D.-H. Oh
Department of Food Science and Biotechnology, College of Agriculture and Life Science, Kangwon National University, Chuncheon 24341, Korea
e-mail: ramachandran865@kangwon.ac.kr

K. Barathikannan
Agricultural and Life Science Research Institute, Kangwon National University, Chuncheon 24341, Korea

R. Chelliah
Department of Food Science and Biotechnology, Kangwon Institute of Inclusive Technology (KIIT), Kangwon National University, Chuncheon 24341, Korea

Saveetha School of Engineering, (SIMATS) University, Tamil Nadu, Chennai 600124, India

M. Rubab
Department of Food Science and Technology University of Management and Technology (UMT), Lahore, Pakistan

S. Sanyal
Department of Electrical and Computer Engineering, Sungkyunkwan University, Suwon 16419, Republic of Korea

A. Husen (ed.), *Secondary Metabolites Based Green Synthesis of Nanomaterials and Their Applications*, Smart Nanomaterials Technology,
https://doi.org/10.1007/978-981-99-0927-8_13

1 Introduction

Oils play a critical role in nutrition and vitamins. Oils continue to be the primary source of essential fatty acids such as arachidonic acids, linoleic and linolenic which enhance food flavor. Triglycerides are the most noticeable oil constituents, and their physical characteristics are influenced by the arrangement and scattering of the fatty acids in the oil at the time. Ninety percentage of the oil produced is of vegetable origin, generated after seed dispensation, and intended for human use. The market is seeing an increase in demand for oils derived from diverse natural sources, particularly for food items. The residual 10% of oil is used to make biodiesel, soap, detergents, biodegradable softeners, animal feed, and other industrial products. There has been an increase in interest in edible oil modification technologies in recent years because of the nutritional and financial significance of these oils. More research is being done on modification technologies to change the properties of the oils and make them appropriate for particular uses. Numerous techniques to expand the safety and excellence of food have been found by researchers [53].

2 The Use of Nanotechnology in the Food Industry

Food qualities like thermal stability, bioavailability and solubility were improved by nanotechnology. According to predictions, nanotechnology will advance the inclusion of health supplements, improve food storage procedures, increase storage period, and enable contamination outlining techniques. Therefore, nanotechnology unquestionably has a big impact on the study of food. One of the technologies that are most likely to transform current approaches to food science then the food manufacturing is nanotechnology. The value of nanotechnology in food systems has been demonstrated by the processing and packaging processes facilitated by nanotechnology [51, 66]. Using different preparation methods, it is possible to produce nanoparticles with a range of physical characteristics, as a result, these particles may be used in food. In the process of encapsulation resulting in minute nanoscale capsules smaller than 1000 nm. According to earlier research, encapsulation is a promising strategy for maintaining the qualities of the natural/native oil throughout time [4].

In addition to its benefits, nanoencapsulation has improved the bioavailability, decreased the processing related side effects, protected from environment, improves the stability, and enhances the functional properties. Additionally, by regulating the distribution of active substances, extending shelf life, and perhaps even preventing the emergence of unfavorable qualities, encapsulation provides a technique to increase biological effectiveness. Above features act as a physical–chemical barrier against pro-oxidant factors like oxygen, free radicals, or ultraviolet light, oil encapsulation may be able to halt or delay oxidation reactions (UV). It also increases the variety of food items intended for enrichment. For instance, the alteration of oil discharge, defense from eco-friendly oxidation responses, improvement of patient compliance and convenience, rise in physical stability, reduction in volatility, decrease in toxicity, and improvement of bioactivity can all be achieved through the use of bioactive oil encapsulation [99].

It also extends the time that food products can be stored. With the help of this technology, ingredients are also protected and bioactive chemical discharge is under control. It specifically enhances the properties of processed foods such as flavor retention, antioxidant protection, shelf life, color, and off-odor. Several methods remain used for encapsulation. Bioactive compounds are often encapsulated using one of the three techniques: The encapsulated agent is surrounded by a barrier structure, contaminated materials are prevented from entering, and the encapsulated agents are positioned to protect them from unintended harm. The first stage of nanoencapsulation is often the creation of nanoemulsions, which are systems composed of aqueous and fatty phases. Emulsifiers remain typically used to emulsify nanoencapsulation [39].

Nanoemulsions also feature small drop sizes and a high surface area. These characteristics, including greater physical stability and increased bioavailability, may make them superior to conventional emulsions. Few of the techniques like emulsion-diffusion and emulsification-solvent evaporation are under investigation for producing oil nanoemulsion and oil nanoencapsulation (NLCs). The objective of this work mainly focused on oil encapsulation in the food industry and taking into account potential challenges like already available food products on the market and pending patent applications. Patent filings and new nanoencapsulated oil products show potential for the usage of oil in numerous manufacturing areas. Additionally, micro- and nanoencapsulation can facilitate (a) a lessening in the core material's vaporization (b) essential solid protection after degradation by a reduction (c) a regulator of the rate of core material discharge (d) alteration of the physical features of the innovative material and (e) protection of essential material after degradation by a decrease [97].

3 The Chemical Makeup of Vegetable Oils

From plant seeds and fruits many vegetable or non-essential or fixed oils are derived. They have an extremely complicated and changeable makeup that varies based on the source, the crop, the season, and the manufacturing techniques. Triglycerides are the primary elements of VO; liable on the basis of the oil, they make up 95 to 98% of the global constituents. At-room-temperature triglycerides are referred to as "oil" in this sentence. Triglycerides are composed of three fatty acid molecules like saturated, mono and poly saturated. Fatty acid length, double bond arrangements and unsaturation levels are important factors which influence the characteristics of VO. Varieties in farming, agronomy, and climate may also result in variations in the fatty acid content of triglycerides. In addition to triglycerides, VO also contains a variety of minor non-triglyceride molecules (less than 5%), which have important biological characteristics and nutritional benefits for the pharmaceutical and nutraceutical industries. They came in two varieties: non-glycerolipids and phospholipids [78].

4 The Chemical Makeup of Essential Oils

Essential oils exist in oily fragrant liquids. Eucalyptus, thyme and salvia oils are obtained from leaves and citrus oil is from zest and fruits. Lavender, rose and jasmine oils from flowers. The EO are typically intricate blends of distinct scent components. They are not exactly oils, but they have poor water solubility. To extract EO as of plant raw materials, cutting-edge techniques have developed during the past few decades. Terpenes, phenylpropanoids, straight-chain compounds, and other groups are the four primary groups that make up plant essential oils.

4.1 Terpenes

The most prevalent substances in plant EO are terpenes. They are created in the cytoplasm of plant cells using the mevalonic acid pathway. Terpene chemicals can be split into two primary groups: oxygenated compounds and terpene hydrocarbons [29].

4.2 Terpenes Hydrocarbon

The isoprene unit connected head-to-tail is the fundamental component of terpenes. Several terpenes can be found in EO depending on the quantity of isoprene units. The two main terpenes present in plant EO are monoterpenes and sesquiterpenes. They are made of two and three units of isoprene, respectively. The structures of these terpenes could be acyclic, cyclic, or aromatic. Depending on the size of their rings, cyclic terpenes can be divided as monocyclic, bicyclic, or tricyclic terpenes. These substances respond quickly to heat and air sources, which causes them to oxidise quickly. Diterpenes are vital components of plant resins and are produced when four isoprene parts are combined head to tail. The extraction technique may change their composition [153].

4.3 Terpenoids

By adding functional groups, terpene can be changed into terpenoid. Monoterpenoid is formed from monoterpene or sesquiterpene. Since "oid" means "derived from" or "like," any molecule having a structure similar to a terpene can be referred to as a "terpenoid." Ex. Carboxylic acids and alcohols [84].

4.4 Phenylpropanoid

These types of compounds are created by the shikimic acid pathway with the combinations of amino acids like phenylalanine, 1-tyrosine and phenylpropene. The flavor and odor were determined by Phenylpropanoids. For example, Eugenol and cinnamaldehyde.

4.5 Straight Chain Compounds

There are no side branches and solely straight-chain compounds in this group. N-heptane and mixtures containing 35 carbon atoms are among them. 3(Z) Hexen-182-1ol is responsible for the green color of the leaf, it is also called as leaf alcohol, while cutting green grass and leaves, it emits a strong grassy-green odor.

4.6 Miscellaneous Group

They are molecules that contain sulfur and nitrogen that are produced when unsaturated fatty acids, lactones, terpenes, and glycosides degrade. 187 indoles and diallyl disulfide are examples of this group of chemicals [87].

5 Limitations in the Medical Practice of Plant Oils

Antique times, EO has remained advocated for a wide range of illnesses and health issues all throughout the world. The pharmacological effects of EO, including its antibacterial, antifungal, antiviral, insecticidal, antioxidant, antidiabetic, anti-inflammatory, antihypertensive, and immunomodulatory characteristics, have been the subject of many investigations. In the medical industry, EO has been used to treat nosocomial infections, clean operating rooms and waiting rooms' air to prevent contamination, and disinfect medical equipment and surfaces. The domains of agriculture and food have seen the application of EO. They are included in food packaging as well as employed as plant and crop protectors, antioxidants, and preservatives in food.

On the other hand, VO is a basis of permanent elements that are in charge of important biological characteristics, especially for our health. Polyphenols, medium chain fatty acids, and tocopherol were also added to some VO to help with their antioxidant and antistress properties. According to recent animal studies, palmitic acid has strong antiviral properties against HIV-1 and HIV-2. Oleic and linoleic acids were thought to be powerful antibacterial substances. Additional experimental research uncovered additional biological advantages of VO, including its hepatoprotective, gastroprotective, anti-inflammatory, analgesic, and antipyretic characteristics, as well as its hypolipidemic, antibacterial, and antiviral activities [106].

The cutaneous way), the oral way, the rectal channel, and the pulmonary way are only a few of the dosage forms in which plant oils have been used (aerosols, nasal drops, sprays). The medicines based on plant oil do, however, come with some significant limitations. Because of their limited solubility in biological fluids, low penetration, and decreased bioavailability, VO oral distribution is constantly difficult because of the likelihood of gastrointestinal tract oxidation. In the case of food-based catalysts, it has been hypothesized, for instance, that the low pH and adsorbed oxygen present in human gastrointestinal juice may enhance the oxidation of VO. Additionally, it was said that the hydroperoxide produced when pancreatic lipase hydrolyzes triacylglycerol that is enclosed in VO is not able to be immersed in its entire method [14].

Additionally, the use of EO is constrained by its toxicity and irritating effects on the gastrointestinal and oesophageal mucosa. If EO is not appropriately or not at all diluted, it may cause allergic reactions in the skin. For instance, using cinnamaldehyde-rich essential oils in excess on a broad area of the skin might cause a serious sensitized reaction and anaphylactic shock, both of which can be fatal. Some oils like cumin can photosensitize people and lead to malignant skin conditions such as hyperpigmentation, burns, itching, and berloque dermatitis. Due to their high volatility and significant risk of destruction upon environmental factors, EO do not require a medicinal vehicle for use. Therefore, it has been suggested that nanoencapsulation is essential for overcoming the aforementioned constraints. Oils are transformed into three different forms such as liquid, solid, or semi-solid, encapsulation techniques enable the retaining of their action for lengthier stages of time. By functioning as barriers between the molecule and the environment, nanocarriers gave protection against them. They also protect the oils from oxidative decay. The greater efficacy, decreased toxicity, more patient compliance, and convenience comes from their capacity to manage oil release.

6 Plant Oils-Loaded Nano-delivery Systems

Intense research is being done on nano-delivery systems because they have several therapeutic advantages, including great drug effectiveness, specificity, and acceptability; (ii) controlled, continued drug release; (iii) lower possibility of harmfulness; (iv) deep tissue transmission and (v) defense of drugs. The two main types of organic nanocarriers that we discussed in this work were lipid-based nanocarriers and polymer-based nanocarriers, which are both experimental methods for encapsulating plant oils [65].

7 Encapsulation in Polymer-Based Nanocarriers

Submicronic core–shell spherical structures known as polymeric nanoparticles (pNP) are created from organic or synthetic polymers. Nanocapsules or nanospheres can be produced, depending on the method used to prepare the nanoparticles. In the matrix-like structure of nanospheres, the medication is uniformly dispersed. However, nanocapsules are vesicular structures with an internal core and a polymeric membrane around the medicine. PNP are currently the focus of considerable research regarding the delivery of plant oils, because it entraps and preserves oils with high stability, biodegradability, and bioavailability. Polymer coated the oil with in the core. The concepts of the several ways have been published for 273 producing pNP, with the most popular ones being nanoprecipitation [143].

8 Spray Drying

Plant oils in micro and nano polymeric particles are encapsulated with spray drying method. The single, continuous conversion of an infeed into a dry powder is the foundation of this process. For the encapsulation of plant oil, oil-in-water emulsion was used with hot air (150–250 °C). In terms of particle size and encapsulation effectiveness (EE%), numerous work studies have revealed with entrapment level of oil is low and diameter was high. A previous study found that the average size of taro starch particles containing almond oil ranged from 1.4 to 31.1 m, with an EE% of 37.5% [86].

A similar EO was encased in chitosan nanoparticles in a different study. In cashew gum, 70% of EE with 335–558 nm size was made by spray drying. Additionally, walnut oil was enclosed in a mixture of maltodextrin and soy protein, and the resulting particles had an EE% of 60% and an average diameter of 4–10 m. In an additional study, a similar EO was enclosed in chitosan nanoparticles, with 70% EE and 335–558 nm particle size. Additionally, walnut oil was enclosed in a mixture of maltodextrin and soy protein, and the resulting particles had an EE% of 60% and an average diameter of 4 to 10 m [70, 71].

The production of unique particles is the spray drying technique's key characteristic. In fact, scanning electron microscopy of microparticles treated with flaxseed oil showed that hollow particles, or particles with concave and shriveled surfaces, had formed. When imagining gum arabic microcapsules with linseed oil. They explained this structure as the result of vacuoles forming inside the particles following the formation of the crust. The amount of this procedure is inversely proportional to the amount of solid in the emulsion. Additionally, spray-dried peony seed oil microparticles studied using confocal scanning laser microscopy revealed that the particles exhibited a full core–shell assembly, a flat surface, and no discernible cracks or breaks. Plant oils generated by spray drying have an in vitro release profile that typically displays a regulated pattern. Depending on the ratio of the two polymers utilized at the time, the amount of Lippiasidoides EO in the alginate: cashew gum units was between 45 and 95%. Because cashew gum has a higher hydrophilic nature than other gums, which allows for more EO diffusion through the polymeric wall, an advanced and quicker discharge rate was attained with a large proportion of cashew gum. Increasing the alginate level, however, resulted in a smaller and more controlled release; after 50 h, just 45% of the oil had been released. This has to do with the growth of the compact network, which ultimately prevented the spread of EO to the outside medium [134].

High stability is provided for encapsulated oils by the particles made throughout the spray drying method. The oxidative stability of almond oil has previously been studied both before and after being encapsulated in taro starch microcapsules. The material's substantially lower porosity was found to offer improved oxidation protection in comparison to unencapsulated almond oil by limiting the oil's exposure to stress situations while being stored. In spray drying process many factors are involved in the entrapment of oils. The impact of wall material type and amount on plant oil-loaded particle characteristics have been the subject of numerous reports. In the past, I also looked at the spray drying method for microencapsulating flaxseed oil. They looked at the effects of maltodextrin combined with various wall components on droplet size, and kinds of altered starch. The maximum viscosity demonstrated by maltodextrin, according to the results, is: Arabic gum emulsion suggests a stronger droplet-movement resistance, which prevents coalescence and results in smaller droplet size [113].

The shelf life of powders is significantly influenced by their moisture content. The moisture content was said to have decreased as the inlet temperature rose. However, higher temperatures (over 175 °C) led to an increase in moisture content because a fast crust formed, which slowed down water evaporation and resulted in particles with increased moisture content. Another factor that changes with the air inlet temperature is the particle size. Mohammed and associates reported in 2017 that the microcapsule setup caused the larger Nigella sativa EO units to be produced at a higher inlet air temperature. In fact, temperatures make it possible for the particles to form earlier, which stops the particles from contracting during drying. For the spray drying of vegetable oils, high temperature is required. This was due to the fact that plant oils only came into contact with hot air for a little period of time; as a result, evaporation only took place at the surface of the oil particles for a period of time between 15 and 30 s, never reaching the inlet temperature of the drying gas. The most popular wall materials include maltodextrin, gums, and chitosan. The nutraceutical and food industries already employ this method extensively because of its advantages, which include simplicity, low production cost and easy production and simple transport. The usage of this technique has a number of limitations, including an absence of consistency, high particle size, poor oil packing, and the potential for oil loss [56].

9 Emulsion Evaporation

The emulsion evaporation approach generates pNPs. In this process, a non-solvent mixture is emulsified at high-shear stress with an organic solution comprised of oil and a polymer. In order to create nanoparticles, the organic solvent is then either evaporated at reduced pressure or heated to ambient temperature while being continuously magnetically stirred. This method has been utilized to include both vegetable and essential oils in pNPs. In comparison to the spray drying method, the emulsion evaporation method yields nanoparticles with a greater EE% and smaller average diameters [105].

In this instance, Minor reported that 399 poly nanoparticles loaded with eugenol and trans-cinnamaldehyde had particle diameters of 179.3 and 173.8 nm and EE%s of 98 and 92%, respectively. Similar outcomes were attained using nanoparticles made of chia mucilage and filled with chia seed oil. These particles had a negative zeta potential (82.8%), a very small particle size (205 nm), and an extraordinarily high EE% (−11.58 mv). Furthermore, poly (lactide-co-glycolide) nanoparticles were made to include clove essential oil (EO), which is high in eugenol and has powerful antibacterial effects. The resultant nanoparticles averaged 237 nm in size, exhibited a −40 mv negative zeta potential, and had an EE% of 93% [152].

10 Nanoprecipitation

A straightforward, speedy, and popular method for creating nanoparticles is nanoprecipitation. The initial report was based on the discovery that when a water-miscible solvent was taken out of a lipophilic solution, polymers began to accumulate at the interface. This technique has been frequently utilized to encapsulate plant oils in pNPs, and lists various research that have done so over the past ten years, encapsulating also crucial or vegetable oils utilizing the nanoprecipitation process. High EE%, narrow dispersion and small particle sizes were all characteristics of the nanoparticles produced by the nanoprecipitation method. Carvacrol-loaded poly (lactide-co-glycolide)-based nanocapsules and Cymbopogon martini EO were prepared by nanoprecipitation method [107].

According to the literature, a variety of parameters have to be taken into account in order to create pNPs with the appropriate physicochemical qualities. The kind and quantity of polymers take a major impact happening the average size of nanoparticles. Creating nanoparticles made of poly (-caprolactone) and loaded with rosemary [5]. According to EO, increasing the amount of polymer results in particles having thicker shells, which increases their mean diameter rather than their number. While the primary ingredients in peppermint, 35% menthol and 27% menthone, produce less consistent, complex, and massive nanoparticles. Contrarily, cinnamon essential oil is primarily composed of cinnamaldehyde, which blocks the action of nanocapsules. 0.1 mg/mL polyvinyl alcohol affects the size of the nanoparticles. Additionally, using span 20 rather than tween 20 produced nanocapsules loaded with rosemary essential oil that were smaller, had a lower polydispersity index, and lost less oil during the solvent evaporation process. The hydrophilic-lipophilic balance (HLB), which is important for the stability of rosemary EO nanoemulsions, was between 12 and 15 when span 20 was present (Table 1 and Fig. 1).

11 General Nanoencapsulation of Colloidal Nanoparticles

Microemulsions and nanoemulsions (d 200 nm) are used as a colloidal delivery system made by low and high-energy methods. There are tiny particles in both systems. The fact that nanoemulsions require a lot less surfactant to create than microemulsions is one of their key advantages. Colloidal particles can be created for a variety of uses, including food, essential oils (EOs), metal, biomedical, medical, sensor, optical, flavoring, beverage, repellant, aroma, and cosmetic items. They can also be utilized for cosmetic purposes. To increase the system's oxidative stability, polyunsaturated fatty acids (PUFA) can be incorporated into colloidal delivery methods, including emulsions [137]. These emulsion-based delivery systems have a high turbidity or opacity because the bulk of their particles have sizes that are equivalent to the wavelengths of light. It is beneficial to employ a transparent delivery method for particular applications, such as some food and beverages. Developed colloidal mixtures of lemon oil, Tween 80, used as a surfactant, and a buffer for soft drinks (pH 2.6). Solid Lipid Nanoparticles (SLNs) (50 nm or as large as 1 m) are used in food and pharma industries, due to its physical and chemical stability, are composed of lipids which is solid at both body and room temperature, low toxic, high encapsulation efficiency, mass production capability, flexibility in the controlled release profile due to the solid matrix, and good organ accessibility.

The tendency of SLNs to crystallize, on the other hand, leads to an extremely constrained region for oil incorporation with a low loading capacity. SLNs are characterized by low encapsulation loads and an explosive risk during storage. SLNs were created by Oehlke et al. [98], utilizing tocopherol and ferulic acid (FA) (Toc).

Table 1 Preparation methods for plant oils encapsulation with their fine applications

Encapsulated oil	Plant (origin)	Wall material	Physicochemical properties			Biological properties	Applications	References
			Size (nm)	Z.pot (mV)	EE (%)			
Nanoprecipitation method								
Costus	*Saussurea lappa*	Eudragit RS 100	145	+45	/	Anticholinesterase, anti-inflammatory	Medicine	[70]
Date seed	*Phoenix dactylifera* L	Eudragit RS 100	207	+59	97	Antidiabetic	Medicine	[69]
Lemongrass	*Cymbopogon citratus* DC	PLA	96.4	/	99	Antifungal	Medicine	[6]
		Cellulose acetate	200	−36	/	Antimicrobial	Medicine	[73]
		PLA	300	−6	/			
Peppermint	*Mentha piperita* L	Cellulose acetate	180	−38	/			
Cinnamon	*Cinnamomum cassia* persl		150	−40	/			
Thyme	*Thymus leptobotrys* L	Eudragit RS 100	144	+80.9	/	Bacteriostatic, fungistatic	Medicine	[30]
	Thymus satureoides L		132	81.6	/			
Pelargonium graveolens	*Pelargonium graveolens* L		113	+80.6	/			
Eugenia caryophyllata	*Eugenia caryophyllata* C		131	+80.7	/			

(continued)

Table 1 (continued)

Encapsulated oil	Plant (origin)	Wall material	Physicochemical properties			Biological properties	Applications	References
			Size (nm)	Z.pot (mV)	EE (%)			
Rosemary	*Rosmarinus officinalis* L	PCL	220	−19.9	99	Antioxidant, antimicrobial	Medicine	[31]
Pine seed	*Pinus pinea* L	Eudragit L 100–55 / Eudragit L100 / Eudragit S100	236–296	/	85–91	Source of PUFA	Medicine	[7]
Free fatty acid	/		260–319	/	83–99			
Babassu	*Attalea speciosa* Mart	PLGA	209.2	−15.8	/	Treatment of benign prostatic hyperplasia	Medicine	[26]
Carvacrol	/	PLGA	209.8	/	26	Antimicrobial	Medicine	[52]
Baccharis dracunculifolia	*Baccharis dracunculifolia* DC	Eudragit RS 100	151.6	+51.7	99.4	Antibacterial	Food	[139]
Oregano	*Origanum vulgare* L	PCL	158–300	−13.8 to −28.5		Antibacterial	Food	[130]
Carvacrol	/	PLA	114.7	+54.7	53.9	Antimicrobial	Food	[94]
Palmorasa	*Cymbopogon martini* Roxb	PCL	282.1	−27.2	99.54	Antioxidant, antimicrobial	Cosmetic	[57]
Geraniol	/		289.3	−26.6	99.88			
Thyme	*Thymus vulgaris* L	Eudragit L 100–55	153.9	−4.11	52.81	Antioxidant	Food	[101]
Sweet orange	*Citrus sinensis* L	Eudragit RS100	57- 208	+39 to +74	56–96	Antimicrobial	Food	[37]
Bergamot	*Citrus bergamia* Risso				28–84			

(continued)

Table 1 (continued)

Encapsulated oil	Plant (origin)	Wall material	Physicochemical properties			Biological properties	Applications	References
			Size (nm)	Z.pot (mV)	EE (%)			
Rosemary	*Rosmarinus officinalis* L	Eudragit EPO	200	/	59	Antioxidant	Cosmetics	[128]
Lavender	*Lavandula dentate* L		200	/	41			
Black seed oil + Indometacin	*Nigella sativa* L	PCL	230–260	− 20 to − 30	Oil = 84 Drug = 70	Anti-inflammatory	Cosmetics	[9]
Argan oil + Indometacin	*Argania spinosa* L		290–350	−40 to −50	65–75			
Pomegranate oil + Silibinin	*Punica granatum* L	Ethylcellulose	/	/	/	Anti-inflammatory for skin damage UVB	Cosmetics	[79]
Rosehip	*Rosa rugosa* Thunb	Eudragit RS100	158	+9.8	/	Accelerate the skin regenerating process	Cosmetics	[21]
Borage oil + Betamethasone dipropionate	*Borago officinalis* L	PCL	210	−16.6	100	Treatment of atopic dermatitis	Cosmetics	[145]
Rice bran	*Oryza sativa* L	PCL	200	−9	/	Sunscreen (treatment of inflammatory disorders of skin)	Cosmetics	[110]
Emulsion evaporation method								
Mastic tree of Chios	*Pistacia lentiscus* L. var. chia	PLA	239–286	−29.1 to −34.5	9–37	Antimicrobial	Cosmetics	[144]

(continued)

Table 1 (continued)

Encapsulated oil	Plant (origin)	Wall material	Physicochemical properties			Biological properties	Applications	References
			Size (nm)	Z.pot (mV)	EE (%)			
Clove	*Syzygium aromaticum* L	PLGA	237.6	−40	93.95	Antimicrobial	Medicine	[104]
Coffee	*Coffea arabica* L	PLA	263	/	112.7	Flavoring agent	Food	[36]
		PHBV	271		94.5			
Ionic gelation method								
Homalomena pineodora	*Homalomena pineodora* L	Chitosan	70	+24.1		Wound healing	Medicine	[114]
Camelthorn	*Alhagi maurorum* L	Chitosan	172	+28.6	/	Antimicrobial	Medicine	[46]
Great morinda	*Morinda citrifolia* L	Chitosan	1006	+43.5	/	Anticancer	Medicine	[108]
Nettle	*Urtica dioica* L	Chitosan	208.3–369.4	/	59.5–68.2	Antimicrobial	Food; medicine	[10]
Tarragon	*Artemisia dracunculus* L	Gelatin + Chitosan	246–505	27.1–37.1	9.8–35.6	Preservative	Food	[149]
Clove	*Syzygium aromaticum* L	Chitosan	268.5	+22.4	/	Antifungal	Food	[45]
Pepper tree	*Schinus molle* L	Chitosan	516.9	+40.2	26.6	Antifungal, anti-aflatoxigenic	Medicine	[75]
Cardamon	*Elettaria cardamomum* L	Chitosan	50–100	>+50	>90	Antimicrobial	Medicine	[54]
	Carum copticum	Chitosan	/	/	36.2	Antioxidant, antimicrobial	Cosmetics; medicine	[32]
Tumeric	*Curcuma longa* L	Alginate + Chitosan	<300	/	71	Anticancer	Medicine	[91]

(continued)

Table 1 (continued)

Encapsulated oil	Plant (origin)	Wall material	Physicochemical properties			Biological properties	Applications	References
			Size (nm)	Z.pot (mV)	EE (%)			
Poppy seed	*Papaver somniferum* L	Gelatin + Arabic gum	<5000	/	76.8	Anticancer, antitumor, antidiabetic	Medicine	[147]
Holy basil	*Ocimum tenuiflorum* L	Gelatin	392.30 0.10^3	/	95.41	Antioxidant	Medicine	[131]
Olive	*Olea europaea* L	Gelatin + Sodium alginate	/	/	/	Source of vitamins and fatty acids	Medicine	[28]
Lemongrass	*Cymbopogon citratus* DC	Alginate + Chitosan	<300	/	86.9	Anticancer	Medicine	[91]
		Chitosan	58	/	/	Antimicrobial	Food	[55]
Spray drying methods								
Soybean	*Glycine max* L	Kafirin + sodium caseinate	10,450	/	/	Source of PUFA	Food	[11, 12]
Linseed	*Linum usitatissimum* L	Alginate + whey protein	/	−44	84	Source of omega 3	Food	[35]
Almond	*Prunus dulcis* L	Taro starch 1600–	$31.1.10^3$	/	37.5	Prevention of colon cancer and cardiovascular diseases	Medicine	[49]
Walnut	*Juglans cinerea* L	Skim milk powder + Tween 80	/	/	/	Source of PUFA, tocopherols and phytosterols	Medicine	[123]

(continued)

Table 1 (continued)

Encapsulated oil	Plant (origin)	Wall material	Physicochemical properties			Biological properties	Applications	References
			Size (nm)	Z.pot (mV)	EE (%)			
Black seed	*Nigella sativa* L	Sodium caseinate + maltodextrin DE10	18,500	/	95.4	Antioxidant	Medicine	[85]
Coffee	*Coffea arabica* L	Arabic gum	$4000–11.10^3$	/	10–30	Antioxidant	Cosmetic	[96]
Linseed	*Macadamia integrifolia*	Sodium caseinate + maltodextrin	9750	/	88.75	Source of MUFA	Medicine	[72]

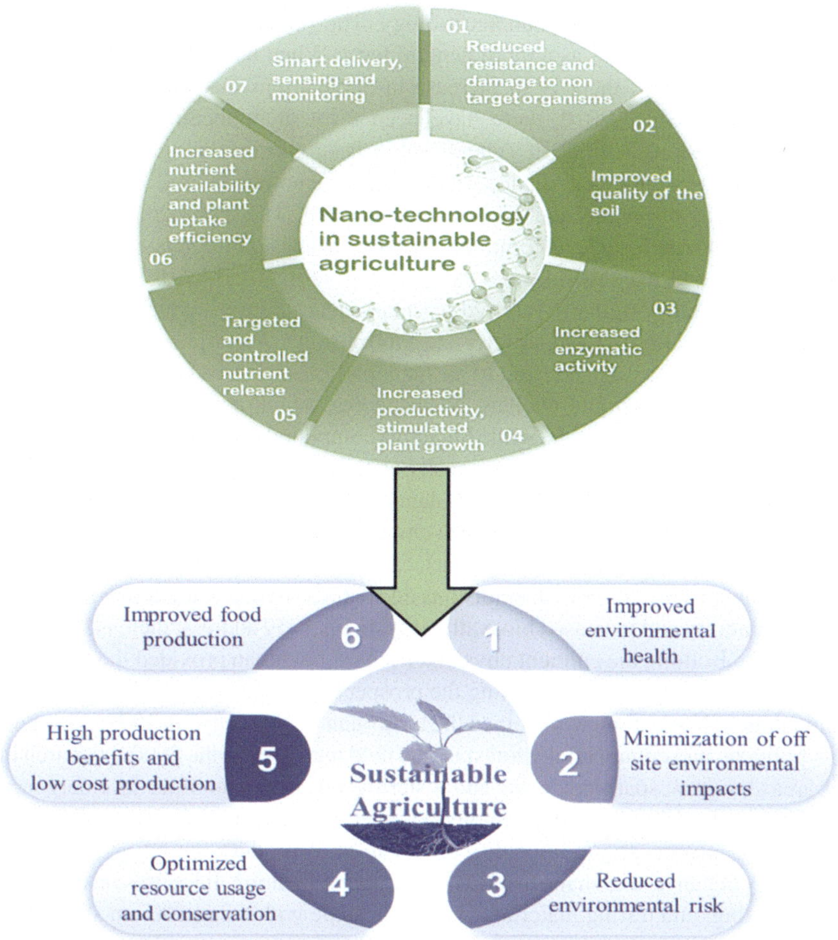

Fig. 1 Utilization of nanotechnology towards plant products on the development of sustainable agriculture

These SLNs are suitable for usage as food additives when a gradual release of the active component would be favorable, according to the scientists' findings.

12 Trends in Oil Nanoencapsulation

Recent searches have been dominated by the four terminologies like "nanoencapsulation," "nanoemulsion," "nanoparticle," and "nanotechnology. However, until the 2000s, articles about the oil and food applications research carried out in the late 1990s accounted for less than 2% of the publications we examined, making this subject a specialized area of nanotechnology. The phrase "nanotechnology" was frequently used in articles as a more general term. When using the combination of these terms and "oil," there is an increase in publications that include the phrase "nanoparticles". Since 2010, there have been a lot of publications about "oils" and "nanoemulsions," either in general subjects or those relating to food. Even though there are significant "nanoparticles" and "nanotechnology," encapsulation is referred to as packaging material or wall material. The terms "nanoencapsulation" and "nanoemulsion" are commonly used in articles on oil nanoencapsulation. Nanoparticles and nanocapsules are two different terms used by different writers. Both terminologies originally referred to "nanoencapsulation," which has come to mean both the creation of nanoparticles and nanocapsules. Nanospheres and nanocapsules are both collectively referred to as "nanoparticles." It was composed of polymeric membrane and liquid nucleus. Nanospheres, which are solid colloidal fragments, are bioactive chemicals, while bioactive compounds are dispersed. Compared to other enclosed systems, cells more easily absorb nanoparticles [22].

A nanoemulsion is the initial stage of nanoencapsulation and is a system made up of oily and aqueous phases that have been emulsified by the addition of an emulsifier. In addition, nanoemulsions have high surface areas and small drop sizes. They may have advantages over traditional emulsions due to these features, including improved physical stability and increased bioavailability. In his 1959 address, there is Plenty of Room at the Bottom, the eminent physicist Richard Feynman provided the first definition of nanotechnology by outlining the prospect of synthesis through direct atom manipulation. Norio Taniguchi coined the term "nanotechnology" in 1974. As scientific, political, and commercial interest in this field increased in the 2000s, research in this area also grew, sparking debate and fostering advancement. Since its emergence as a field in the 1980s, nanotechnology has seen a surge in scholarly publications and popular awareness. Additionally, the creation of products based on nanoscale technology began to take shape in terms of commercialization. It focuses on creating, characterizing, and testing materials at the nanoscale with 1 and 100 nm sizes. The smallest particle size exhibits unusual and unexpected features with respect to the growing surface area, which opens up a wide range of potential technological applications. Nanotechnology can improve food's macroscale characteristics like flavor, texture, industrial processes, and color intensity as well as strategies for bioactive compounds, water solubility, and thermal and storage stability. The big food corporations have developed plans for using nanotechnologies in functional foods using their own research divisions [129].

13 Application Status of Oil Nanoencapsulation

The various researchers reported the terms "food," "oil," "nanoencapsulation," "nanoemulsion," "nanoparticle," or "nanotechnology" in abstracts to quantify the expansion of the food discipline. According to the trends the majority of the expansion in the field of food nanotechnology took place around 2010 as a result of the extensive research on nanotechnology that was done in the late 1990s and the creation of food materials for the nanoparticle process [34].

The industrial applications for oil nanoencapsulation. In the next ten years, the value of food-related nanotechnology products, which currently total US$1 billion (mainly made up of packaging innovations, prototypes, and nanoparticle coatings for health-promoting items), might rise to US$20 billion. Numerous reviews have given positive feedback to the research teams and businesses and other institutions that have been at the forefront of the food nanotechnology field. Despite various analyses covering the investment in food nanotechnology and the novel applications of nanotechnology for primary production, there aren't any studies that cover oil nanoencapsulation when considering oil as the encapsulating ingredient. Numerous evaluations on the use of nanotechnology, and most of them concentrate on applications in the food industry. Some of the applications of the microencapsulation process in the food industry have been described as follows: the core material is protected, evaporation was reduced, easier handling and controlled release. Oil nanoencapsulation is appropriate for these purposes as well [34].

Chia seed oil and high-oleic palm oil (HOPO) may serve various purposes, according to earlier studies. The first study looked for the best HOPO-derived nanoemulsion microfluidization, manufacturing, and storage settings, while the second study proposed viable remedies to protect the oil from lipid oxidation and improve solubility and stability, demonstrates the viability of food microencapsulation; fish oil in microcapsules used as a functional material. Fish oil's disagreeable flavor is concealed by the microencapsulating method, and this bread is already commercially available. A logical development of the technique is the use of compounds and nanoencapsulation in the food business. As was already mentioned, oil nanoencapsulation encourages the decrease in oxidation by creating protective barriers throughout the nanoencapsulation process. Fats, oils, and lipid-based meals deteriorate mostly through oxidation reactions, which lower the nutritional content and sensory quality of the item. They observed that despite using nanoemulsions for encapsulating offers potential, although fatty acids are required for human nutrition but cannot be generated by the human body. Therefore, humans must eat food to obtain them. PUFA containing oils are linked to a decreased risk of developing cancer as well as chronic disorders like cardiovascular or neurological conditions. Bioactive oils are frequently utilized for their nutritional advantages, however, one of their main drawbacks is the loss of active ingredients during storage, due to their sensitivity to environmental factors. Some of these things might be bad for biological tissues. This oxidation destroys the oil's properties and nutritional value while also giving it a foul flavor and odor. The additional active components in these oils may have tissue-regenerative, antiviral, antibacterial, anticancer, and/or anti-inflammatory properties [34].

Oils contain tocopherols and polyphenols, both of which have significant antioxidant properties. Antioxidants, therefore, differ in composition and properties depending on the type of oil. As a result, the oils from grape seed, argan, sunflower, and olive have significant antioxidant content. Additionally, labile compounds including sterols, carotenoids, xanthophyll, flavonols, and monoterpenes might affect an oil's nutritional value and health advantages. Furthermore, EOs contain combinations of physiologically active compounds and provide distinctive molecular templates and possibly bioactive molecules. EOs are made up of volatile secondary metabolites, which also have antifungal, antibacterial, antioxidant, anti-inflammatory, antiviral, and anticancer activities. Some of these oils are highly prone to oxidation even though it has been discovered that they have antimicrobial properties (thyme and lemongrass oil).EOs are categorized as naturally occurring, bioactive compounds that are thought to be ideal for use in preventing the spread of foodborne diseases. But due to their high volatility and possibility for considerable alterations in the sensory aspects of food products, the incorporation has challenges. The antibacterial properties of oregano, carvacrol, thymol, and -terpinene have all been used in food preparation. The use of many essential oils has improved the food quality and functional properties [34].

According to earlier experiments, thyme essential oil was nanoencapsulated and added to beef burgers. They discovered that the encapsulating procedure lengthened the thyme essential oil's shelf life and reduced the amount of active components that vaporized at the start of storage. Additionally, the antioxidant and antibacterial action of the oil may be maintained or even increased throughout the course of refrigerated storage due to the gradual discharge of thyme essential oil during storage. Thyme oil enhanced the sensory quality and acceptability of food products, and there were positive changes in the oxymyoglobin level and redness of the burgers compared to the controls. According to studies, using essential oils as natural preservatives in food can increase food quality and safety while also replacing dangerous synthetic food preservatives. The encapsulated eucalyptus was used to determine the antibacterial properties of the loaded nanoparticles on Listeria monocytogenes and Salmonella enteritidis bacteria. The scientists found that the bactericidal action of nanoparticles was more effective against gram-positive than gram-negative bacteria. The heteropolysaccharide cashew gum (CG), which has an Arabic gum-like structure, is extracted from the exudate of the Anacardium occidentale tree, which is common in northeastern Brazil. Because of its ability to interact with water, cashew gum can function as an adhesive, emulsifier, and stabilizer, making it a viable alternative to gum arabic, which is more expensive (Fig. 2).

14 Nanoencapsulation Techniques Used in Various Oils

This chapter looked at research that used nanoencapsulated oils in the food industry, and was made to show the technologies, wall materials, and nanoencapsulated oils used. The various techniques utilized to make nanocapsules containing oils include emulsion-diffusion technique. The methods are comparable in general and share some features with one another. In the case of emulsion-diffusion, a polymer is deposited around the droplets as a result of dilution, but in the case of emulsification-solvent evaporation, an emulsion is created using a polymer solution and an aqueous phase. At the conclusion of both methods, the solvent is evaporated. Microfluidizers

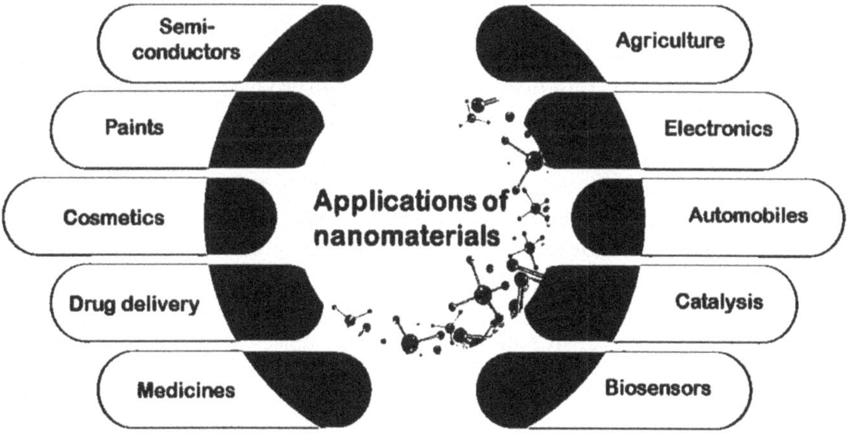

Fig. 2 Nanomaterials and its various field of application

are used in high-shear homogenization, a type of high-energy emulsification that produces mechanical shear. To operate, a liquid jet is split into two sections. Each component fits through a little gap. Typically, emulsions larger than 1 m in diameter are first created using other techniques, and microfluidizer size is decreased [77].

Using the spray dryer method, the active component is dissolved or disseminated in a biopolymer solution. The dispersion is atomized in a heated air chamber, which also rapidly evaporates the solvent, to produce a dry particle containing the active ingredient embedded in a porous wall material. The supercritical fluid extraction of emulsions (SFEE) technique uses supercritical carbon dioxide to quickly remove the organic solvent from an oil-in-water emulsion that already contains a bioactive chemical and its coated polymer (CO_2). When the solvent is removed, both chemicals precipitate, leaving water with suspended particles. Additionally, fish oil is commonly used in oil nano and microencapsulation. Additionally used are HOPO, rice bran oil, sunflower oil, roasted coffee oil, chia oil, thyme oil, garlic essential oil, and chia oil. Based on the use and type of oil that needs to be nanoencapsulated, different types of wall materials have been employed to manufacture oil nanocapsules. In contrast to oil microencapsulation, oil nanoencapsulation frequently does not involve the mixing of wall materials [34].

Because the particles must be at least 1000 nm in size, the wall material is typically used alone. The interaction between the fewest possible components in the formulation of the nanocapsules is also improved due to the bigger contact surface. Most prevalent coating materials used in the operations are biodegradable polymers. Oil nanoencapsulation frequently uses polycaprolactone, whey protein, sodium caseinate, chia seed mucilage, maltodextrin, or modified starch as wall materials. Some authors failed to mention the type of wall material they utilized in their study because of the spontaneous emulsification method that was employed. The size of the needed nanoparticles, their solubility and stability in water, and other parameters are taken into consideration while selecting the wall material [124].

The majority of the polymers utilized in wall building are poly(-caprolactone) (PCL). With the aid of cationic or anionic, coordination, or radical catalysts, the cyclic monomer caprolactone is ring-opened polymerized to produce PCL, a polymer. The molecular weight and crystallinity of this polymer, which has a semicrystalline structure, are directly correlated. It can change chemical properties, due to its melting point of 59–64 °C, solubility in inorganic solvents, and good mix compatibility. Additionally, PCL is a semicrystalline, synthetic, biocompatible, and totally biodegradable polymer. The Food and Medicine Administration has given its approval for drug delivery (FDA). Because of slow breakdown, PCL is most appropriate for long-term distribution or controlled delivery in the digestive system is required. PCL has the advantages of being inexpensive, incredibly in vitro stable, and hydrophobic. Fish oil is commonly enclosed in PCL during the emulsion-diffusion process, and emulsions are extracted using supercritical fluid. Whey protein can be used to nanoencapsulate bioactive substances because of their biocompatibility and biodegradability [62].

Whey protein was used to create stable nanocapsules with a diameter of 163 nm using HOPO and microfluidization. After synthesis, important elements including size, polydispersity index (PDI), and zeta potential are employed to fundamentally define the oil nanoparticles. Due to their capacity to alter the physicochemical and pharmacological behaviors of the enclosed components, nanocapsules' size and size dispersion are crucial. Laser diffraction (LD) and a Coulter counter are two techniques that can be used to measure nanoparticle size, commonly known as the mean diameter or z-average. Particle size distribution and destabilization were controlled by Dynamic light scattering (DLS), which is the technique that is most frequently used. This technique is suitable for characterizing particles smaller than 1 m, but it is not very accurate when used with substantial size discrepancies. Oils that have been nanoencapsulated typically have a diameter of 100 nm or less. Nanoencapsulated materials with the size between 163 and 2268 nm by microfluidization method [20]. Size dispersion is indicated by the PDI, an indicator that gauges particle homogeneity in suspension, values between 0.1 and 0.25 signify a limited distribution, while values higher than 0.5 suggest a wide distribution. The electrical potential of the nanoparticle as affected by the elements of the nanocapsule and the medium in which they are distributed is known as the "Zeta potential," and it is shown by suspended particles, macromolecules, and material surfaces. This measure is widely used to assess suspension stability in colloidal dispersions.

Zeta potential values of more than 30 mV and less than 30 mV encourage high stability and the prevention of particle aggregation. Most of the research we looked at here produced outcomes in this range (30 and 30 mV). Because of the properties of the wall material, the zeta potential of nanoencapsulated oils is typically varied. When chia seed mucilage was used as the wall material, researchers examined the zeta potential of the encapsulated chia oil and discovered that it was 11.58 1.87 mV. Using an FTIR spectrometer, high spectral resolution data are simultaneously collected across a wide spectrum range. For example, oil nanoparticle peaks were observed with wavelengths between 400 and 500 and 4000 cm^1 with a resolution of 4 cm^1 [140].

15 Incorporation of Nanoencapsulated Oils into the Food Market and Patent Application

Food currently possesses both micro and nanoscale structures thanks to technology. Margarine included fat crystals that were even smaller than 10 m in size and water droplets that were less than 10 m in size. Beverages like tea, beer, and coffee have been found to contain particles smaller than 100 nm. Foods like eggs or soy were found to have protein structures of around 300 nm, whereas foods like milk were found to contain bigger oil particles of about 800 nm. All fresh and processed food had nanoscale structures, which forced the body to adapt over time to deal with them. The use of nanotechnology in business has not received much research. Furthermore, no oil nanoencapsulated products were found in commerce. An article that stresses this information is accessible, although there are a number of oil microencapsulation products on the market. This result could be explained by the fact that nanotechnology is a relatively new and difficult field to work in.

Additionally, there are significant concerns about the market adoption of this technology due to the regulatory gap in nanotechnology. Numerous ongoing EU research initiatives on toxicology, among other areas of nano-safety, are focused on the regulation of nanotechnology. The active EU projects are the NanoLyse project, which intends to create analytical methods for the identification of designed nanoparticles in food, and the NanoReTox project, which focuses on the impacts of exposure to engineered nanoparticles on human health and the environment. The Environmental Protection Agency (EPA) and Food and Drug Administration (FDA) in the United States, as well as the Health and Consumer Protection Directorate of the European Commission, have started to address the potential risks posed by nanoparticles. There are currently no specific rules governing the manufacturing, handling, or labeling of either artificial nanoparticles or the goods and materials that include them. Although there is no explicit law regarding nanoparticles, there are several food sector patent filings that have been filed in various nations. The subject of the Argentine patent application WO2018029626 was chia oil in an edible nanoemulsion [89].

Edible chia oil nanoemulsions for desserts and liquids that are transparent, including jellies and juices. Cinnamon oil was the subject of a Korean patent application (KR20160005182), which sought to prevent the growth and spread of harmful food germs. In addition to food industry, it will be used in pharma and cosmetic industries. It is possible to utilize this mustard oil to make food and medicine more bacterially resistant. A palm oil nanoemulsion is described in the Chinese patent CN103750050 that addresses problems with current young animal feed. The resulting nanoemulsion had a high oil content, a small dispersion granularity, good stability, and higher digestibility for cattle due to the procedure's simplicity, ease of control over the reaction process, quick production time, low equipment investment costs, and low production costs. Oil nanoencapsulation patent filings are on the rise, which suggests that several developments have been developed and attests to the technology's widespread use (Fig. 3).

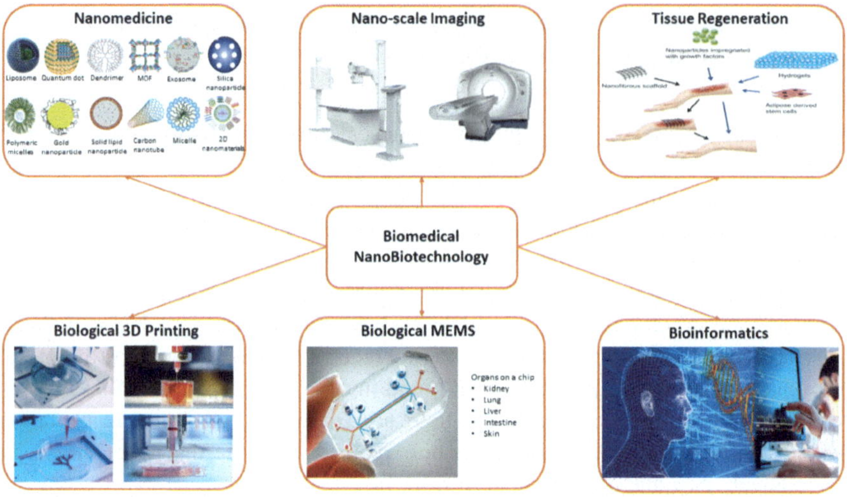

Fig. 3 Application of nanotechnology towards green synthase nanoparticles on Bio-medication application

16 Utilizing Lipid-Based Nanocarriers to Encase

16.1 Liposomes

Bangham et al. [13] developed the first liposomes for drug delivery with lipid bilayer and aqueous compartments. They can be separated into three groups based on their size and lamellarity. Small unilamellar vesicles (SUV), which range in size from 20 to 100 nm, large unilamellar vesicles (LUV), which are larger than 100 nm, and multilamellar vesicles (MLV), which contain six layers, make up the first three categories. Phospholipids such as phosphatidylcholine, phosphatidylethanolamine, phosphatidylserine, phosphatidylglycerol, and phosphatidylinositol typically make up the liposome bilayer. Cholesterol may also be added to the lipid bilayer to increase fluidity and stability and decrease the permeability (Table 2).

16.2 Thin Lipid Film Hydration

Thin lipid film hydration, also known as thin membrane hydration, is the first method used to create liposomes. To achieve size reduction and particle consistency, liposome size reduction methods like sonication or extrusion were used. The thin hydration technique was widely employed to create liposomes that contained plant oils [42].

According to various research studies, the liposomes made using this technology have a modest average diameter and a spherical form. An earlier study found that

Table 2 Preparation methods for plant oils encapsulation in liposomes with their fine applications

Encapsulated oil	Plant (origin)	Wall material	Physicochemical properties			Biological properties	Applications	References
			Size (nm)	Z.pot (mV)	EE (%)			
Thin hydration method								
Myristica	*Myristica fragrans* Houtt	Soybean phospholipid	/	/	/	Antimicrobial, antioxidant	Food	(Zhu et al. 2020)
Laurel	*Laurelia sempervirens* Tul	L-α-PCH	200	/	/	Antibacterial, antioxidant	Food	[146]
Cardamon	*Elettaria cardamomum* L	Phospholipid	71.8–147.9	−17.4 to −10.7	<60	Antimicrobial, antioxidant	Food	[58]
Chrysanthemum	*Chrysanthemum pacificum* L	Soybean phospholipid	132.4	−37.6	42.7	Antibacterial	Food	[74]
Garlic	*Allium sativum* L	Soybean PCH	115	−12.36	86.00	Antibacterial, antioxidant	Food	[92]
Sterol	/	PCH	< 200	−9 to −14	>89	Antioxidant	Food	[103]
Rosemary	*Rosmarinus officinalis* L	L-α-PEA dioleoyl	79	−2.77	45	Anticancer	Medicine	[117]
		Phospholipon 90G	≈200	−20 to −35	65	Antioxidant; antibacterial Anti-inflammatory	Medicine	[111]
Sage	*Salvia triloba* L				57			
Eucalyptus	*Eucalyptus globulus* L'Her	Soybean phospholipid	/	/	97.35	Antimicrobial	Medicine	[136]
Cineole	/	Egg yolk PCH	128	6.8	67.7	Permeation enhancers	Medicine	[116]
Limonene	/		102	6.1	71.0			
Cinnamon	*Cinnamomum cassia* persl	Soybean phospholipid	144.3	−54.7	29.2	Antibacterial	Medicine	[23]

(continued)

Table 2 (continued)

Encapsulated oil	Plant (origin)	Wall material	Physicochemical properties			Biological properties	Applications	References
			Size (nm)	Z.pot (mV)	EE (%)			
Bergamot	*Monarda didyma* L	Phospholipid	185.1	−2.95	75.0	Anticancer	Medicine	[18]
Bergapten	/							
Macassar kernels	*Brucea javanica* L	Phospholipid	108.2	−57.0	92.4	Anticancer	Medicine	[24]
Zanthoxylum tingoassuiba	*Zanthoxylum tingoassuiba* L 1,2-Dipalmitoyl	PCH	210–3630	/	68.5–79.2	Anticancer	Medicine	[27]
DHA	/	1,2-Dipalmitoyl PCH	99	−15.7	81.4	Anticancer; anti-inflammatory	Medicine	[3]
EPA	/	1,2-Dipalmitoyl PCH	/	/	56.9	Anticancer; anti-inflammatory	Medicine	[44]
EPA + DHA	/	Soybean phospholipid	<100	−48.4	/	Anticancer; anti-inflammatory	Medicine	[109]
Zataria multiflora	*Zataria multiflora* Boiss	Soybean PCH	99.9	/	18.4	Antimicrobial	Medicine Cosmetics	[59]
Ajwain	*Trachyspermum copticum* L	Soybean PCH	170.8	/	85.9	Antimicrobial	Cosmetics	[133]
Solvent injection method								
Black seed	*Nigella sativa* L	Egg yolk PCH	50–900	/	34–87	Analgesic	Medicine	[115]

(continued)

Table 2 (continued)

Encapsulated oil	Plant (origin)	Wall material	Physicochemical properties			Biological properties	Applications	References
			Size (nm)	Z.pot (mV)	EE (%)			
Galangal	*Alpinia officinarum* L	Phospholipid	61.9	−16	/	Antimicrobial; antioxidant	Medicine	[93]
		Phospholipid	125.6	−35.5	35.6	Antimicrobial; antioxidant	Medicine	[148]
Clove	*Syzygium aromaticum* L	Phospholipon 90H	/	/	54.2–59.7	Anti-inflammatory	Medicine	[121]
		Phospholipid	/	/	/	Anti-inflammatory	Medicine	[120]
Eugenol	/	Lipoid S100	74.6	/	24.2	Anticancer	Medicine	[103]
		Phospholipon 90H	/	/	63.1–63.5	Anti-inflammatory	Medicine	[121]
		Phospholipid	/	/	/	Anti-inflammatory	Medicine	[120]
Coix seed oil + B-carotene	*Coix lacryma-jobi* L	Egg yolk PCH	156.7- 193.1	−47.8 to −32.4	80.8–83.8	Anticancer	Medicine	(Bai, Zheng et al. 2019)
Linseed oil + quercetin	*Linum usitatissimum* L	Soybean phospholipid	262.2	/	/	Source of PUFA	Food	[50]
Linoleic acid	/	Soybean PCH	266	/	83.3	/	Food	[141]

EPA: Eicosapentaenoic acid; DHA: Docosahexaenoic acid; PUFA: Polyunsaturated fatty acids; PCH: phosphatidylcholine; PEA: phosphatidylethanolamine clove EO-loaded liposomes were extremely small with a size of 149.2 nm, had a high entrapment efficiency (20.41%), and contained 5.0 mg/mL of clove oil. In order to lessen the volatility of salvia EO, a similar process to that described in the paper was used to encapsulate the oil. According to images captured by atomic force microscopy, the resulting liposomes were spherical in shape and ranged from 100 to 700 nm on average [68].

For bergamot EO and its portion, the generated liposomes displayed a high EE% of 75.0 and 77.0%, respectively. The diameters of EO-loaded and empty liposomes created using the thin hydration method have been compared by a number of writers. Loaded liposomes were about 75 nm in size, but empty liposomes were typically 142 nm in size. However, studies claimed that there was no difference in the size of the liposomes between MLVs and SUVs that had been empty and those that had been filled with Zanthoxylum tingoassuiba EO. Additionally, lipid hydration approach was used for nerolidol- and cyclodextrin/nerolidol-loaded liposomes development. Similar activity was seen when cinnamon essential oil (EO) was encapsulated in thinly hydrated soy lecithin/cholesterol-based liposomes. The concentration of cinnamon oil ranges from 3.0 to 6.0 mg/mL while the vesicle size increased steadily from 133 to 157 nm. When the amounts of cardamom, thyme, and chrysanthemum essential oils were raised, similar results were also observed. The EO concentration had an impact on liposome size in addition to that. The impact of the preparation process was studied by creating garlic EO nano-phytosomes using a thin hydration technique. For instance, nano-phytosomes are a new lipid-based vesicular delivery system in which nutraceuticals and phosphatidylcholine molecules form a stable combination by the establishment of stable hydrogen bonds [70, 71].

In comparison to the smaller particles produced by the probe sonication approach, the homogenized particles had an average size of about 161 and 135 nm. The zeta potential values of all the produced phytosomes were lower (−12.36 to −3.86 mV) than expected and compared to prior reports for soybean phosphatidylcholine-made nano-vesicular systems (about 26 mV) [81].

16.3 Solvent Injection

The lipid is thought to be first dissolved into an organic phase, such as ethanol or ether, and then injected into an aqueous solution, causing the liposomes to scatter, according to the solvent injection method, which was initially introduced in 1973 [16]. The organic solvent was then evaporated at lower pressure after that. This method is frequently applied for encapsulating plant oils. The primary advantage of solvent injection is the ability to manufacture submicron liposomes (under 100 nm) with a narrow size distribution without the need for extrusion.

Furthermore, it was shown that the ethanol injection method was more scaleable than the thin hydration method. In addition, several EO liposomes with eugenol and clove oil were produced in the lab on a small and big scale (membrane contactor). The results showed that the mean sizes of DCL and DCL2 produced at a large scale and loaded with clove and eugenol were equivalent to those found at a laboratory scale. Additionally, it was discovered that the ethanol injection technique was scalable in comparison to the thin hydration method. This was explained by the fact that eugenol and lipidic bilayers only had a limited amount of interaction due to the aromatic cycle's complete integration into the hydrophobic cavity of hydroxypropyl—cyclodextrin. Regarding the zeta potential or polydispersity index, none of the liposomes that were studied could be distinguished from one another.

16.4 Solid Lipid Nanoparticles and Nanostructured Lipid Carriers

Due to their capacity to combine the benefits of polymeric and liposomal nanoparticles, solid lipid nanoparticles (SLN), a new class of lipid-based carriers, have emerged. Solid lipid nanoparticles are suspensions of nanoscale solid lipid particles dispersed in an aqueous media, and they were first created in 1990. Surfactants and a lipid matrix, which solidifies into a lipid at room temperature, make up their composition. Triacylglycerols, fatty acids, and acyl glycerols are widely used as building blocks to create the lipid core. Fatty acids include stearic, palmitic, decanoic, and benzoic acids (Table 3) [88].

These lipids remain exceptional in terms of purity, cost, safety, and biodegradability. Phospholipids, poloxamer, pluronic® 954 F127, bile salts, tween 80, and polyvinyl alcohol are typical surfactants in contrast. To boost the loading capacity of SLN, the second generation of lipid nanoparticles, also known as nanostructured lipid carriers (NLC), was developed in 1999. The matrix of an NLC is composed of both solid and liquid lipid oil. These carriers may provide more space for the loading of bioactive compounds, resulting in a greater concentration of the active molecules trapped and a lower possibility of expulsion from the lipid phase. The improvement in entrapment efficiency is the primary advantage of using SLN or NLC as nanocarriers for the delivery of plant oil. Oils have strong miscibility with the lipid matrix because they are frequently lipophilic, and as a result, they can entrap a lot of oil. The improvement in entrapment efficiency is the primary advantage of using SLN or NLC as nanocarriers for the delivery of plant oil. Oils have strong miscibility with the lipid matrix because they are frequently lipophilic, and as a result, they can entrap a lot of oil [135] (Fig. 4).

16.5 High-Pressure Homogenization

Plant oils were frequently encapsulated in SLN using the high-pressure homogenization process because it yields small, very homogeneous particles. High-pressure liquids (100–2000 bar) were driven through a small hole in order to disrupt the

Table 3 Preparation methods for plant oils encapsulation in solid lipid nanoparticles and nanostructured lipid carriers with their fine applications

Encapsulated oil	Plant (origin)	Wall material	Physicochemical properties			Biological properties	Applications	References
			Size (nm)	Z.pot (mV)	EE (%)			
Solid lipid nanoparticles								
High-pressure homogenization								
Carvacrol	/	Cocoa butter	/	/	/	Antioxidant	Medicine	[17]
Eugenol	/	Caprylic triglyceride Stearic acid	332	−18.7	91.8	Anticancer	Medicine	[40]
Sclareol	/	Compritol	128	−29.3	/	Treatment of metabolic disorder	Medicine	[19]
Frankincense + Myhrr oil	*Commiphora myrrha* L	Compritol888 ATO	113.3	−16.8	80–60	Antitumor	Medicine	[126]
Chamomile	*Matricaria chamomilla* L	Stearic acid	542	−35.9	/	Wound healing	Cosmetics	[38]

(continued)

Table 3 (continued)

Encapsulated oil	Plant (origin)	Wall material	Physicochemical properties			Biological properties	Applications	References
			Size (nm)	Z.pot (mV)	EE (%)			
Gac fruit EO	*Momordica Cochinchinensis* L	Emulgade	200	/	90	Antioxidant; Anti-aging Skin tanning, skin care and skin protection	Cosmetics	[76]
Copaiba oil	*Copaifera officinalis* L	Cetyl palmitate	126.06	/	/	Antifungal	Cosmetics	[132]
Copaiba oil + Allantoin			118.63					
Eugenol	/	Hydrogenated palm oil Coconut oil	164.5–172.9	/	/	Antifungal	Food	[80]
Thymol	/		166.4–170.5	/	/			
Cinnamaldehyde	/		198.0–202.1	/	/			
Ultrasonication method								
Coconut	*Cocos nucifera* L	Stearic acid	608	−47.5	99.9	Moisturizing lotion	Cosmetics	[95]
Eugenol	/	Stearic acid Caprylic triglyceride	87.7	−19.2	98.5	Antifungal	Medicine	[40]
		Stearic acid/Compritol	363–423	/	/			[41]

(continued)

Table 3 (continued)

Encapsulated oil	Plant (origin)	Wall material	Physicochemical properties			Biological properties	Applications	References
			Size (nm)	Z.pot (mV)	EE (%)			
Linalool	/	Myristyl myristate Cetyl esters Cetyl palmitate	90–130	−4	>80	Anticancer	Medicine	[112]
Clove	*Syzygium aromaticum* L	PrecirolRATO 5 Stearic acid Glyceryl monostearate	397.2	−20.9	/	Antimicrobial	Medicine	[33]
β-elemene	/	Glyceryl monostearate Glyceryl tristearate	128.4	−3.07	/	Antitumor	Medicine	[47]
Zataria multiflora	*Zataria multiflora* Boiss	Glyceryl monostearate PrecirolRATO 5	255.5	−37.8	84	Antifungal	Medicine	[90]
		Precirol ATO 5 Stearic acid Glyceryl monostearate	220–486	−37.8 to −27.2	85.3–95.2	Fungicide Pesticide	Medicine Food	
Nanostructured lipid carriers								
High-pressure homogenization								
Cinnamon	*Cinnamomum cassia* persl	Cocoa butter Almond oil	100–120	−14.9 to −17.5	>82.1	Antimicrobial, antioxidant	Food	[15]

(continued)

Table 3 (continued)

Encapsulated oil	Plant (origin)	Wall material	Physicochemical properties			Biological properties	Applications	References
			Size (nm)	Z.pot (mV)	EE (%)			
Sunflower Crambe Canola	*Helianthus annuus* L. *Crambe maritima* L. *Brassica napus* L	Sunflower oil Crambe oil Canola oil	48.23 – 342.10	−22.3 to −29.7	/	Source of phytosterols	Food	[118]
Sucupira	*Pterodon emarginatus* L	ImwitorR 900 K Tripalmitin	148.1	/	99.9	Antidiabetic	Medicine	[142]
Clove	*Syzygium aromaticum* L	Beeswax Coconut oil	121–367	/	≈60	Cholinesterase enzyme inhibition	Medicine	[25]
Ridolfia segetum	*Ridolfia segetum* L	PrecirolRATO 5	143	16.3	100	Antioxidant; anti-inflammatory	Medicine	[82]
B-elemene	/	Glycerol monostearate Glyceryl monolinoleate Oleoyl macro-golglycerides	138.9	−20.2	82.11	Antitumor	Medicine	[125]
Thymoquinone	/	Hydrogenated palm oil Olive oil LipoidRS100	75	−31	/	Antiulcer	Medicine	[1]
Olive	*Olea europaea* L	Hydrogenated palm oil	61.14	−25.4	/	Source of PUFA	Medicine	[48]

(continued)

Table 3 (continued)

Encapsulated oil	Plant (origin)	Wall material	Physicochemical properties			Biological properties	Applications	References
			Size (nm)	Z.pot (mV)	EE (%)			
Grape seed + Laurel	*Vitis vinifera* L	n-Hexadecyl palmitate Glycerol monostearate MyritolR318	<150	/	/	Antitumor; antioxidant	Medicine	[67]
Squaw mint	*Mentha pulegium* L	PrecirolRATO 5 MiglyolR812	40–250	−10 to −15	/	Wound healing	Cosmetics	[61]
Rosemary	*Rosmarinus officinalis* L	PrecirolRATO 5 MiglyolR812	100–250	−15.7	92.1	Wound healing	Cosmetics	[60]
Peppermint	*Mentha piperita* L	PrecirolRATO 5 MiglyolR812	40–250	−10 to −15	/	Wound healing	Cosmetics	[43]
Pomegranate seed	*Punica granatum* L	Glycerol monostearate Emulgade	108–145	34.4 to −44.3	51–70	Antioxidant UVA protector	Cosmetics	[8]
Wheat germ	*Triticum aestivum* L							
Blackcurrant seed	*Ribes nigrum* L							

(continued)

Table 3 (continued)

Encapsulated oil	Plant (origin)	Wall material	Physicochemical properties			Biological properties	Applications	References
			Size (nm)	Z.pot (mV)	EE (%)			
Sesame	*Sesamum indicum* L							
Carrot root	*Daucus carota* subsp. Sativus							
Raspberry seed	*Rubus idaeus* L							
Rice bran	*Oryza sativa* L							
Ultrasonication method								
Carvacrol + Astaxanthin	/	Lecithin Beeswax	58.8	−36	Asta = 95 Car = 88	Antioxidant Antibiofilm	Medicine	[122]
Houttuynia cordata	*Houttuynia cordata* Thunb	Glyceryl behenate	171	−19.3	90.20	Treatment of pneumonia, respiratory and urinary tract infections	Medicine	[150]
Blackcurrant	*Ribes nigrum* L	Myverol RX GMS 95P Chemal EO-20	/	/	/	Source of PUFA	Medicine	[64]

(continued)

Table 3 (continued)

Encapsulated oil	Plant (origin)	Wall material	Physicochemical properties			Biological properties	Applications	References
			Size (nm)	Z.pot (mV)	EE (%)			
Blackberry	*Rubus fruticosus* L							
Raspberry	*Rubus idaeus* L							
Strawberry	*Fragaria Ananassa* Duchesne							
Plum	Prunus domestica L							
Eucalyptus	*Eucalyptus globulus* L	Olive oil	50–60	−22.07	101.7	Wound healing	Cosmetics	[119]
Peppermint	*Mentha piperita* L	Miglyol-812 Precirol-ATO 5	40–250	−10 to −15	93.2	Wound healing	Cosmetics	[43]
Thymol	/	Illipe butter Calendula oil	107.7	−11.6	89.1	Anti-inflammatory Antipsoriatic	Cosmetics	[102]

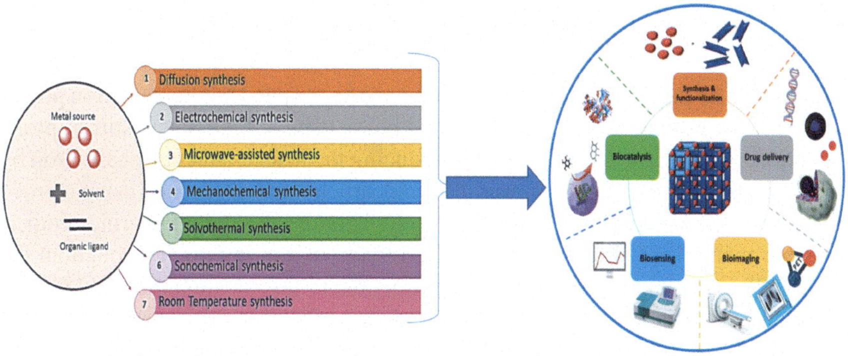

Fig. 4 Utilization of different methods for nanomaterial synthesis based on organic ligand

particles down to the submicron range (micron size). This caused fluid to accelerate over a very small distance while moving at a high speed of more than 1000 km/h. High-pressure homogenization, a process that includes raising the lipid's temperature by 5–10° over its melting point, is used to encapsulate plant oils. The next phase includes homogenization, either warm or cold. In warm homogenization, a pre-emulsion is made by stirring the plant oil-lipid liquify at a similar temperature as a hot aqueous surfactant solution. After the resulting pre-emulsion was driven over a piston-gap homogenizer and cooled to room temperature to permit the lipid to crystallize and produce SLN, a heated oil-in-water (O/W) nanoemulsion was produced. Cold homogenization was used to cool the oil-lipid melt, resulting in solid lipid microparticles that were later mixed with a cold aqueous surfactant solution to form a pre-suspension. To create SLN 990, this pre-suspension is then homogenized. There take already been studies published that examine the use of a high-pressure homogenization technique to encapsulate plant oils [2].

Numerous investigations on the size of SLN made with this technology have shown that it has a minor average diameter, good stability, and a high EE%. For instance, SLN containing 1:1 ratios of frankincense and myrrh EO was produced using the heat homogenization technique. The generated SLN was assessed to have an EE% of 80.60%, an average mean size of 113.3 nm, and a zeta potential of –16.8 mV. The same method was used to make both the copaiba oil with and without the allantoin form of the SLN, which is widely used in cosmetology for its keratolytic, moisturizing, calming, and wound healing properties. The dynamic light scattering investigation revealed that the copaiba oil-loaded SLN and the copaiba oil: allantoin-co-loaded SLN had average particle sizes of 126.06 nm and 118.63 nm, respectively.

16.6 Ultrasonication/High-Speed Homogenization

SLN can also be produced via ultrasonication or a high-speed homogenization procedure. It was combined with the melted lipid phase using a high-speed stirrer, probe sonication, or magnetic stirring while introducing the heated aqueous phase drop by drop. The obtained pre-emulsion was ultrasonically treated in one step using a probe sonicator and a water bath. To avoid re-crystallization, the manufacturing temperature was maintained at least 5 °C overhead the lipid melting point throughout the process. Plant oils are often encapsulated using the ultrasonication method [83].

Tween 80 and poloxamer 188 were used as surfactants to form the multiflora EO-loaded SLN, while Precirol® 1105 1106 ATO 5 (Glyceryl monostearate) served as the lipidic matrix. With a high EE% of 84%, a zeta potential of −37.8 mV, and a particle size of approximately 255.5 nm, the resultant SLN was produced. The same procedure remained used to produce cardamom EO loaded-NLC, which was made with tween 80 as a surfactant, cocoa butter as a solid lipid, and olive oil as a liquid lipid. The resultant particles were less than 100 nm in size and had a smooth, rounded surface. The particles were stable during a 70-day storage period at room temperature and 4 °C, just like NLC, which was made using lecithin as an emulsifier and palm stearin as a solid lipid.

17 Application Agriculture Field

The usage of synthetic chemicals in the agriculture field cause serious problem for environment and human health. If synthetic fungicides are used excessively, they may also cause additional negative side effects. There have been reports of the development of resistance in pathogen populations for various fungi. Additionally, it has been highlighted that during the past few years, consumer concerns regarding crops carrying fungicidal residues have increased. Environmentally friendly fungicides are preferred instead of chemical fungicides. Antifungal substances like EO that are GRAS (generally regarded as safe) have been mentioned in reports. Due to its broad range of efficacy against insect pests and plant pathogenic fungi, EO has been used as an eco-friendly fungicide. However, it is challenging to replace synthetic chemicals since EOs demonstrate poor stability during the dispensation steps. For this purpose, nanoencapsulation has been planned. Conferring to the works, pNPs are frequently used for EO encapsulation in the agricultural industry. Few articles have examined the use of lipid-based nanocarriers in agriculture over the past ten years. To stop the germination of seeds from invasive alien species, rosemary EO was initially encapsulated in liposomes. The potency of rosemary essential oil was increased by incorporating it into liposomes, which kept the volatile components from leaving and carried them to the seed. Additionally, it was discovered that loaded SLN containing Pelargonium graveolens EO was more stable in outdoor conditions and had a better larvicidal effect against Phthorimaea operaculella than free oil did [151].

18 Food Field

Consumer demands for eating wholesome, minimally processed meals have significantly increased in recent years. Functional foods are essential for boosting health and preventing degenerative illnesses including diabetes, cancer, and heart disease, which cause nearly two-thirds of global fatalities. Fatty acid enriched foods have advantages due to their health benefits. Consuming omega-fatty acids are linked to a lower chance of developing cancer, cardiovascular disease, and inflammatory and autoimmune diseases. However, during processing and storage, polyunsaturated fatty acids are very vulnerable to oxidation, which lowers their nutritional value and sensory quality. Investigations have been made into the antibacterial mechanism of Eos-loaded nanocarriers. The Staphylococcus aureus-fighting properties of liposomes containing clove essential oil. It was discovered that clove oil release from liposomes was activated by pore-forming toxins, which have a specific antibacterial effect for Staphylococcus aureus. In fact, the liposomes did not fuse with biological membranes after the bacteria discharged their toxins; instead, clove oil was generated. 99.87% population reduction in Staphylococcus aureus was seen after 24 h of liposome therapy. Since Escherichia coli could not secrete toxins, clove oil was unable to kill them, which contributed to the population decline observed after liposome therapy [63].

It was shown that EO-loaded nanocarriers maintained their antioxidant and antibacterial capabilities for a long period. The physicochemical characteristics of freshly cut bananas in contact with free rosemary EO, rosemary oil with mucilage, and rosemary oil liposomes were assessed in this context after 9 days of storage at 5 °C. According to sensory research, liposomal formulations were more effective at improving the quality and extending the shelf life of freshly cut bananas. Similar to how cardamom EO is converted into NLC and nanoliposomes, this considerably preserves the antioxidant and antibacterial properties for a long time. After 30 days, the NLC revealed MIC values for both bacteria that were approximately 1100 g/ml lower than those for cardamom oil emulsion. The MIC values for Staphylococcus aureus and Escherichia coli were around 4400 and 2200 g/ml.

The biological activity of the plant oils contained in nanocarriers was influenced by the operating circumstances and particle makeup. In 2017, Mohammed and his colleagues demonstrated the correlation between strong antioxidant efficacy of Nigella sativa EO-loaded microparticles produced by spray drying technology. Results demonstrated that Gram negative bacteria and fungi might be killed by SLN formulations, especially those that contained stearic acid rather than glycerol monostearate.

19 Medicinal Field

19.1 Antimicrobial Activity

Everywhere in the world, there is a growing concern about how to treat bacterial infections. Due to their propensity to build biofilms and the rise of multi-drug resistant bacteria, all host defensive mechanisms and large doses of antibiotics are easily tolerated by these germs. Scientists from all over the world are searching for new antimicrobial agents to combat this problem, such as formulations based on plant oils, which are economical and eco-friendly antibacterial agents. Then, nanotechnology is coupled with these agents. In this regard, rheological measurements were used to examine how carvacrol-loaded poly (lactide-co-glycolide) nanocapsules affected Staphylococcus epidermidis biofilms' viscoelastic characteristics. Results showed that carvacrol-loaded nanoparticles might decrease the mechanical stability and flexibility of prefabricated biofilms, allowing antimicrobial agents to enter the biofilms' deep core. *Escherichia coli* and *Candida albicans* tested strains had greater antibiofilm effects, according to the results. The produced nanoparticles' cytotoxicity was further investigated, and the results showed that human cells grew well and with high cell viability [52].

19.2 Larvicidal Activity

Lippiasidoides EO possesses larvicidal efficacy against Aedes (Stegomyia aegypti) larvae, the dengue vector that causes a number of diseases and occasionally even fatalities in tropical countries like Brazil, according to several studies [127]. To improve the larvicidal action of Angico gum, chitosan nanoparticles encompassing LippiaSidoides EO were created. Twenty third-instar Stegomyia aegypti larvae were utilized to investigate the in vivo activity. The findings demonstrated a clear correlation between oil loading and larval mortality. The trial oil: angico gum-chitosan 1:20, for instance, showed the lowest loading (3.3%) and the lowest mortality (52% after 24 h and 60% after 48 h). The sample oil angico gum-chitosan: 1:10 demonstrated 85 and 92% mortality after 24 and 48 h, respectively. Similar findings were made by Abreu et al. when they investigated the larvicidal activity of LippiaSidoides EO loaded-cashew gum: chitosan nanoparticles [100].

19.3 Antitumor Activity

The anticancer efficacy of frankincense and myrrh oil-loaded SLN was assessed on H22 tumor-bearing mice. The findings demonstrated that SLN was more effective at inhibiting tumor growth than suspension and cyclodextrin inclusion. Oil suspension, oil with cyclodextrin, and oil with SLN had similar inhibition rates of 31.23, 34.81, and 43.66%. Similar results were seen when comparing liposomal formulations containing bergamot EO to their free drug counterparts at various doses and time intervals. These formulations showed a higher reduction in the viability of human SH-SY5Y neuroblastoma cells and a greater increase in cell death [70, 71].

19.4 Cosmetic Field

One of the most intricate physiological processes is wound healing. In reality, improving tissue regeneration and preventing infections are crucial to reducing pain, discomfort, and scarring. Numerous chemical agents have been suggested for wound treatment, but their applications are constrained by their drawbacks and adverse effects. Due to their anti-inflammatory, antibacterial, and antioxidant qualities, plant oils have recently drawn a lot of attention from academics. In this case, it is being looked into whether NLC containing peppermint EO can treat an infected wound. For the in vivo study, Pseudomonas aeruginosa and Staphylococcus aureus were injected into two 5 mm full-thickness wounds that were produced in each animal. The outcomes showed that peppermint EO-NLC could hasten wound healing in the infected wound model by lowering the overall bacterial count. Comparatively, administering Ligusticum chuanxiong EO-loaded liposomes to hypertrophic scars in a rabbit ear model led to a drop in TGF-b1, MMP-1, collagen I, and collagen III levels and an increase in apoptotic cells, caspase-3 and -9 apoptotic markers [70, 71].

19.5 Textiles

Consumers today are fascinated by functional textiles because of its significant advantages in enhancing the inherent qualities of textile products and promoting new functionalities like flame retardancy, UV resistance, photocatalysis, antibacterial, antibiotic, and wrinkle recovery to the fabrics. Consumers of textiles in the medical industry are growing more aware of the potentially dangerous effects that bacteria may have on common hospital supplies including curtains, bed sheets, pillowcases, gowns, and masks. To examine their antibacterial efficacy against Staphylococcus aureus, gelatin and arabic gum-based moxa oil microcapsules were grafted onto cotton fabric in this case. The outcomes showed a strong antibacterial effect. Alginate nanoparticles infused with Ocimum sanctum oil also exhibited outstanding antibacterial properties and great washability. Bacillus cereus, Pseudomonas aeruginosa, and Staphylococcus aureus all displayed 100% inhibition, with the exception of Escherichia coli (90% suppression after 20 washes). Staphylococcus aureus, Escherichia coli, Bacillus cereus, and Pseudomonas aeruginosa each had a lower bacterial 1518 inhibition after 30 washings, dropping to 95, 87, 98, and 98%, respectively [138].

20 Conclusion and Perspectives

The rising demand for eco-friendly products, particularly essential and vegetable oils, that have similar efficacy as or greater efficacy than chemical drugs over the past 20 years has commended investigators to focus the maximum of their efforts on evolving novel methods to extend the shelf life of these bioactive agents. Plant oils are actually highly vulnerable to outside influences when used as such. Because of their nanoscale size range, these bioactive compounds can be effectively encapsulated to improve their physical stabilities, protect them from oxidation and other detrimental environmental factors including light, heat, oxygen, and humidity, and increase their effectiveness. In this bibliographical essay, we concentrate on recent advances in lipid- and polymer-based nanocarriers, including solid lipid nanoparticles, liposomes, and polymeric nanoparticles, for the development of fresh ideas for the active use of plant oils. Each carrier offers a quantity of profits and drawbacks. For instance, compared to liposomes, which have a wider range of applications due to their extremely comparable chemical composition to that of the skin layers, solid lipid nanoparticles and polymeric nanoparticles both display superior oil stability and continued release.

The aforementioned investigations found that although being safer than polymeric nanoparticles since they do not include organic solvents, solid lipid nanoparticles were used less frequently to encapsulate both essential and vegetable oils. Oil loss and encapsulation efficiency were lost due to high temperature. Numerous methods are available for nanocarrier preparation, but the selection of appropriate method plays a crucial role for selecting potential carrier features accountable for the best achievement of the ultimate goal. The aforementioned nanotechnological systems offer valuable insights for usage in a number of sectors, including textiles, medicine, cosmetics, food, and agriculture. These days, there is a lot of interest in the connection between plant oils and active chemicals. Lipid core nanocapsules and lipid carriers with nanostructures are potential carriers for encasing both plant oil and manufactured medicines. For hydrophobic medications to have a high loading capacity and physical stability, it is crucial to use plant oils properly. Additionally, the oil may support or even improve the biological effects of these medications. As a result, research in the area of plant oil encapsulation can significantly advance the development of innovative approaches to the treatment and identification of serious chronic diseases.

References

1. Abdelwahab SI, Sheikh BY, Taha MM, How CW, Abdullah R, Yagoub U, El-Sunousi R, Eid EE (2013) Thymoquinone-loaded nanostructured lipid carriers: preparation, gastroprotection, in vitro toxicity, and pharmacokinetic properties after extravascular administration. Int J Nanomedicine 8:2163–2172
2. Akhavan S, Assadpour E, Katouzian I, Jafari SM (2018) Lipid nano scale cargos for the

protection and delivery of food bioactive ingredients and nutraceuticals. Trends Food Sci Technol 74:132–146
3. Alaarg A, Jordan NY, Verhoef JJ, Metselaar JM, Storm G, Kok RJ (2016) Docosahexaenoic acid liposomes for targeting chronic inflammatory diseases and cancer: an in vitro assessment. Int J Nanomedicine 11:5027–5040
4. Alfadul SM, Elneshwy AA (2010) Use of nanotechnology in food processing, packaging and safety–review. Afr J Food, Agric, Nutr Dev 10(6)
5. Alizadeh-Sani M, Moghaddas Kia E, Ghasempour Z, Ehsani A (2021) Preparation of active nanocomposite film consisting of sodium caseinate, ZnO nanoparticles and rosemary essential oil for food packaging applications. J Polym Environ 29(2):588–598
6. Antonioli G, Fontanella G, Echeverrigaray S, Delamare APL, Pauletti GF, Barcellos T (2020) Poly (lactic acid) nanocapsules containing lemongrass essential oil for postharvest decay control: in vitro and in vivo evaluation against phytopathogenic fungi. Food Chem 326:126997
7. Averina E, Allémann E (2013) Encapsulation of alimentary bioactive oils of the Baikal Lake area into pH-sensitive micro-and nanoparticles. LWT-Food Sci Technol 53(1):271–277
8. Badea G, Lăcătuşu I, Badea N, Ott C, Meghea A (2015) Use of various vegetable oils in designing photoprotective nanostructured formulations for UV protection and antioxidant activity. Ind Crops Prod 67:18–24
9. Badri W, Miladi K, Eddabra R, Fessi, H, Elaissari A (2015) Elaboration of nanoparticles containing indomethacin: argan oil for transdermal local and cosmetic application. J Nanomater
10. Bagheri R, Ariaii P, Motamedzadegan A (2021) Characterization, antioxidant and antibacterial activities of chitosan nanoparticles loaded with nettle essential oil. J Food Meas Charact 15(2):1395–1402
11. Bai C, Zheng J, Zhao L, Chen L, Xiong H, McClements DJ (2019) Development of oral delivery systems with enhanced antioxidant and anticancer activity: coix seed oil and β-carotene coloaded liposomes. J Agric Food Chem 67(1):406–414
12. Bai X, Li C, Yu L, Jiang Y, Wang M, Lang S, Liu D (2019) Development and characterization of soybean oil microcapsules employing kafirin and sodium caseinate as wall materials. Lwt 111:235–241
13. Bangham AD, Standish MM, Watkins JC (1965) Diffusion of univalent ions across the lamellae of swollen phospholipids. J Mol Biol 13(1):238-IN27
14. Barar FSK (2000) Essentials of pharmacotherapeutics. S. Chand Publishing
15. Bashiri S, Ghanbarzadeh B, Ayaseh A, Dehghannya J, Ehsani A, Ozyurt H (2020) Essential oil-loaded nanostructured lipid carriers: the effects of liquid lipid type on the physicochemical properties in beverage models. Food Biosci 35:100526
16. Batzri S, Korn ED (1973) Single bilayer liposomes prepared without sonication. Biochimica et Biophysica Acta (BBA)-Biomembranes 298(4):1015–1019
17. Carvalho FO, Silva ÉR, Nunes PS, Felipe FA, Ramos KPP, Ferreira LAS, Lima VNB, Shanmugam S, Oliveira AS, Guterres SS, Camargo EA, Cravalho Olivera TV, de Albuquerque Júnior RLC, de Lucca Junior W, Quintans-Júnior LJ, Araújo AAS (2020) Effects of the solid lipid nanoparticle of carvacrol on rodents with lung injury from smoke inhalation. Naunyn Schmiedebergs Arch Pharmacol 393(3):445–455
18. Celia C, Trapasso E, Locatelli M, Navarra M, Ventura CA, Wolfram J, Carafa M, Morittu VM, Britti D, Di Marzio L, Paolino D (2013) Anticancer activity of liposomal bergamot essential oil (BEO) on human neuroblastoma cells. Colloids Surf B 112:548–553
19. Cerri GC, Lima LCF, Lelis DdF, Barcelos LdS, Feltenberger JD, Mussi SV, Monteiro-Junior RS, Santos RASd, Ferreira LAM, Santos SHS (2019) Sclareol-loaded lipid nanoparticles improved metabolic profile in obese mice. Life Sci 218:292–299
20. Chaudhary S, Kumar S, Kumar V, Sharma R (2020) Chitosan nanoemulsions as advanced edible coatings for fruits and vegetables: composition, fabrication and developments in last decade. Int J Biol Macromol 152:154–170
21. Contri RV, Kulkamp-Guerreiro IC, da Silva SJ, Frank LA, Pohlmann AR, Guterres SS (2016) Nanoencapsulation of rose-hip oil prevents oil oxidation and allows obtainment of gel and film topical formulations. AAPS PharmSciTech 17(4):863–871

22. Crucho CI, Barros MT (2017) Polymeric nanoparticles: a study on the preparation variables and characterization methods. Mater Sci Eng, C 80:771–784
23. Cui H, Li W, Li C, Vittayapadung S, Lin L (2016) Liposome containing cinnamon oil with antibacterial activity against methicillin-resistant Staphylococcus aureus biofilm. Biofouling 32(2):215–225
24. Cui Y, Wu Z, Liu X, Ni R, Zhu X, Ma L, Liu J (2010) Preparation, safety, pharmacokinetics, and pharmacodynamics of liposomes containing brucea javanica oil. AAPS PharmSciTech 11(2):878–884
25. de Meneses AC, Marques EBP, Leimann FV, Gonçalves OH, Ineu RP, de Araújo PHH, de Oliveira D, Sayer C (2019) Encapsulation of clove oil in nanostructured lipid carriers from natural waxes: preparation, characterization and in vitro evaluation of the cholinesterase enzymes. Colloids Surf A 583:123879
26. de Sousa VP, Crean J, de Almeida Borges VR, Rodrigues CR, Tajber L, Boylan F, Cabral LM (2013) Nanostructured systems containing babassu (Orbignya speciosa) oil as a potential alternative therapy for benign prostatic hyperplasia. Int J Nanomed 8:3129
27. Detoni CB, de Oliveira DM, Santo IE, Pedro AS, El-Bacha R, da Silva Velozo E, Ferreira D, Sarmento B, de Magalhães Cabral-Albuquerque EC (2012) Evaluation of thermal-oxidative stability and antiglioma activity of Zanthoxylum tingoassuiba essential oil entrapped into multi- and unilamellar liposomes. J Liposome Res 22(1):1–7
28. Devi N, Hazarika D, Deka C, Kakati D (2012) Study of complex coacervation of gelatin A and sodium alginate for microencapsulation of olive oil. J Macromol Sci Part A 49(11):936–945
29. Dhifi W, Bellili S, Jazi S, Bahloul N, Mnif W (2016) Essential oils' chemical characterization and investigation of some biological activities: a critical review. Medicines 3(4):25
30. El Asbahani A, Miladi K, Addi EA, Bitar A, Casabianca H, El Mousadik A, Hartmann D, Jilale A, Renaud F, Elaissari A (2015) Antimicrobial activity of nano-encapsulated essential oils: Comparison to non-encapsulated essential oils. J Colloid Sci Biotechnol 4(1):39–48
31. Ephrem E, Greige-Gerges H, Fessi H, Charcosset C (2014) Optimisation of rosemary oil encapsulation in polycaprolactone and scale-up of the process. J Microencapsul 31(8):746–753
32. Esmaeili A, Asgari A (2015) In vitro release and biological activities of Carum copticum essential oil (CEO) loaded chitosan nanoparticles. Int J Biol Macromol 81:283–290
33. Fazly Bazzaz BS, Khameneh B, Namazi N, Iranshahi M, Davoodi D, Golmohammadzadeh S (2018) Solid lipid nanoparticles carrying Eugenia caryophyllata essential oil: the novel nanoparticulate systems with broad-spectrum antimicrobial activity. Lett Appl Microbiol 66(6):506–513
34. Ferreira CD, Nunes IL (2019) Oil nanoencapsulation: development, application, and incorporation into the food market. Nanoscale Res Lett 14(1):1–13
35. Fioramonti SA, Stepanic EM, Tibaldo AM, Pavón YL, Santiago LG (2019) Spray dried flaxseed oil powdered microcapsules obtained using milk whey proteins-alginate double layer emulsions. Food Res Int 119:931–940
36. Freiberger EB, Kaufmann KC, Bona E, de Araújo PHH, Sayer C, Leimann FV, Gonçalves OH (2015) Encapsulation of roasted coffee oil in biocompatible nanoparticles. LWT-Food Sci Technol 64(1):381–389
37. Froiio F, Ginot L, Paolino D, Lebaz N, Bentaher A, Fessi H, Elaissari A (2019) Essential oils-loaded polymer particles: preparation, characterization and antimicrobial property. Polymers 11(6):1017
38. Gad HA, Abd El-Rahman FAA, Hamdy GM (2019) Chamomile oil loaded solid lipid nanoparticles: a naturally formulated remedy to enhance the wound healing. J Drug Deliv Sci Technol 50:329–338
39. Galanakis CM (2021) Functionality of food components and emerging technologies. Foods 10(1):128
40. Garg A, Singh S (2011) Enhancement in antifungal activity of eugenol in immunosuppressed rats through lipid nanocarriers. Colloids Surf B 87(2):280–288

41. Garg A, Singh S (2013) Targeting of eugenol-loaded solid lipid nanoparticles to the epidermal layer of human skin. Nanomedicine 9(8):1223–1238
42. Ghanbarzadeh S, Valizadeh H, Zakeri-Milani P (2013) Application of response surface methodology in development of sirolimus liposomes prepared by thin film hydration technique. BioImpacts: BI 3(2):75
43. Ghodrati M, Farahpour MR, Hamishehkar H (2019) Encapsulation of Peppermint essential oil in nanostructured lipid carriers: in-vitro antibacterial activity and accelerative effect on infected wound healing. Colloids Surf A 564:161–169
44. Hadian Z, Sahari MA, Moghimi HR, Barzegar M (2014) Formulation, characterization and optimization of liposomes containing eicosapentaenoic and docosahexaenoic acids; a methodology approach. Iran J Pharm Res 13(2):393–404
45. Hasheminejad N, Khodaiyan F, Safari M (2019) Improving the antifungal activity of clove essential oil encapsulated by chitosan nanoparticles. Food Chem 275:113–122
46. Hassanshahian M, Saadatfar A, Masoumipour F (2020) Formulation and characterization of nanoemulsion from Alhagi maurorum essential oil and study of its antimicrobial, antibiofilm, and plasmid curing activity against antibiotic-resistant pathogenic bacteria. J Environ Health Sci Eng 18(2):1015–1027
47. He H, Yao J, Zhang Y, Chen Y, Wang K, Lee RJ, Yu B, Zhang X (2019) Solid lipid nanoparticles as a drug delivery system to across the blood-brain barrier. Biochem Biophys Res Commun 519(2):385–390
48. How CW, Rasedee A*, Abbasalipourkabir R (2013) Characterization and cytotoxicity of nanostructured lipid carriers formulated with olive oil, hydrogenated palm oil, and polysorbate 80. IEEE Trans NanoBioscience 12(2):72–78
49. Hoyos-Leyva J, Bello-Perez L, Agama-Acevedo J, Alvarez-Ramirez J, Jaramillo-Echeverry L (2019) Characterization of spray drying microencapsulation of almond oil into taro starch spherical aggregates. Lwt 101:526–533
50. Huang J, Wang Q, Chu L, Xia Q (2020) Liposome-chitosan hydrogel bead delivery system for the encapsulation of linseed oil and quercetin: preparation and in vitro characterization studies. LWT 117:108615
51. Husen A, Siddiqi KS (2022) Advances in smart nanomaterials and their applications. Elsevier Inc., 50 Hampshire St., 5th Floor, Cambridge, MA 02139, USA
52. Iannitelli A, Grande R, Di Stefano A, Di Giulio M, Sozio P, Bessa LJ, ... Cellini L (2011) Potential antibacterial activity of carvacrol-loaded poly (DL-lactide-co-glycolide)(PLGA) nanoparticles against microbial biofilm. Int J Mol Sci 12(8):5039–5051
53. Izquierdo MS, Montero D, Robaina L, Caballero MJ, Rosenlund G, Ginés R (2005) Alterations in fillet fatty acid profile and flesh quality in gilthead seabream (Sparus aurata) fed vegetable oils for a long term period. Recovery of fatty acid profiles by fish oil feeding. Aquaculture 250(1–2):431–444
54. Jamil B, Abbasi R, Abbasi S, Imran M, Khan SU, Ihsan A, Javed S, Bokhari H, Imran M (2016) Encapsulation of cardamom essential oil in chitosan nano-composites: in-vitro efficacy on antibiotic-resistant bacterial pathogens and cytotoxicity studies. Front Microbiol 7:1580
55. Jiang Y, Lan W, Sameen DE, Ahmed S, Qin W, Zhang Q, Chen H, Dai J, He L, Liu Y (2020) Preparation and characterization of grass carp collagen-chitosan-lemon essential oil composite films for application as food packaging. Int J Biol Macromol 160:340–351
56. Joana Gil-Chávez G, Villa JA, Fernando Ayala-Zavala J, Basilio Heredia J, Sepulveda D, Yahia EM, González-Aguilar GA (2013) Technologies for extraction and production of bioactive compounds to be used as nutraceuticals and food ingredients: an overview. Compr Rev Food Sci Food Saf 12(1):5–23
57. Jummes B, Sganzerla WG, da Rosa CG, Noronha CM, Nunes MR, Bertoldi FC, Barreto PLM (2020) Antioxidant and antimicrobial poly-ε-caprolactone nanoparticles loaded with Cymbopogon martinii essential oil. Biocatal Agric Biotechnol 23:101499
58. Keivani Nahr F, Ghanbarzadeh B, Hamishehkar H, Kafil HS, Hoseini M, Moghadam BE (2019) Investigation of physicochemical properties of essential oil loaded nanoliposome for enrichment purposes. LWT 105:282–289

59. Khatibi SA, Misaghi A, Moosavy MH, Akhondzadeh Basti A, Mohamadian S, Khanjari A (2018) Effect of nanoliposomes containing Zataria multiflora Boiss. essential oil on gene expression of Shiga toxin 2 in Escherichia coli O157:H7. J Appl Microbiol 124(2):389–397
60. Khezri K, Farahpour MR, Mounesi Rad S (2019) Accelerated infected wound healing by topical application of encapsulated Rosemary essential oil into nanostructured lipid carriers. Artif Cells Nanomedicine Biotechnol 47(1):980–988
61. Khezri K, Farahpour MR, Mounesi Rad S (2020) Efficacy of Mentha pulegium essential oil encapsulated into nanostructured lipid carriers as an in vitro antibacterial and infected wound healing agent. Colloids Surf A 589:124414
62. Kokai LE, Lin YC, Oyster NM, Marra KG (2009) Diffusion of soluble factors through degradable polymer nerve guides: controlling manufacturing parameters. Acta Biomater 5(7):2540–2550
63. Kolanowski W, Jaworska D, Weißbrodt J (2007) Importance of instrumental and sensory analysis in the assessment of oxidative deterioration of omega-3 long-chain polyunsaturated fatty acid-rich foods. J Sci Food Agric 87(2):181–191
64. Krasodomska O, Paolicelli P, Cesa S, Casadei MA, Jungnickel C (2016) Protection and viability of fruit seeds oils by nanostructured lipid carrier (NLC) nanosuspensions. J Colloid Interface Sci 479:25–33
65. Kyriakoudi A, Spanidi E, Mourtzinos I, Gardikis K (2021) Innovative delivery systems loaded with plant bioactive ingredients: formulation approaches and applications. Plants 10(6):1238
66. Kumar A, Choudhary A, Kaur H, Mehta S, Husen A (2021) Metal-based nanoparticles, sensors and their multifaceted application in food packaging. J Nanobiotechnol 19(256):1–25
67. Lacatusu I, Mitrea E, Badea N, Stan R, Oprea O, Meghea A (2013) Lipid nanoparticles based on omega-3 fatty acids as effective carriers for lutein delivery. preparation and in vitro characterization studies. J Funct Foods 5(3):1260–1269
68. Lai A, Saleem Q, Macdonald PM (2015) Centerband-only-detection-of-exchange 31 P nuclear magnetic resonance and phospholipid lateral diffusion: theory, simulation and experiment. Phys Chem Chem Phys 17(38):25160–25171
69. Lammari N, Froiio F, Louaer M, Cristiano MC, Bensouici C, Paolino D, Louaer O, Meniai AH, Elaissari A (2020) Poly (ethyl acrylate-co-methyl Methacrylate-co-trimethylammoniethyl methacrylate chloride)(Eudragit RS100) Nanocapsules as Nanovector Carriers for Phoenix dactylifera L. Seeds Oil: a Versatile Antidiabetic Agent. Biomacromol 21(11):4442–4456
70. Lammari N, Louaer O, Meniai AH, Fessi H, Elaïssari A (2021) Plant oils: From chemical composition to encapsulated form use. Int J Pharm 601:120538
71. Lammari N, Demautis T, Louaer O, Meniai AH, Casabianca H, Bensouici C, Devouassoux G, Fessi H, Bentaher A, Elaïssari A (2021) Nanocapsules containing Saussurea lappa essential oil: formulation, characterization, antidiabetic, anti-cholinesterase and anti-inflammatory potentials. Int J Pharm 593:120138
72. Laohasongkram K, Mahamaktudsanee T, Chaiwanichsiri S (2011) Microencapsulation of Macadamia oil by spray drying. Procedia Food Sci 1:1660–1665
73. Liakos IL, Iordache F, Carzino R, Scarpellini A, Oneto M, Bianchini P, Grumezescu AM, Holban AM (2018) Cellulose acetate-essential oil nanocapsules with antimicrobial activity for biomedical applications. Colloids Surf B 172:471–479
74. Lin L, Gu Y, Sun Y, Cui H (2019) Characterization of chrysanthemum essential oil triple-layer liposomes and its application against Campylobacter jejuni on chicken. LWT 107:16–24
75. López-Meneses A, Plascencia-Jatomea M, Lizardi-Mendoza J, Fernández-Quiroz D, Rodríguez-Félix F, Mouriño-Pérez R, Cortez-Rocha M (2018) Schinus molle L. essential oil-loaded chitosan nanoparticles: preparation, characterization, antifungal and anti-aflatoxigenic properties. LWT 96:597–603
76. Mai C, Le TTT, Diep T, Le THN, Nguyen DT, Bach LG (2018) Development of solid lipid nanoparticles of Gac (Momordica cocochinensis Spreng) oil by nano-emulsion technique. Asian J Chem 30:293–297
77. Manikkavasagan A (2014) Formulation and evaluation of polymeric nanoparticles of felodipine. Doctoral dissertation, College of Pharmacy, Madras Medical College, Chennai.

78. Manthey NW, Cardona F, Aravinthan T, Wang H, Cooney T (2010) Natural fibre composites with QLD based fibres and vegetable oils. Incorporating sustainable practice in mechanics and structures of materials. ACMSM21, Melbourne, Victoria, pp 469–473
79. Marchiori MCL, Rigon C, Camponogara C, Oliveira SM, Cruz L (2017) Hydrogel containing silibinin-loaded pomegranate oil based nanocapsules exhibits anti-inflammatory effects on skin damage UVB radiation-induced in mice. J Photochem Photobiol B 170:25–32
80. McDaniel A, Tonyali B, Yucel U, Trinetta V (2019) Formulation and development of lipid nanoparticle antifungal packaging films to control postharvest disease. J Agric Food Res 1:100013
81. Mehnert W, Mäder K (2012) Solid lipid nanoparticles: production, characterization and applications. Adv Drug Deliv Rev 64:83–101
82. Miranda M, Cruz MT, Vitorino C, Cabral C (2019) Nanostructuring lipid carriers using Ridolfia segetum (L.) Moris essential oil. Mater Sci Eng C 103:109804
83. Modarres-Gheisari SMM, Gavagsaz-Ghoachani R, Malaki M, Safarpour P, Zandi M (2019) Ultrasonic nano-emulsification—a review. Ultrason Sonochem 52:88–105
84. Moghaddam M, Mehdizadeh L (2017) Chemistry of essential oils and factors influencing their constituents. In Soft chemistry and food fermentation. Academic Press, pp 379–419
85. Mohammed NK, Tan CP, Abd Manap Y, Alhelli AM, Hussin ASM (2017) Process conditions of spray drying microencapsulation of Nigella sativa oil. Powder Technol 315:1–14
86. Mohammed NK, Tan CP, Manap YA, Muhialdin BJ, Hussin ASM (2020) Spray drying for the encapsulation of oils—a review. Molecules 25(17):3873
87. Morsy NFS (2017) Chemical structure, quality indices and bioactivity of essential oil constituents. Active ingredients from aromatic and medicinal plants. pp 175–206
88. Müller RH, Mäder K, Gohla S (2000) Solid lipid nanoparticles (SLN) for controlled drug delivery–a review of the state of the art. Eur J Pharm Biopharm 50(1):161–177
89. Murray CW, Egan SK, Kim H, Beru N, Bolger PM (2008) US food and drug administration's total diet study: dietary intake of perchlorate and iodine. J Eposure Sci Environ Epidemiol 18(6):571–580
90. Nasseri M, Golmohammadzadeh S, Arouiee H, Jaafari MR, Neamati H (2016) Antifungal activity of Zataria multiflora essential oil-loaded solid lipid nanoparticles in-vitro condition. Iran J Basic Med Sci 19(11):1231–1237
91. Natrajan D, Srinivasan S, Sundar K, Ravindran A (2015) Formulation of essential oil-loaded chitosan–alginate nanocapsules. J Food Drug Anal 23(3):560–568
92. Nazari M, Ghanbarzadeh B, Samadi Kafil H, Zeinali M, Hamishehkar H (2019) Garlic essential oil nanophytosomes as a natural food preservative: its application in yogurt as food model. Colloid Interface Sci Commun 30:100176
93. Neng L, Tian G, Wei Z, Jihua L, Yupo C, Lijing L (2016) Preparation and characterization of galanga essential oil liposomes. In: Proceedings of the 2016 international conference on advances in energy, environment and chemical science, Atlantis Press
94. Niza E, Božik M, Bravo I, Clemente-Casares P, Lara-Sanchez A, Juan A, Klouček P, Alonso-Moreno C (2020) PEI-coated PLA nanoparticles to enhance the antimicrobial activity of carvacrol. Food Chem 328:127131
95. Noor NM (2013) The effect of virgin coconut oil loaded solid lipid particles (VCO-SLPs) on skin hydration and skin elasticity. Jurnal Teknologi 62(1)
96. Nosari AB, Lima JF, Serra OA, Freitas LAP (2015) Improved green coffee oil antioxidant activity for cosmetical purpose by spray drying microencapsulation. Rev Bras 25:307–311
97. Nuruzzaman MD, Rahman MM, Liu Y, Naidu R (2016) Nanoencapsulation, nano-guard for pesticides: a new window for safe application. J Agric Food Chem 64(7):1447–1483
98. Oehlke K, Behsnilian D, Mayer-Miebach E, Weidler PG, Greiner R (2017) Edible solid lipid nanoparticles (SLN) as carrier system for antioxidants of different lipophilicity. PLoS ONE 12(2):e0171662
99. Pateiro M, Gómez B, Munekata PE, Barba FJ, Putnik P, Kovačević DB, Lorenzo JM (2021) Nanoencapsulation of promising bioactive compounds to improve their absorption, stability, functionality and the appearance of the final food products. Molecules 26(6):1547

100. Paula HC, Sombra FM, Abreu FO, Paul R (2010) Lippia sidoides essential oil encapsulation by angico gum/chitosan nanoparticles. J Braz Chem Soc 21:2359–2366
101. Piña-Barrera AM, Álvarez-Román R, Báez-González JG, Amaya-Guerra CA, Rivas-Morales C, Gallardo-Rivera CT, Galindo-Rodríguez SA (2019) Application of a multisystem coating based on polymeric nanocapsules containing essential oil of Thymus vulgaris L. to increase the shelf life of table grapes (Vitis vinifera L.). IEEE Trans Nanobiosci 18(4):549–557
102. Pivetta TP, Simões S, Araújo MM, Carvalho T, Arruda C, Marcato PD (2018) Development of nanoparticles from natural lipids for topical delivery of thymol: Investigation of its anti-inflammatory properties. Colloids Surf, B 164:281–290
103. Poudel A, Gachumi G, Wasan KM, Dallal Bashi Z, El-Aneed A, Badea I (2019) Development and characterization of liposomal formulations containing phytosterols extracted from canola oil deodorizer distillate along with tocopherols as food additives. Pharmaceutics 11(4):185
104. Priyadarshini BM, Antipina MN, Fawzy AS (2018) Formulation and characterisation of poly (lactic-co-glycolic acid) encapsulated clove oil nanoparticles for dental applications. IET Nanobiotechnol 12(3):311–317
105. Pulingam T, Foroozandeh P, Chuah JA, Sudesh K (2022) Exploring various techniques for the chemical and biological synthesis of polymeric nanoparticles. Nanomaterials 12(3):576
106. Rabail R, Shabbir MA, Sahar A, Miecznikowski A, Kieliszek M, Aadil RM (2021) An intricate review on nutritional and analytical profiling of coconut, flaxseed, olive, and sunflower oil blends. Molecules 26(23):7187
107. Rai R, Alwani S, Badea I (2019) Polymeric nanoparticles in gene therapy: new avenues of design and optimization for delivery applications. Polymers 11(4):745
108. Rajivgandhi G, Saravanan K, Ramachandran G, Li J-L, Yin L, Quero F, Alharbi NS, Kadaikunnan S, Khaled JM, Manoharan N, Li W-J (2020) Enhanced anti-cancer activity of chitosan loaded Morinda citrifolia essential oil against A549 human lung cancer cells. Int J Biol Macromol 164:4010–4021
109. Rasti B, Jinap S, Mozafari MR, Yazid AM (2012) Comparative study of the oxidative and physical stability of liposomal and nanoliposomal polyunsaturated fatty acids prepared with conventional and Mozafari methods. Food Chem 135(4):2761–2770
110. Rigo LA, da Silva CR, de Oliveira SM, Cabreira TN, da Silva CdB, Ferreira J, Beck RCR (2015) Nanoencapsulation of rice bran oil increases its protective effects against UVB radiation-induced skin injury in mice. Eur J Pharm Biopharm 93:11–17
111. Risaliti L, Kehagia A, Daoultzi E, Lazari D, Bergonzi MC, Vergkizi-Nikolakaki S, Hadjipavlou-Litina D, Bilia AR (2019) Liposomes loaded with Salvia triloba and Rosmarinus officinalis essential oils: in vitro assessment of antioxidant, antiinflammatory and antibacterial activities. J Drug Deliv Sci Technol 51:493–498
112. Rodenak-Kladniew B, Islan GA, de Bravo MG, Durán N, Castro GR (2017) Design, characterization and in vitro evaluation of linalool-loaded solid lipid nanoparticles as potent tool in cancer therapy. Colloids Surf B 154:123–132
113. Romero-Hernandez HA, Sánchez-Rivera MM, Alvarez-Ramirez J, Yee-Madeira H, Yañez-Fernandez J, Bello-Pérez LA (2021) Avocado oil encapsulation with OSA-esterified taro starch as wall material: physicochemical and morphology characteristics. LWT 138:110629
114. Rozman NAS, Tong WY, Leong CR, Anuar MR, Karim S, Ong SK, Yusof FAM, Tan W-N, Sulaiman B, Ooi ML (2020) Homalomena pineodora essential oil nanoparticle inhibits diabetic wound pathogens. Sci Rep 10(1):1–11
115. Rushmi ZT, Akter N, Mow RJ, Afroz M, Kazi M, de Matas M, Rahman M, Shariare MH (2017) The impact of formulation attributes and process parameters on black seed oil loaded liposomes and their performance in animal models of analgesia. Saudi Pharm J 25(3):404–412
116. Saffari M, Hoseini Shirazi F, Moghimi HR (2016) Terpene-loaded liposomes and isopropyl myristate as chemical permeation enhancers toward liposomal gene delivery in lung cancer cells; a comparative study. Iran J Pharm Res 15(3):261–267
117. Salari S, Salari R (2019) Nanoliposomal system of rosemary essential oil made by specific human cell phospholipids and evaluation of its anti-cancer properties. Appl Nanosci 9(8):2085–2089

118. Santos VdS, Braz BB, Silva AÁ, Cardoso LP, Ribeiro APB, Santana MHA (2019) Nanostructured lipid carriers loaded with free phytosterols for food applications. Food Chem 298:125053
119. Saporito F, Sandri G, Bonferoni MC, Rossi S, Boselli C, Icaro Cornaglia A, Mannucci B, Grisoli P, Vigani B, Ferrari F (2018) Essential oil-loaded lipid nanoparticles for wound healing. Int J Nanomed 13:175–186
120. Sebaaly C, Jraij A, Fessi H, Charcosset C, Greige-Gerges H (2015) Preparation and characterization of clove essential oil-loaded liposomes. Food Chem 178:52–62
121. Sebaaly C, Charcosset C, Stainmesse S, Fessi H, Greige-Gerges H (2016) Clove essential oil-in-cyclodextrin-in-liposomes in the aqueous and lyophilized states: from laboratory to large scale using a membrane contactor. Carbohyd Polym 138:75–85
122. Shakeri M, Razavi SH, Shakeri S (2019) Carvacrol and astaxanthin co-entrapment in beeswax solid lipid nanoparticles as an efficient nano-system with dual antioxidant and anti-biofilm activities. LWT 107:280–290
123. Shamaei S, Seiiedlou SS, Aghbashlo M, Tsotsas E, Kharaghani A (2017) Microencapsulation of walnut oil by spray drying: effects of wall material and drying conditions on physicochemical properties of microcapsules. Innov Food Sci Emerg Technol 39:101–112
124. Shchukina EM, Graham M, Zheng Z, Shchukin DG (2018) Nanoencapsulation of phase change materials for advanced thermal energy storage systems. Chem Soc Rev 47(11):4156–4175
125. Shi F, Yang G, Ren J, Guo T, Du Y, Feng N (2013) Formulation design, preparation, and in vitro and in vivo characterizations of β-Elemene-loaded nanostructured lipid carriers. Int J Nanomed 8:2533–2541
126. Shi F, Zhao JH, Liu Y, Wang Z, Zhang YT, Feng NP (2012) Preparation and characterization of solid lipid nanoparticles loaded with frankincense and myrrh oil. Int J Nanomed 7:2033–2043
127. Silva WJ, Dória GAA, Maia RT, Nunes RS, Carvalho GA, Blank AF, ... Cavalcanti SCH (2008) Effects of essential oils on Aedes aegypti larvae: alternatives to environmentally safe insecticides. Bioresour Technol 99(8):3251–3255
128. Silva-Flores PG, Pérez-López LA, Rivas-Galindo VM, Paniagua-Vega D, Galindo-Rodríguez SA, Álvarez-Román R (2019) Simultaneous GC-FID quantification of main components of Rosmarinus officinalis L. and Lavandula dentata essential oils in polymeric nanocapsules for antioxidant application. J Anal Methods Chem
129. Solans C, Solé I (2012) Nano-emulsions: formation by low-energy methods. Curr Opin Colloid Interface Sci 17(5):246–254
130. Souza JR, Bonfim KS, Lorevice MV, Correa DS, Mattoso LHC, de Moura MR (2020) Antibacterial properties of oregano essential oil encapsulated in poly (ε-caprolactone) nanoparticles. Adv Sci Eng Med 12(7):864–869
131. Sutaphanit P, Chitprasert P (2014) Optimisation of microencapsulation of holy basil essential oil in gelatin by response surface methodology. Food Chem 150:313–320
132. Svetlichny G, Külkamp-Guerreiro IC, Cunha SL, Silva FEK, Bueno K, Pohlmann AR, Fuentefria AM, Guterres SS (2015) Solid lipid nanoparticles containing copaiba oil and allantoin: development and role of nanoencapsulation on the antifungal activity. Die Pharmazie—Int J Pharm Sci 70(3):155–164
133. Tabatabai MB, Mirjalili M, Yazdiyan F, Hekmatimoghaddam S (2019) Antibacterial activity and cytotoxicity of nanoliposomic and nanoniosomic essential oil of trachyspermum copticum. Proc Natl Acad Sci, India Sect B: Biol Sci 89(3):1109–1116
134. Taheri A, Jafari SM (2019) Gum-based nanocarriers for the protection and delivery of food bioactive compounds. Adv Coll Interface Sci 269:277–295
135. Tamjidi F, Shahedi M, Varshosaz J, Nasirpour A (2013) Nanostructured lipid carriers (NLC): a potential delivery system for bioactive food molecules. Innov Food Sci Emerg Technol 19:29–43
136. Tang J, Ge Y (2017) Development and evaluation of novel eucalyptus essential oil liposomes/chitosan composite sponges for medical use. Fibers Polym 18(3):424–433

137. Tapiero H, Ba GN, Couvreur P, Tew KD (2002) Polyunsaturated fatty acids (PUFA) and eicosanoids in human health and pathologies. Biomed Pharmacother 56(5):215–222
138. Thiagamani SMK, Rajini N, Siengchin S, Rajulu AV, Hariram N, Ayrilmis N (2019) Influence of silver nanoparticles on the mechanical, thermal and antimicrobial properties of cellulose-based hybrid nanocomposites. Compos B Eng 165:516–525
139. Timbe PPR, da Motta AdS, Isaía HA, Brandelli A (2020) Polymeric nanoparticles loaded with Baccharis dracunculifolia DC essential oil: preparation, characterization, and antibacterial activity in milk. J Food Process Preserv 44(9):e14712
140. Ushikubo FY, Furukawa T, Nakagawa R, Enari M, Makino Y, Kawagoe Y., ... Oshita S (2010) Evidence of the existence and the stability of nano-bubbles in water. Colloids Surfs A PhysChemical Eng Asp361(1–3):31–37
141. Vélez MA, Perotti MC, Hynes ER, Gennaro AM (2019) Effect of lyophilization on food grade liposomes loaded with conjugated linoleic acid. J Food Eng 240:199–206
142. Vieira R, Severino P, Nalone LA, Souto SB, Silva AM, Lucarini M, Durazzo A, Santini A, Souto EB (2020) Sucupira oil-loaded nanostructured lipid carriers (NLC): lipid screening, factorial design, release profile, and cytotoxicity. Molecules 25(3):685
143. Vrignaud S, Benoit JP, Saulnier P (2011) Strategies for the nanoencapsulation of hydrophilic molecules in polymer-based nanoparticles. Biomaterials 32(33):8593–8604
144. Vrouvaki I, Koutra E, Kornaros M, Avgoustakis K, Lamari FN, Hatziantoniou S (2020) Polymeric nanoparticles of Pistacia lentiscus var. chia essential oil for cutaneous applications. Pharmaceutics 12(4):353
145. Weber J, Funk NL, Motta MH, Guedes AM, Visintainer APC, Tedesco SB, Da Silva CdB (2016) Association of borage oil and betamethasone dipropionate in lipid-core nanocapsules: characterization, photostability and in vitro irritation test. J Nanosci Nanotechnol 16(2):1354–1362
146. Wu Z, Zhou W, Pang C, Deng W, Xu C, Wang X (2019) Multifunctional chitosan-based coating with liposomes containing laurel essential oils and nanosilver for pork preservation. Food Chem 295:16–25
147. Yang X, Gao N, Hu L, Li J, Sun Y (2015) Development and evaluation of novel microcapsules containing poppy-seed oil using complex coacervation. J Food Eng 161:87–93
148. Yupo C, Wei Z, Jihua L, Xiaobing H, Yuan Y, Lijing L (2016) Preparation and stability evaluation of pectin coated galanga essential oil liposomes. In: Proceedings of the 2016 2nd international conference on architectural, civil and hydraulics engineering (ICACHE 2016), Atlantis Press
149. Zhang H, Liang Y, Li X, Kang H (2020) Effect of chitosan-gelatin coating containing nano-encapsulated tarragon essential oil on the preservation of pork slices. Meat Sci 166:108137
150. Zhao Y, Chang Y-X, Hu X, Liu C-Y, Quan L-H, Liao Y-H (2017) Solid lipid nanoparticles for sustained pulmonary delivery of Yuxingcao essential oil: preparation, characterization and in vivo evaluation. Int J Pharm 516(1):364–371
151. Zaker M (2016) Natural plant products as eco-friendly fungicides for plant diseases control-A review. Agric 14(1):134–141
152. Zhang LQ, Zhang ZG, Fu Y, Xu Y (2015) Research progress of trans-cinnamaldehyde pharmacological effects. Zhongguo Zhong yao za zhi= Zhongguo zhongyao zazhi= China journal of Chinese materia medica 40(23):4568–4572
153. Zuzarte M, Salgueiro L (2015) Essential oils chemistry. In: Bioactive essential oils and cancer. Springer, Cham, pp 19–61

Tree Bark and Their Role in Nanomaterials Synthesis and Applications

Avtar Singh, Payal Malik, Anupama Parmar, Rohini, and Harish Kumar Chopra

Abstract Nanomaterials have become a ubiquitous part of modern life and are recognized as a fascinating class of materials with a broad range of uses such as in catalysis, medicine, agriculture, electronics, personal care and other areas. Nanomaterials are significantly different from their bulk materials and known to have a high surface area and outstanding electrical, optical, and mechanical properties. The desired properties of nanomaterials can be tailored by rational designing of nanoparticles (NPs). The properties of nanomaterials can be tuned by precise control over the shape, size of the NPs, and reaction conditions. In this regard, capping agents have a profound effect on the shape and size of NPs. In the recent past, an extensive range of capping/stabilizing agents has been tested in the synthesis of nanomaterials. Various types of extracts obtained from plants have emerged as cost-effective green capping and reducing agents which can fine-tune the properties of nanomaterials. In this chapter, advances in the bark extract mediated synthesis of nanomaterials with a special focus on metal and metal oxide NPs will be discussed. In the first section, the compositions of different bark extracts have been discussed, followed by various green syntheses of metal and metal oxide NPs using tree bark extracts. The next section covers the impact of reducing and capping agents such as polysaccharides, lignin, tannins, suberin etc. on the properties of nanomaterials. In the last section, challenges and future perspectives of the field are discussed.

Keywords Nanomaterials · Bark extract · Green synthesis · Reducing agents · Capping agents

A. Singh · P. Malik · H. K. Chopra (✉)
Department of Chemistry, Sant Longowal Institute of Engineering and Technology, Distt. Sangrur (Pb.), Longowal 148106, India
e-mail: hk67@rediffmail.com

A. Parmar
PG Department of Chemistry, M. M. Modi College, Patiala 147001, India

Rohini
Department of Chemistry, Baba Farid College, Bathinda 151001, India

A. Husen (ed.), *Secondary Metabolites Based Green Synthesis of Nanomaterials and Their Applications*, Smart Nanomaterials Technology,
https://doi.org/10.1007/978-981-99-0927-8_14

Abbreviations

AAS	Atomic absorption spectroscopy
AgNPs	Silver nanoparticles
AuNPs	Gold nanoparticles
BET	Brunauer, Emmett and Teller
CuONPs	Copper oxide nanoparticles
DLS	Dynamic light scattering
EDX	Energy-dispersive X-ray spectroscopy
FTIR	Fourier transform infrared spectroscopy
NPs	Nanoparticles
SEM	Scanning electron microscopy
SPR	Surface plasmon resonance
TEM	Transmission electron microscopy
TiO_2NPs	Titanium oxide nanoparticles
UV-Vis	Ultraviolet-visible
ZnONPs	Zinc oxide nanoparticles

1 Introduction

Nanotechnology has gained much consideration from the academic and industrial researchers because of the fascinating properties and myriad applications of nanomaterials in drug delivery, DNA analysis and gene therapy, anticancer agents, antimicrobial agents, biosensing, catalysis, food science, forestry, agricultural sectors etc. [4, 10, 14, 23, 24, 26, 27-29, 38, 39, 60]. Nanoparticles (NPs) exhibit different properties than the bulk materials due to their nano-scale size, increased surface area and diverse morphologies and appeared as neoteric candidates in biomedicines, catalysis etc. [19, 62, 69]. Conventional chemical synthesis of the NPs poses several problems such as toxicity, requirement of costly reducing and capping agents, environmental issues [25, 52]. To overcome these limitations and for the production of biocompatible NPs, several greener alternatives have been formulated, the most common being the implementation of plant extracts as a reducing and stabilizing agent [13, 22, 37].

The leaf, fruit, bark and root extracts of various plants consist of wide variety of natural products such as polyphenols, flavonoids, carbohydrates, terpenoids, and alkaloids (Fig. 1) and these natural products can be used as the eco-friendly, cost effective and greener reducing or stabilizing agents for the nanoparticle synthesis. The phytochemical assisted synthesis of NPs reduces the risk of chemical toxicity and also offers additional benefit of synergistic combination of nanoparticle cores with the capping layer containing biological active compounds and offers an alternative to the NPs obtained from conventional chemical methods.

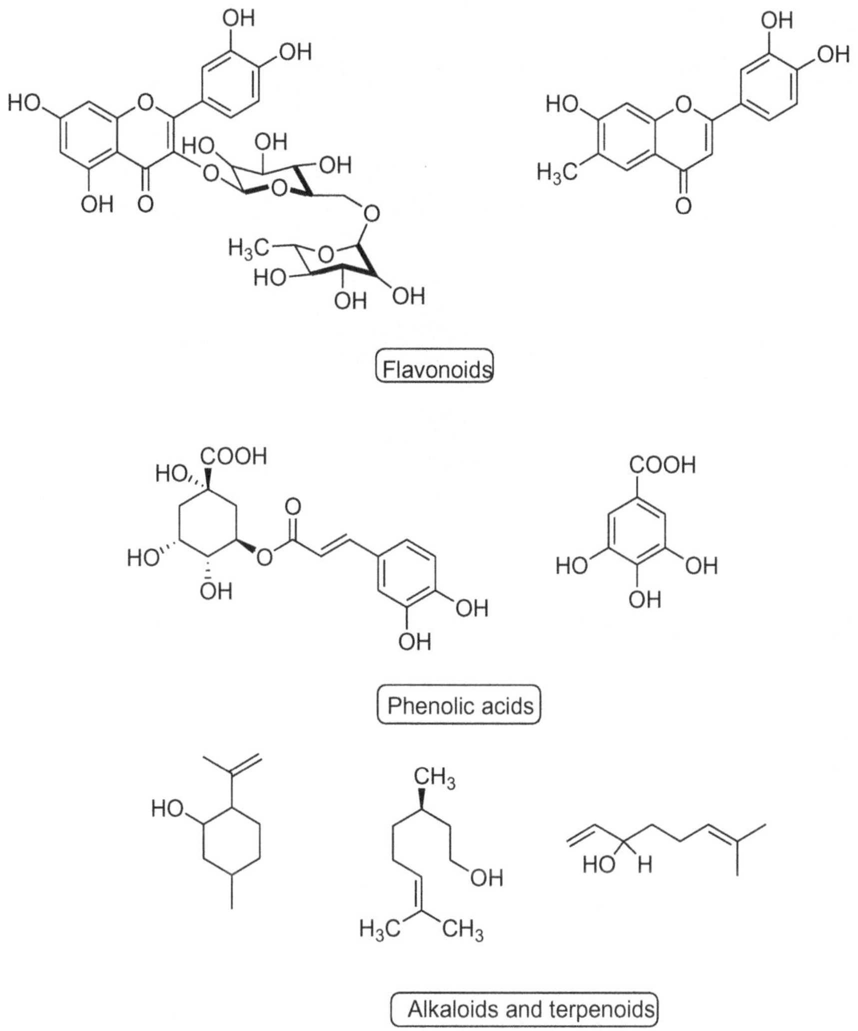

Fig. 1 Major phytochemicals present in tree barks

Plant bark is abundant, easy to process and consists of several natural reducing and stabilizing compounds, so, the bark extract of several plants have been successfully employed in metal and metal oxide NPs without adding any additional reducing/capping agents (Fig. 2).

Herein, we have discussed a variety of tree bark extracts which have been used in the synthesis of metal and metal oxide NPs with special focus on coinage metals. Due to the availability of plethora of research and literature reports on this topic, we have discussed the literature published in the past decade.

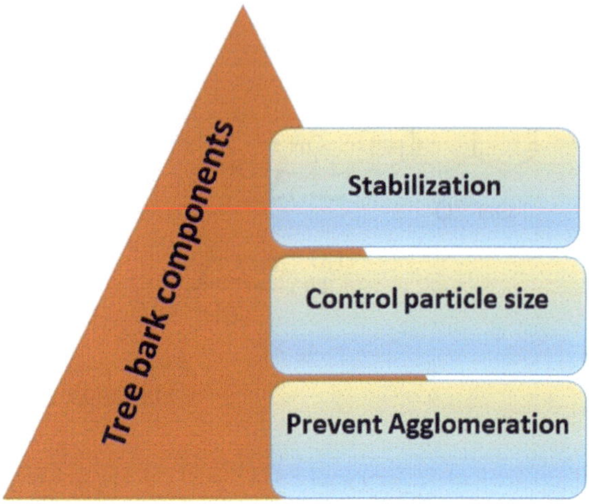

Fig. 2 Role of tree bark extract in the synthesis of nanoparticles

The chapter gives an overview of the phytochemical profile of various bark extracts, the effect of bark concentration and other reaction parameters on the morphology and size distribution of the metal NPs, important applications of these NPs such as antimicrobial activity, anticancer activity, catalytic potential and biosensing.

2 Metal NPs

2.1 Silver NPs (AgNPs)

The utilization of bark extracts as a reducing and stabilizing agent to synthesize the silver NPs have been summarized in this section. Silver is one of the most explored metals for the synthesis of zero valent NPs. The AgNPs have potential pharmacological, biosensing, catalytic and nanomedicines applications [15, 46, 63, 68, 72].

Shetty et al. [61] synthesized the biogenic AgNPs, from the bark extract of an important medicinal plant, *Alstonia scholaris*. The synthesized particles were characterized by using UV, FTIR, XRD, TEM and particle size analysis. Also, the antifungal activity of these NPs was tested against several fungal species. Moreover, these NPs were found to be effective against the bacteria present in domestic water supply pipelines.

Iravani and Zolfaghari [32] exploited the bark of *Pinus eldarica* in the green synthesis of spherical AgNPs. In this study, the authors have thoroughly observed the effect of the amount of extract used, substrate concentration, temperature, and pH on the synthesis of these NPs. Temperature and pH were found to be two important parameters in the NPs synthesis; it was observed that the production increases and particle size decreases at high temperature. Also, the large sized NPs were obtained at an acidic pH and small sized, more dispersed NPs were produced at alkaline pH. Almalah et al. [3] reported the eco-friendly synthesis of AgNPs from using the *Cinnamomum zylinicum* bark extract as a reducing agent. The formation of the spherical NPs has been confirmed by the UV–Vis spectroscopy, FTIR, SEM and TEM etc. The antimicrobial potential of the NPs was assessed against several multi drug resistance bacterial strains and results showed that biogenic silver NPs were effective against some Gram-negative and Gram-positive bacteria.

Moyo et al. examined [45] *Afzelia quanzensis* bark extract to synthesize the AgNPs of 10–80 nm size; XRD analysis confirmed the crystalline fcc structure. FTIR analysis showed that the presence of carboxyl (–C=O) and amine (N–H) groups in the bark extract were mainly responsible for the reduction of Ag (I) ions to Ag (0) NPs. The synthesized AgNPs were found to have excellent stability in water and exhibit antibacterial potential against several bacterial cultures. Similarly, [1] synthesized the biogenic AgNPs from *Terminalia arjuna* bark extract; the synthesized NPs showed good antibacterial activity against *Escherichia coli.*

Yugandhar et al. [75] synthesized and characterized AgNPs from the bark extract of *Syzygium alternifolium* obtained from south east ghat. The synthesis of NPs was confirmed by a broad peak at 448 nm; The FTIR study reflects that the phenols and proteins present in the bark extract act as the stabilizing and capping agent to prevent agglomeration. Further, the antifungal and antibacterial study was also carried out against several fungal and bacterial strains respectively.

Patil and Kim [49] described the potency of *Ficus benghalensis* and *Azadirachta indica* bark extracts in the nanoparticle production in the absence of any other reducing or capping agent. The synthesized NPs were characterized by UV–Vis spectroscopy, ATR-FTIR, DLS, Fe-SEM and AFM. The AgNPs synthesized by this method show good antibacterial activity as well as anticancer activity against MG-63 osteosarcoma cell line in a concentration dependent manner. Bhakya et al. [11] investigated the potential of stem bark extracts of *Helicteres isora* as the reducing and capping agent for the production of AgNPs. The synthesized NPs were found to have numerous biological applications like antioxidant activity, antibacterial activity and the anticancer activity was assessed on oral carcinoma (KB) cell line although the AgNPs which were found to be cytotoxic at a high concentration and prolonged use.

Pattanayak et al. [51] have used *Butea monosperma* bark extract as a reducing as well as capping agent in the synthesis of AgNPs by. The size and shape of these NPs was observed by HRTEM, DLS and XRD studies. Further, the antibacterial activity was tested against both types of Gram bacteria; also, the NPs were found to be potent anticancer agents against the KG-1A cell line. Similarly, Das et al. [16] have also examined the synthesis and antimicrobial properties of the AgNPs synthesized from the bark extract of *Butea monosperma.* Arya et al. [6] reported the microwave assisted synthesis of AgNPs using the *Prosopis juliflora* bark collected from Rajasthan, India. The physicochemical characterization of the NPs was performed by UV–Vis, FTIR,

EDAX and FESEM etc. The AgNPs thus synthesized were found to have antibacterial activity; in vitro anticancer activity against A549 cells and also used in the catalytic reduction of 4-nitrophenol. The same group has reported the biosynthesis of AgNPs from the bark extract of *Acacia nilotica* and investigated the numerous potent applications of these NPs viz. antibacterial, anticancer, antibiofilm and wound healing etc. Ramanathan et al. [56] evaluated the antimicrobial potential of AgNPs synthesized by using the bark extract of *Solanum trilobatum*. Both the Disc diffusion and well diffusion methods were carried out for the inhibition of microbial species and it was demonstrated that the antimicrobial potential of these bark extract capped AgNPs is far better than the bark extract itself and chemically synthesized AgNPs. This study clearly shows that the biosynthesized AgNPs exhibit enhanced biological activity and can be a better substitute in medical applications [33]. Reported the utilization of aqueous and ethanolic bark extracts of *Acacia ferruginea* in the AgNPs synthesis under sunlight. These NPs were characterized by a variety of analytical techniques viz. UV–Vis spectroscopy, FTIR, XRD, FE-SEM, EDS, HR-TEM, DLS and Zeta potential. The antioxidant activity of these NPs was tested by the DPPH method and the antioxidant potential of the aqueous bark mediated AgNPs was found to be higher than that of ethanolic bark mediated NPs. Also, both types of NPs were found to have moderate antibacterial activity.

Chandra Sekhar et al. [14] reported the environment friendly biosynthesis of AgNPs from the bark extract of *Syzygium cumini*. The synthesized NPs exhibit enhanced antibacterial activity against *Escherichia coli* and *Bacillus subtilis* as compared to the plant extract and the fresh $AgNO_3$ solution [35]. Synthesized and characterized the AgNPs by adding the *Garcinia mangostana* bark extract to the silver nitrate. The AgNPs were found to be effective bactericidal agents against several human pathogenic bacteria; and also show the larvicidal activity against the larvae of the dengue vector *A. aegypti*.

Bharathi et al. reported [12] the phyto-synthesis of AgNPs using the *Diospyros montana* stem bark extract. The phytochemical screening of the methanolic bark extract confirms the presence of alkaloids, saponins, polyphenols, proteins etc. which can act as the reducing as well as the capping agent for the nanoparticle formation. The antioxidant activity of these NPs was determined by DPPH and H_2O_2 free radical scavenging methods and the antibacterial potential is also tested against few Gram positive and Gram negative bacterial strains [70]. Described the microwave mediated green synthesis of AgNPs using the underbark extract of *Handroanthus impetiginosus*. The antimicrobial study shows that these NPs are effective against Gram positive bacteria; also the cytotoxicity study reveals that the NPs are not toxic against HaCaT mammalian cells [44]. Utilized the bark extract of *Elaeagnus angustifolia* for the green synthesis of AgNPs with potential antibacterial application. Tahir et al. [74] reported the Thespesia populnea bark extract mediated biosynthesis of AgNPs. After characterization, these NPs were tested for their biological and catalytic potential. The investigations showed that the green NPs have antimicrobial activity against several fungal and bacterial cultures and cytotoxic viability against the SK-MEL cell line. In photocatalysis, the NPs were found to show 90% degradation of methylene blue dye under sunlight.

Recently, [42] described the synthesis of AgNPs using the *Picea abies* and *Pinus nigra* bark extracts as reducing as well as capping agents. Both types of NPs exhibited antimicrobial and antimitotic activity; the biological potential of these NPs is greater than as compared to the bark extracts used for their synthesis (Table 1).

2.2 Gold NPs (AuNPs)

Gold NPs are reported for their antibacterial potential against many pathogens, anticancer application, antioxidant activity and some other important physiological effects [21, 30, 31]. Also, the AuNPs are being used in separation science, catalysis and sensing applications [2, 48, 50, 58]. The present section comprises a brief review about the synthesis, size, morphology and applications of AuNPs synthesized using the bark extract of some medicinal plants [17] used the *Saraca indica* bark extract for the biogenic production of polyshaped AuNPs. The synthesized NPs showed a good catalytic efficacy in the reduction of 4-nitrophenol to 4-aminophenol at room temperature. Similarly, [43] reported the synthesis of AuNPs under very mild conditions using the bark extract of *Mimusops elengi* without the use of any other reducing or capping agent. The formation of the AuNPs was established by surface plasmon resonance band at 536–541 nm. These NPs were used as efficient catalysts for the reduction of nitrophenols to the corresponding amino compounds in aqueous medium.

Also, Guo et al. [20] designed the controllable synthesis of AuNPs from the aqueous bark extract of *Eucommia ulmoides*. These NPs were used for the catalytic degradation of azo dyes; reactive yellow 179 and congo red in the presence of $NaBH_4$.

Suganthy et al. [65] reported the phytosynthesis of triangular and spherical shaped AuNPs from the bark of the *Terminalia arjuna*. This synthesis method is mild, efficient and can be used for the large scale synthesis of the NPs. The characterization of the NPs was done by using FTIR, XRD, FESEM, EDX, HRTEM, DLS, and zeta potential analysis etc. The synthesized NPs were biocompatible and non-toxic in nature and found to have potential biological applications viz. antioxidant, anticholinesterase, and antiamyloidogenic etc. Pinto et al. [53] examined the effect of various reaction parameters on the size and morphology of AuNPs synthesized by employing *Eucalyptus globules* extract. The observations suggested that temperature and reaction time has no marked effect on the shape and size of the NPs. The morphology is slightly dependent on the pH of the reaction medium and size of the NPs can be controlled by varying the extract concentration; as decrease in the extract concentration leads to the increase in the size of the AuNPs [18] used the cinnamon bark extract as a reducing and stabilizing effect for the greener synthesis of AuNPs. After characterization these NPs were used as biosensors for the detection of bovine serum albumin protein using fluorescence. It was observed that the fluorescence intensity of a mixture of NPs and eosin Y dye was increased in the presence of bovine serum albumin protein [64] described the synthesis of AuNPs using the bark extract of the well recognized medicinal plant *Panax ginseng*. The synthesized

Table 1 Size, morphology and applications of AgNPs synthesized from various bark extracts

Plant	Size/shape	Applications	References
Alstonia scholaris	50 nm/Spherical	Antimicrobial	[61]
Pinus eldarica	10–40 nm/Spherical	–	[32]
Cinnamomum Zylinicum	10–80 nm/Spherical	Antibacterial	[3]
Afzelia quanzensis	10–80 nm/Spherical	Antibacterial	[45]
Terminalia arjuna	30–50 nm/Spherical	Antibacterial	[1]
Syzygium alternifolium	4–48 nm/Spherical	Antimicrobial	[75]
Ficus benghalensis and *Azadirachta indica*	40–50 nm/Spherical	Antibacterial, Anticancer	[49]
Helicteres isora	16–95 nm/Spherical	Antioxidant, Antibacterial, Anticancer	[11]
Butea monosperma	18–50 nm/Spherical	Antibacterial, Anticancer	[51]
Butea monosperma	81 nm/Spherical	Antimicrobial	[16]
Prosopis juliflora	10–50 nm/Spherical	Antibacterial, Anticancer, Catalytic	[6]
Acacia nilotica	20–50 nm/Spherical	Antibacterial, Anticancer, Antibiofilm, Wound healing	[7]
Solanum trilobatum	50-100 nm/Spherical	Antimicrobial	[56]
Acacia ferruginea	14–32 nm/Spherical	Antioxidant, Antibacterial,	[33]
Syzygium cumini	10–20 nm/Spherical	Antimicrobial	[14]
Garcinia mangostana	65 nm/Spherical	Antimicrobial Larvicidal	[35]
Diospyros montana	28 nm/Spherical	Antioxidant, Antibacterial,	[12]
Handroanthus impetiginosus	13.4 nm/Spherical	Antimicrobial	[70]
Elaeagnus angustifolia	65–90 nm/Spherical	Antibacterial	[44]
Thespesia populnea	40–50 nm/Spherical	Antimicrobial, Anticancer, Photocatalytic	[74]
Picea abies and *Pinus nigra*	78.48 and 77.66 nm/Spherical	Antimicrobial, Antimitotic	[42]

NPs were fully characterized by using UV–Vis, FTIR, EDAX, FESEM, XRD etc. and were used as antimicrobial agents against the pathogenic bacteria like *Bacillus anthracis, Vibrio parahaemolyticus, Staphylococcus aureus, Escherichia coli,* and *Bacillus cereus.* Bahram et al. [9] reported the greener alternative for the synthesis of AuNPs using willow tree bark extract without the addition of any other chemical reagents. The synthesized NPs were found to be a very effective colorimetric spectrophotometric sensor for the detection of cysteine. It was observed that these NPs were selective towards cysteine over the other amino acids [54] described the eco-friendly synthesis of AuNPs by using the bark extract of *Mammea suriga.* The effect of various reaction parameters on the size and morphology of the NPs was thoroughly studied and the synthesized NPs were tested for their antibacterial potential against some Gram-positive (*B. subtilis and S. aureus*) and Gram-negative (*E. coli and P. aeruginosa*) bacterial strains [36] described the phytochemical assisted synthesis of AuNPs by using the bark extract of *Pterocarpus santalinus.* These AuNPs were also used as effective antibacterial agents against the Gram positive and Gram negative strains.

Rodríguez-León et al. [57] synthesized the AuNPs using the bark extract of *Mimosa tenuiflora.* The cytotoxicity was assessed in human umbilical vein endothelial cells and these NPs were found to have moderate toxicity. The catalytic potential of these NPs was tested in the degradation of methylene blue dye and 50% degradation was achieved in 190s. Velammal et al. [55] reported the bioreductive synthesis of spherical AuNPs using the *Plumbago zeylanica* bark extract. In vitro cytotoxicity and DNA binding capability of these NPs were assessed and these NPs were found to have moderate toxic nature. Also, the AuNPs were found to exhibit the antioxidant and antimicrobial activity (Table 2).

3 Metal Oxide NPs

3.1 Titanium Oxide (TiO_2) and Copper Oxide (CuO) NPs

Tahir et al. [47] developed a simple environmentally benign method for the synthesis of TiO_2 NPs using the extract of cinnamon powder. XRD analysis confirmed the presence of anatase phase TiO_2 with a crystallite size of 70.1 nm. SEM revealed that the TiO_2 NPs are spherical shapes with a size range of 70–150 nm. Band gap analysis suggested that there is an increment of the band gap of TiO_2 NPs in visible light absorption and as compared to the bulk material; therefore NPs can be used in solar cells (Table 3).

Devi et al. [76] synthesized and characterized biocompatible CuO NPs with the stem bark of *Syzygium alternifolium* and explored their antimicrobial and anticancer activities. FTIR studies established that phenols and primary amines act as stabilizing agents for the NPs. The crystallographic nature of CuO NPs with an average size of 17.2 nm was confirmed by XRD results. TEM results revealed that biosynthesized

Table 2 Size, morphology and applications of AuNPs synthesized from various bark extracts

Plant	Size/Shape	Applications	References
Saraca indica	15–23 nm/Spherical	Catalytic	[17]
Mimusops elengi	9–14 nm/Spherical	Catalytic	[43]
Eucommia ulmoides	16–18 nm/Spherical	Catalytic degradation	[20]
Terminalia arjuna	3–70 nm/Spherical	Antioxidant, Anticholinesterase, Antiamyloidogenic	[65]
Eucalyptus globules	Variable size/mostly spherical	–	[53]
Cinnamon bark	35 nm/Spherical	Sensors	[18]
Panax ginseng	10–40 nm/Spherical	Antimicrobial	[64]
Willow tree bark	15 nm/Spherical	Sensors	[9]
Mammea suriga	23 nm/Spherical	Antibacterial	[54]
Pterocarpus santalinus	13–26 nm/Spherical	Antibacterial	[36]
Mimosa tenuiflora	20-200 nm/Polyshaped	Catalytic	[57]
Plumbago zeylanica	10–25 nm/Spherical	DNA binding, Antioxidant, Antimicrobial	[55]

Table 3 Size, morphology and applications of TiO_2 and CuO NPs synthesized from various bark extracts

Plant name	Shape/size (in nm) of NPs	Applications	References
Cinnamon powder	Spherical/50–100	Solar cells	[47]
Syzygium alternifolium	Spherical,/5–13	Antimicrobial, anticancer	[76]
Rubia cordifolia	Irregular spherical shapes/50.72	Antimicrobial, photodegradation	[73]

NPs are non-agglomerated spherical in shape with particle size ranging from 5–13 nm. Among the used microbes these CuO NPs exhibited the highest inhibitory activity against *E. coli* and *T. harzianum*. The authors examined the anticancer activity of CuO NPs, and they displayed significant anticancer activity against MDA-MB-231 human breast cancer lines.

Recently, Sekar et al. [73] reported bark extract mediated (*Rubia cordifolia*) synthesis of CuO NPs and their biomedical and environmental applications were explored. From FTIR analysis, it was confirmed that the functional groups act as a reducing as well as stabilizing agent in the formation of NPs. FE-SEM.

DLS results analysis revealed that the CuO NPs are spherical in shape with a size of 50.72 nm. CuO NPs displayed high antibacterial activity against *E.coli*, *P. aeruginosa*, *S. aureus* and *B. cereus*p pathogenic bacteria. In addition, to investigate the photocatalytic activity of the CuONPs MB and crystal violet dyes were used, and the results revealed that these NPs exhibited degradation potentials, 81.84% and 64.0% degradation was observed in MB and crystal violet respectively.

3.2 *Zinc Oxide (ZnO) NPs*

David et al. [67] have synthesized ZnO NPs using $Zn(NO_3)_2$ and stem bark extract of *Boswellia ovalifoliolata* and explored their antimicrobial activities. On the basis of FT-IR analysis it was confirmed that primary and secondary amine functionalities are responsible for the stabilization of ZnO NPs. XRD analysis results confirmed the hexagonal wurzite structure of NPs. SEM and TEM results revealed the irregular hexagonal morphology of agglomerated ZnO NPs which have an average size of 66.25 nm. These NPs exhibited high antimicrobial activity against *B. cereus*, *S. aureus*, *E. coli*, *K. pneumoniae*, and *S. typhi*. Moreover, biosynthesized ZnO NPs displayed significant cytotoxic against MDA-MB 231 and MCF-7 breast cancer cell lines (Table 4).

Later, Babujanarthanam et al. [66] have used *Alstonia scholaris* bark extract for the synthesis of ZnO NPs. The NPs were characterized using various analytical techniques. TEM analysis confirmed the formation of spherical shaped agglomerated NPs with the size range of 30–20 nm. The authors investigated the anti-microbial activity of the synthesized ZnO NPs using gram-positive and negative bacteria, and these were found to be potent anti-bacterial and anti-fungal agents with low cytotoxicity.

Rajendran et al. [59] reported *Terminalia arjuna* bark mediated synthesis of ZnO NPs using different processing methods *i.e.* sonication, wet-chemical and hydrothermal. To analyze morphology, particle size, and surface area of ZnO NPs XRD, SEM, TEM and BET were performed. It was observed that the processing method has a profound effect on NPs size and surface area. In all the cases spherical NPs were obtained, however, their size and surface area varied with the processing

Table 4 Size, morphology and applications of ZnO NPs synthesized from various bark extracts

Plant name	Shape/size (in nm) of NPs	Applications	References
Boswellia ovalifoliolata	Irregular hexagonal/66.25	Antimicrobial, anticancer	[67]
Alstonia scholaris	Spherical/30–20	Antimicrobial	[66]
Terminalia arjuna	Spherical/21–45	Antimicrobial	[59]
Albizia lebbeck	Hexagonal/66.25	Antimicrobial, anticancer	[71]
Kalopanax septemlobus	Flower-like/500	Photodegradation	[41]
Cinnamomum verum	Cone/45	Antimicrobial	[5]
Acacia caesia	Hexagonal/32	Antimicrobial, photodegradation	[8]
Amygdalus scoparia	Spherical/29	Antimicrobial, anticancer, antidiabetic	[34]

method. In the hydrothermal method, smallest NPs of 21 nm with a surface area of 217 m^2 g^{-1} were obtained. Hydrothermally synthesized ZnO NPs showed remarkable antimicrobial action against *S. aureus* and *E. coli* due to their high surface area.

In 2019, [71] described a simple, environmentally benign route for the synthesis of ZnO employing *Albizia lebbeck* stem bark extract as a stabilizer. ZnO NPs were characterized by UV–vis, FTIR, XRD, SEM, and EDX. XRD analysis confirmed the presence of a hexagonal wurzite structure. SEM results revealed the presence of irregular hexagonal morphology of the NPs and the average size was found to be 66.25 nm; agglomeration was also observed. ZnO NPs displayed antibacterial activity against *B. cereus, S. aureus, E. coli, K. pneumoniae,* and *S. typhi.* Moreover, biosynthesized ZnO NPs are significantly cytotoxic on MDA-MB 231 and MCF-7 breast cancer cell lines.

Later, Aquisman et al. [40] synthesized ZnO NPs using fresh aqueous extract of leaf and stem-bark of *Carica papaya* and compared their anti-microbial efficiency for pathogenic bacteria i.e. *S. aureus, E. coli, A. baumanni, E. aquaticum,* and *K. pneumonia.* FTIR and AAS analysis results confirmed that the bark and leaf extract acts as a capping agent during the synthesis of ZnO NPs. The biosynthesized ZnO NPs have agglomerated petal-like morphologies and size ranges between 54 and 97 nm. The anti-microbial studies revealed that the ZnO NPs exhibited high inhibition activity against the all the five pathogenic bacteria (vide supra), the leaf extract-mediated ZnO NPs displayed better antimicrobial activity.

Yang et al. [41] prepared *Kalopanax septemlobus* bark extract mediated synthesis of ZnO NPs and investigated their efficiency in the degradation of methylene blue dye (MB). FE-TEM results confirmed that the bark extract produced flower-like ZnO NPs. These NPs act as a photocatalyst for the degradation of MB and degraded the dye ~97% within 30 min under UV light irradiation.

In 2020, Gowtham et al. [5] used *Cinnamomum verum* bark extract for the synthesis of ZnO-NPs and their efficacy against *S. aureus* and *E. coli.* was evaluated. PXRD and TEM analysis confirmed the hexagonal wurtzite phase of ZnO and cone-like shape of the ZnO NPs with a size of ~45 nm. Anti-microbial studies demonstrated that these NPs inhibit the growth of both *E. coli* and *S. aureus* with a MIC of 125 μg mL^{-1} and 62.5 μg mL^{-1} respectively.

Nair et al. [8] reported a green synthesis ZnO NPs using the bark extract of *Acacia caesia* (L.) *Willd.* The presence of NPs was confirmed by XRD and SEM, and the analysis revealed that hexagonal NPs were synthesized with an average crystalline size of 32.32 nm. FTIR analysis revealed the presence of the functional groups which confirms that the extract is acting as a capping agent. The photocatalytic activity of ZnO NPs was investigated using MB 92.2% dye degraded within 40 min. In addition, these NPs displayed significant antibacterial, antifungal and anti-inflammatory activity.

Kiani et al. [34] used *Amygdalus scoparia* stem bark extract for the biosynthesis of ZnO NPs and explored their anti-microbial and anticancer activities. SEM and TEM results confirmed the formation of agglomerated spherical ZnO NPs with an average size of 29 nm. To evaluate the antimicrobial activities of ZnO NPs the authors have selected *E. coli*, *E. aerigenes*, *S. aureus*, *P. oryzae*, *F. thapsinum*, and *F. semitectum*. These NPs exhibited remarkable inhibitory activity against all the microbes. In addition, ZnO NPs exhibited remarkable inhibitory effects on cancer cell lines. These ZnO NPs were proved to be efficient anti-diabetic agents.

4 Conclusions

The above discussed literature showed the potential of different tree bark extracts in the greener synthesis of various metal and metal oxide NPs. Bark-obtained phytochemicals provide a viable alternative to the existing reducing/capping agents. The results established that these phyto-constituents profoundly affect the size and morphology of the synthesized NPs. These biosynthesized NPs exhibited remarkable antimicrobial, anticancer and photocatalytic properties. However, agglomeration is one of the major limitations associated with these biosynthesized NPs. In addition, the mechanisms of bark-assisted synthesis of NPs have not yet been established, and only scarce reports are available in the literature. There is a need to carry out extensive mechanistic investigations on the synthesis of NPs. The tree barks are less explored as compared to the leaves or other parts; therefore, other tree bark should be explored for the synthesis of NPs. We anticipate that the bark-mediated NPs synthesis will emerge as an interesting area of research due to its environmental viability and interesting biological activities.

References

1. Ahmed Q, Gupta N, Kumar A, Nimesh S (2017) Antibacterial efficacy of silver nanoparticles synthesized employing *Terminalia arjuna* bark extract. Artif Cell Nanomed Biotechnol 45(6):1192–1200
2. Ahmed S, Ikram S (2016) Biosynthesis of gold nanoparticles: a green approach. J Photochem Photobiol B: biology 161:141–153
3. Almalah HI, Alzahrani HA, Abdelkader HS (2019) Green synthesis of silver nanoparticles using *cinnamomum zylinicum* and their synergistic effect against multi-drug resistance bacteria. J Nanotechnol Res 1(3):95–107
4. Ansari MA, Murali M, Prasad D, Alzohairy MA, Almatroudi A, Alomary MN, Niranjana SR (2020) *Cinnamomum verum* bark extract mediated green synthesis of ZnO nanoparticles and their antibacterial potentiality. Biomolecules 10(2):336
5. Ansari MA, Murali M, Prasad D, Alzohairy MA, Almatroudi A, Alomary MN, Udayashankar AC, Singh SB, Asiri SM, Ashwini BS, Gowtham HG (2020) *Cinnamomum verum* bark extract mediated green synthesis of ZnO nanoparticles and their antibacterial potentiality. Biomolecules 10(2):336
6. Arya G, Kumari RM, Gupta N, Kumar A, Chandra R, Nimesh S (2018) Green synthesis of silver nanoparticles using *Prosopis juliflora* bark extract: reaction optimization, antimicrobial and catalytic activities. Artif Cell Nanomed Biotechnol 46(5):985–993
7. Arya G, Kumari RM, Pundir R, Chatterjee S, Gupta N, Kumar A, Nimesh S (2019) Versatile biomedical potential of biosynthesized silver nanoparticles from *Acacia nilotica* bark. J Appl Biomed 17(2):115–124

8. Ashwini J, Aswathy TR, Rahul AB, Thara GM, Nair AS (2021) Synthesis and characterization of zinc oxide nanoparticles using *Acacia caesia* bark extract and its photocatalytic and antimicrobial activities. Catalysts 11(12):1507
9. Bahram M, Mohammadzadeh E (2014) Green synthesis of gold nanoparticles with willow tree bark extract: a sensitive colourimetric sensor for cysteine detection. Anal Methods 6(17):6916–6924
10. Baranwal A, Mahato K, Srivastava A, Maurya PK, Chandra P (2016) Phytofabricated metallic nanoparticles and their clinical applications. RSC Adv 6(107):105996–106010
11. Bhakya S, Muthukrishnan S, Sukumaran M, Grijalva M, Cumbal L, Benjamin JF, Kumar TS, Rao MV (2016) Antimicrobial, antioxidant and anticancer activity of biogenic silver nanoparticles–an experimental report. RSC Adv 6(84):81436–81446
12. Bharathi D, Diviya Josebin M, Vasantharaj S, Bhuvaneshwari V (2018) Biosynthesis of silver nanoparticles using stem bark extracts of *Diospyros montana* and their antioxidant and antibacterial activities. J Nanostr Chem 8(1):83–92
13. Burlacu E, Tanase C, Coman NA, Berta L (2019) A review of bark-extract-mediated green synthesis of metallic nanoparticles and their applications. Molecules 24(23):4354
14. Chandra Sekhar EC, Rao KK, Rao KMS, Alisha SB (2018) A simple biosynthesis of silver nanoparticles from *Syzygium cumini* stem bark aqueous extract and their spectrochemical and antimicrobial studies. J Appl Pharma Sci 8(1):73–79
15. Chung IM, Park I, Seung-Hyun K, Thiruvengadam M, Rajakumar G (2016) Plant-mediated synthesis of silver nanoparticles: their characteristic properties and therapeutic applications. Nanoscale Res Lett 11(1):1–14
16. Das M, Smita SS (2018) Biosynthesis of silver nanoparticles using bark extracts of *Butea monosperma* (Lam.) Taub. and study of their antimicrobial activity. Appl Nanosci 8(5):1059–1067
17. Dash SS, Majumdar R, Sikder AK, Bag BG, Patra BK (2014) *Saraca indica* bark extract mediated green synthesis of polyshaped gold nanoparticles and its application in catalytic reduction. Appl Nanosci 4(4):485–490
18. ElMitwalli OS, Barakat OA, Daoud RM, Akhtar S, Henari FZ (2020) Green synthesis of gold nanoparticles using cinnamon bark extract, characterization, and fluorescence activity in Au/eosin Y assemblies. J Nanoparticle Res 22(10):1–9
19. Gour A, Jain NK (2019) Advances in green synthesis of nanoparticles. Artif Cells Nanomed Biotechnol 47(1):844–851
20. Guo M, Li W, Yang F, Liu H (2015) Controllable biosynthesis of gold nanoparticles from a *Eucommia ulmoides* bark aqueous extract. Spectrochim Acta Part A Mol Biomol Spectrosc 142:73–79
21. Husen A (2017) Gold nanoparticles from plant system: synthesis, characterization and their application. In: Ghorbanpourn M, Manika K, Varma A (eds) Nanoscience and plant–soil systems, vol. Springer International Publishing AG, Gewerbestrasse 11, 6330 Cham pp 455–479
22. Husen A (2019) Natural product-based fabrication of zinc oxide nanoparticles and their application. In: Husen A, Iqbal M (eds) N*anomaterials* and plant potential. Springer International Publishing AG, Gewerbestrasse 11, 6330 Cham, pp 193–291. https://doi.org/10.1007/978-3-030-05569-1_7
23. Husen A (2020a) Introduction and techniques in nanomaterials formulation. In: Husen, A, Jawaid M (eds) Nanomaterials for agriculture and forestry applications. Elsevier Inc. 50 Hampshire St., 5th Floor, Cambridge, MA 02139, USA, pp 1–14
24. Husen A (2021) Harsh environment and plant resilience (Molecular and Functional Aspects). Springer Nature Switzerland AG, Gewerbestrasse 11, 6330 Cham, Switzerland https://doi.org/10.1007/978-3-030-65912-7
25. Husen A, Siddiqi KS (2014) Phytosynthesis of nanoparticles: Concept, controversy and application. Nanoscale Res Lett 9(229):1–24
26. Husen A, Siddiqi KS (2022) Advances in smart nanomaterials and their applications. Elsevier Inc., 50 Hampshire St., 5th Floor, Cambridge, MA 02139, USA

27. Husen A (2020b) Interactions of metal and metal-oxide nanomaterials with agricultural crops: an overview. In: Husen A, Jawaid M (eds) Nanomaterials for agriculture and forestry applications. Elsevier Inc. 50 Hampshire St., 5th Floor, Cambridge, MA 02139, USA, pp 167–197. https://doi.org/10.1016/B978-0-12-817852-2.00007-X
28. Husen A (2020c) Carbon-based nanomaterials and their interactions with agricultural crops (Eds. Husen A, Jawaid M) Elsevier Inc. 50 Hampshire St., 5th Floor, Cambridge, MA 02139, USA, pp 199–218. https://doi.org/10.1016/B978-0-12-817852-2.00008-1
29. Husen A (2022) engineered nanomaterials for sustainable agricultural production, soil improvement and stress management. Elsevier Inc., 50 Hampshire St., 5th Floor, Cambridge, MA 02139, USA
30. Husen A, Iqbal M (2019). Nanomaterials and plant potential. Springer International Publishing AG, Gewerbestrasse 11, 6330 Cham, Switzerland. https://doi.org/10.1007/978-3-030-05569-1
31. Iravan S (2011) Green synthesis of metal nanoparticles using plants. Green Chem 13(10):2638–2650
32. Iravani S, Zolfaghari B (2013). Green synthesis of silver nanoparticles using *Pinusel darica* bark extract. BioMed Res Int
33. Jeevitha M, Ravi PV, Subramaniyam V, Pichumani M, Sripathi SK (2022). Biomimetic syntheses of silver nanoparticles using *A. ferruginea* bark extracts and tenable approaches for developing anti-infectives. Appl. Nanosci. 1–16
34. Jobie FN, Ranjbar M, Moghaddam AH, Kiani M (2021) Green synthesis of zinc oxide nanoparticles using *Amygdalus scoparia Spach* stem bark extract and their applications as an alternative antimicrobial, anticancer, and anti-diabetic agent. Adv Powder Technol 32(6):2043–2052
35. Karthiga P, Rajeshkumar S, Annadurai G (2018) Mechanism of larvicidal activity of antimicrobial silver nanoparticles synthesized using *Garcinia mangostana* bark extract. J Cluster Sci 29(6):1233–1241
36. Keshavamurthy M, Srinath BS, Rai VR (2018). Phytochemicals-mediated green synthesis of gold nanoparticles using *Pterocarpus santalinus* L. (Red Sanders) bark extract and their antimicrobial properties. Particulate Sci Technol 36(7):785–790
37. Koduru JR, Kailasa SK, Bhamore JR, Kim KH, Dutta T, Vellingiri K (2018) Phytochemical-assisted synthetic approaches for silver nanoparticles antimicrobial applications: a review. Adv Colloid Interface Sci 256:326–339
38. Kumar A, Choudhary A, Kaur H, Guha S, Mehta S, Husen A (2022) Potential applications of engineered nanoparticles in plant disease management: a critical update. Chemosphere 295:133798
39. Kumar A, Choudhary A, Kaur H, Mehta S, Husen A (2021) Metal-based nanoparticles, sensors and their multifaceted application in food packaging. J Nanobiotechnol 19(256):1–25. https://doi.org/10.1186/s12951-021-00996-0
40. Kwabena DE, Wee BS, Fun CS, Kok KY, Assim ZB, Aquisman AE (2019) Comparative evaluation of antibacterial efficacy of biological synthesis of ZnO nanoparticles using fresh leaf extract and fresh stem-bark of *Carica papaya*. Nano Biomed Eng 11(3):264–271
41. Lu J, Batjikh I, Hurh J, Han Y, Ali H, Mathiyalagan R, Yang DC (2019) Photocatalytic degradation of methylene blue using biosynthesized zinc oxide nanoparticles from bark extract of *Kalopanax septemlobus*. Optik 182:980–985
42. Macovei I, Luca SV, Skalicka-Woźniak K, Sacarescu L, Pascariu P, Ghilan A, Doroftei F, Ursu EL, Rimbu CM, Horhogea CE, Lungu C (2021) Phyto-functionalized silver nanoparticles derived from conifer bark extracts and evaluation of their antimicrobial and Cytogenotoxic effects. Molecules 27(1):217–224
43. Majumdar R, Bag BG, Ghosh P (2016) *Mimusops elengi* bark extract mediated green synthesis of gold nanoparticles and study of its catalytic activity. Appl Nanosci 6(4):521–528
44. Mortazavi-Derazkola S, Yousefinia A, Naghizadeh A, Lashkari S, Hosseinzadeh M (2021) Green synthesis and characterization of silver nanoparticles using *Elaeagnus angustifolia* bark extract and study of Its antibacterial effect. J Polymers Environ 29(11):3539–3547
45. Moyo M, Gomba M, Nharingo T (2015) *Afzelia quanzensis* bark extract for green synthesis of silver nanoparticles and study of their antibacterial activity. Int J Ind Chem 6(4):329–338

46. Mustapha T, Misni N, Ithnin NR, Daskum AM, Unyah NZ (2022) A review on plants and microorganisms mediated synthesis of silver nanoparticles, role of plants metabolites and applications. Int J Environ Res Pub Health 19(2):674
47. Nabi G, Raza W, Tahir MB (2020) Green synthesis of TiO_2 nanoparticle using cinnamon powder extract and the study of optical properties. J Inorg and Organomet Polym Mater 30(4):1425–1429
48. Nadeem M, Abbasi BH, Younas M, Ahmad W, Khan T (2017) A review of the green syntheses and anti-microbial applications of gold nanoparticles. Green Chem Lett Rev 10(4):216–227
49. Nayak D, Ashe S, Rauta PR, Kumari M, Nayak B (2016) Bark extract mediated green synthesis of silver nanoparticles: evaluation of antimicrobial activity and antiproliferative response against osteosarcoma. Mater Sci Engg: C 58:44–52
50. Patil MP, Kim GD (2017) Eco-friendly approach for nanoparticles synthesis and mechanism behind antibacterial activity of silver and anticancer activity of gold nanoparticles. Appl Microbiol Biotechnol 101(1):79–92
51. Pattanayak S, MdMasud RM, Dipanwita M, Chakraborty S, Dash SK, Chattopadhyay S, Roy S, Chattopadhyay D, Chakraborty M (2017) *Butea monosperma* bark extract mediated green synthesis of silver nanoparticles: characterization and biomedical applications. J Saudi Chem Soc 21:673–684
52. Peralta-Videa JR, Zhao L, Lopez-Moreno ML, de la Rosa G, Hong J, Gardea-Torresdey JL (2011) Nanomaterials and the environment: a review for the biennium 2008–2010. J Haz Mater 186(1):1–15
53. Pinto RJ, Lucas JM, Morais MP, Santos SA, Silvestre AJ, Marques PA, Freire CS (2017) Demystifying the morphology and size control on the biosynthesis of gold nanoparticles using *Eucalyptus globulus* bark extract. Ind Crops Prod 105:83–92
54. Poojary MM, Passamonti P, Adhikari AV (2016) Green synthesis of silver and gold nanoparticles using root bark extract of *Mammea suriga*: characterization, process optimization, and their antibacterial activity. Bio Nano Sci 6(2):110–120
55. PriyaVelammal S, Devi TA, Amaladhas TP (2016) Antioxidant, antimicrobial and cytotoxic activities of silver and gold nanoparticles synthesized using *Plumbago zeylanica* bark. J Nanostr Chem 6(3):247–260
56. Ramanathan S, Gopinath SC, Anbu P, Lakshmipriya T, Kasim FH, Lee CG (2018) Eco-friendly synthesis of *Solanum trilobatum* extract-capped silver nanoparticles is compatible with good antimicrobial activities. J Molecular Str 1160:80–91
57. Rodríguez-León E, Rodríguez-Vázquez BE, Martínez-Higuera A, Rodríguez-Beas C, Larios-Rodríguez E, Navarro RE, Iñiguez-Palomares RA (2019) Synthesis of gold nanoparticles using *Mimosa tenuiflora* extract, assessments of cytotoxicity, cellular uptake, and catalysis. Nanoscale Res Lett 14(1):1–16
58. S Kumar J, Rajeshkumar S, Kumar SV (2017). Phyto-assisted synthesis, characterization and applications of gold nanoparticles–a review. Biochem Biophy Rep 11:46–57
59. Saha R, Karthik S, Balu KS, Suriyaprabha R, Siva P, Rajendran V (2018) Influence of the various synthesis methods on the ZnO nanoparticles property made using the bark extract of *Terminalia arjuna*. Mater Chem Phy 209:208–216
60. Sharma G, Nim S, Alle M, Husen A, Kim JC (2022) Nanoparticle-mediated delivery of flavonoids for cancer therapy: prevention and treatment. In: Kim JC, Alle M, Husen A (eds) Smart nanomaterials in biomedical applications. Nanotechnology in the Life Sciences. Springer, Cham. https://doi.org/10.1007/978-3-030-84262-8_3
61. Shetty P, Supraja N, Garud M, Prasad TNVKV (2014) Synthesis, characterization and antimicrobial activity of *Alstonia scholaris* bark-extract-mediated silver nanoparticles. J Nanostr Chem 4(4):161–170
62. Shreyash N, Bajpai S, Khan MA, Vijay Y, Tiwary SK, Sonker M (2021) Green synthesis of nanoparticles and their biomedical applications: a review. ACS Appl Nano Mater 4(11):11428–11457
63. Siddiqi KS, Husen A, Rao RAK (2018) A review on biosynthesis of silver nanoparticles and their biocidal properties. J Nanobiotechnol 16(14):1–28

64. Singh P, Kim YJ, Wang C, Mathiyalagan R, Yang DC (2016) The development of a green approach for the biosynthesis of silver and gold nanoparticles by using *Panax ginseng* root extract, and their biological applications. Artif Cell Nanomed Biotechnol 44(4):1150–1157
65. Suganthy N, Sri Ramkumar V, Pugazhendhi A, Benelli G, Archunan G (2018) Biogenic synthesis of gold nanoparticles from *Terminalia arjuna* bark extract: assessment of safety aspects and neuroprotective potential via antioxidant, anticholinesterase, and antiamyloidogenic effects. Environ Sci Pollut Res 25(11):10418–10433
66. Supraja N, Prasad TNVKV, Gandhi AD, Anbumani D, Kavitha P, Babujanarthanam R (2018) Synthesis, characterization and evaluation of antimicrobial efficacy and brine shrimp lethality assay of *Alstonia scholaris* stem bark extract mediated ZnONPs. Biochem Biophy Rep 14:69–77
67. Supraja N, Prasad TNVKV, Krishna TG, David E (2016) Synthesis, characterization, and evaluation of the antimicrobial efficacy of *Boswellia ovalifoliolata* stem bark-extract-mediated zinc oxide nanoparticles. Appl Nanosci 6(4):581–590
68. Tarannum N, Gautam YK (2019) Facile green synthesis and applications of silver nanoparticles: a state-of-the-art review. RSC Adv 9(60):34926–34948
69. Thakur PK, Verma V (2021) A Review on green synthesis, characterization and anticancer application of metallic nanoparticles. Appl Biochem Biotechnol 193(7):2357–2378
70. Tormena RPI, Rosa EV, Mota BDFO, Chaker JA, Fagg CW, Freire DO, Martins PM, da Silva ICR, Sousa MH (2020). Evaluation of the antimicrobial activity of silver nanoparticles obtained by microwave-assisted green synthesis using *Handroanthus impetiginosus* (Mart. ex DC.) Mattosunderbark extract. RSC Adv 10(35):20676–20681
71. Umar H, Kavaz D, Rizaner N (2019) Biosynthesis of zinc oxide nanoparticles using *Albizia lebbeck* stem bark, and evaluation of its antimicrobial, antioxidant, and cytotoxic activities on human breast cancer cell lines. Internat J Nanomed 14:87
72. Vanlalveni C, Lallianrawna S, Biswas A, Selvaraj M, Changmai B, Rokhum SL (2021) Green synthesis of silver nanoparticles using plant extracts and their antimicrobial activities: A review of recent literature. RSC Adv 11(5):2804–2837
73. Vinothkanna A, Mathivanan K, Ananth S, Ma Y, Sekar S (2022) Biosynthesis of copper oxide nanoparticles using *Rubia cordifolia* bark extract: characterization, antibacterial, antioxidant, larvicidal and photocatalytic activities. Environ Sci Pollut Res 29:1–12
74. Yahya Tahir M, Ahmad A, A Alothman, A, Mushab MSS, Ali S (2022) Green synthesis of silver nanoparticles using *Thespesia populnea* Bark extract for efficient removal of methylene blue (MB) degradation via photocatalysis with antimicrobial activity and for anticancer activity. Bioinorg Chem Appl
75. Yugandhar P, Haribabu R, Savithramma N (2015). Synthesis, characterization and antimicrobial properties of green-synthesised silver nanoparticles from stem bark extract of *Syzygium alternifolium* (Wt.) Walp. Biotech 5(6):1031–1039
76. Yugandhar P, Vasavi T, Uma P, Devi M (2017) Bioinspired green synthesis of copper oxide nanoparticles from *Syzygium alternifolium* (Wt.) Walp: characterization and evaluation of its synergistic antimicrobial and anticancer activity. Appl Nanosci 7:417–427

Green and Cost-Effective Nanomaterials Synthesis from Aquatic Plants and Their Applications

Yogita Abhale, Trupti Pagar, Rajeshwari Oza, Kun-Yi Andrew Lin, Alejandro Perez Larios, Canh Minh Vu, and Suresh Ghotekar

Abstract Aquatic bionanotechnology is one of the foremost exciting fields of research in advanced science and engineering. Although there are numerous approaches to producing diverse nanomaterials (NMs), there is growing interest in developing cost-effective, high-yield, non-toxic, and environmentally acceptable technologies. Cost-effectiveness, a decrease in the use of noxiuos chemicals, and a wealth of resource availability are the key benefits of greener synthesis. Numerous biological organisms have been employed over the past 10 years to advance new, environmentally friendly, cost-effective techniques for synthesizing NMs. Plant material-mediated NMs have a higher rate of synthesis and greater stability. Aquatic plants have not been substantially studied for the fabrication of metal/metal oxide NMs

Y. Abhale
Department of Chemistry, Government College Daman (Veer Narmad South Gujarat University, Surat), UT of DNH & DD, Daman 396210, India

T. Pagar
Department of Chemistry, G.M.D Arts, B.W Commerce and Science College, Savitribai Phule Pune University, Sinnar 422103, Maharashtra, India

R. Oza
Department of Chemistry, S.N. Arts, D.J.M. Commerce and B.N.S. Science College (Autonomous), Savitribai Phule Pune University, Sangamner 422605, Maharashtra, India

K.-Y. A. Lin
Department of Environmental Engineering and Innovation and Development Center of Sustainable Agriculture, National Chung Hsing University, 250 Kuo-Kuang Road, Taichung, Taiwan

A. P. Larios
Research Laboratory in Nanomaterials, Water and Energy, Engineering Department, University of Guadalajara, Campus Los Altos, Tepatitlán de Morelos, 47600 Jalisco, Mexico

C. M. Vu
Advanced Institute of Science and Technology, The University of Da Nang, Da Nang 550000, Vietnam

S. Ghotekar (✉)
Department of Chemistry, Smt. Devkiba Mohansinhji Chauhan College of Commerce and Science, Dadra and Nagar Haveli (UT), University of Mumbai, Silvassa 396 230, India
e-mail: ghotekarsuresh7@gmail.com

A. Husen (ed.), *Secondary Metabolites Based Green Synthesis of Nanomaterials and Their Applications*, Smart Nanomaterials Technology,
https://doi.org/10.1007/978-981-99-0927-8_15

compared to terrestrial resources. The purpose of the current chapter is to speculatively discuss the significance of using abundant aquatic resources and their assisted NPs fabrication, particularly aquatic plant-mediated NMs production, to get beyond the drawbacks of chemical and physical approaches. Finally, recent developments in more environmentally benign metal/metal oxide NMs synthesis, as well as those particles' size, morphology, distribution, and potential uses as antimicrobial, antioxidant, anticancer agents, and bio-catalytic candidates, are briefly discussed along with the future direction of research in this area.

Keywords Green synthesis · Nanomaterials · Aquatic plants · Applications

1 Introduction

Nanotechnology, one of today's cutting-edge, multidisciplinary domains, is presently getting a lot of attention and has a big impact on agriculture, the environment, and pharmacology [11, 27, 29-31, 54, 76]. Nanoparticle (NP) research has received a lot of attention because of its extremely small size (in nm) and high surface-to-volume ratio, which produce chemical and physical alterations in their features compared to other substances with the same chemical composition [23, 28, 61]. Additionally, NPs are employed in environmental remediation procedures such as removing heavy toxic metals and pollutants and detecting contamination [41, 44, 80]. Due to their nanoscale size, NPs are well suited for use in a wide range of industries, including food industry, chemistry, energy, medical, environment, forestry and agriculture sectors [8, 21, 31, 32, 41, 56, 57, 77]. Top-down and bottom-up approaches for NP amalgamation are the two prime categories (Fig. 1) [69]. The "top-to-bottom" approach involves reducing the size of the relevant bulk material using a range of techniques, including sputtering, milling, grinding, thermal/laser ablation, etc. While the "bottom-up" approaches include electrochemical procedures, chemical reductions involve the self-assembly of atoms to new nuclei, which grow into nanosized particles [11]. Because chemical and physical fabrication approaches have significant drawbacks like poor surface formation, high energy requirements, high cost of production, low production rate, and the use of noxious reducing agents, it is imperative to develop an eco-benign approach for synthesizing NPs [12, 13, 54]. Benefits of the biological synthesis process include the employment of physiologically active molecules, such as phenols, enzymes, biopolymers, and flavonoids, which operate as reducing and capping agents, large-scale manufacturing potential, and less energy use [21, 19, 43]. Diverse biological sources, including microbial biomass and plant parts, have been applied in the manufacturing process. Nevertheless, microbes-assisted fabrication has major downsides, such as maintaining cell cultures and working in an aseptic condition. On the other hand, plant-assisted bio-fabrication has been found to be the most profitable approach [26, 63, 67].

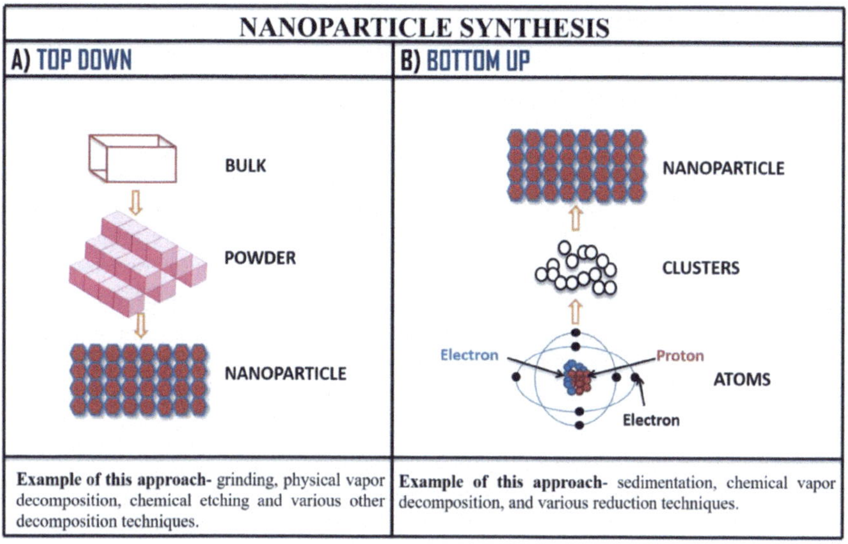

Fig. 1 Top-down and bottom-up approach of NPs synthesis (Reproduced from ref. [69], Open access)

Due to the unusual convergence of contemporary biotechnology and the nanotechnology areas, aquatic bionanotechnology is one of today's most intriguing technological fields [9, 10, 14]. Diverse aquatic entities (actinomycetes, cyanobacteria, bacteria, fungus, macro- and microalgae, soft corals, sponges, seaweeds, and plants) play fascinating roles in the environmentally benign manufacture of NPs [78]. These organisms could be used as an effective bio-reducing, capping, and stabilizing agents in one-pot synthesis of NMs, because they are frequently rich in a variety of biological constituents, including alkaloids, proteins, carbohydrates, polypeptides, polysaccharides, polyphenols, polymers, and pigments [75].

Diverse type of the NPs capable of being sourced from aquatic plants, such as Ag, Ag/AgCl, Au, Cu, Fe_3O_4, and ZnO NPs may be explored in biomedical applications including bio-labeling and bio-imaging, sensors, cancer therapy, antioxidant, antibacterial, antiviral, antidiabetic, and antifouling efficacies [49, 75, 78]. This chapter effectively assesses the effectiveness of NPs biogenically fabricated from aqueous extracts or active bio-compounds originating from aquatic plants with their characteristics, and their stupendous industrial and biomedical uses.

2 Green Synthesis of NMs

2.1 *Using Microorganisms*

Recently, interest has grown in the benefits of microorganism-based NP synthesis over conventional chemical syntheses [22] These features include the ability to survive harsh conditions, energy-efficient synthesis at temperature, consumption of less toxic chemicals and byproducts, an abundance of natural resources, and simple proportions. Microorganisms that create NPs either externally or intracellularly include bacteria, fungi, and yeast. These procedures involve the enzymatic reduction of metal ions, producing NPs with a limited size distribution that is well-dispersed [21, 36]. As naturally occurring capping agents, peptides, proteins, and DNA offer stability and prevent the agglomeration of NPs. Contrary to the extracellular approach, which involves enzymatic reduction of metal ions electrostatically linked to the microorganism's cell wall or surface, the intracellular mechanism uses metal ions that diffuse into the cell, interacting with enzymes to produce NPs [59].

2.2 *Using Plants*

Plants have developed a variety of strategies to survive in diverse environments and ecological situations, some of which take the form of secondary metabolites. More than 250 phytochemicals, including polyphenols, tannins, resveratrol, isoflavones, and anthocyanidins, are found in plant species ranging from algae to angiosperms. They differ structurally because of changes in the quantity and positions of methyl, hydroxyl, isoprenoid, and glycosidic groups, among other things [11, 21]. Plant extracts and their secondary metabolites are frequently utilized to synthesize NPs that adhere to the principles of green chemistry [33–35, 41]. Compared to microbial approaches, plant-assisted bio-fabrication of NPs is more desirable because it does not require the complex process of maintaining the cell culture. This further supports the simple scaling up of the production of NPs [54]. The successful utilization of numerous plant extracts for producing metal NPs employed in biomedical, veterinary, pharmaceutical, and environmental applications have occurred despite the large number of studies devoted to this topic [11, 12, 19, 20]. These studies on the fabrication of metallic NPs are unique in terms of morphology, size, fabrication parameters, and stability. However, there has not been much advancement in employing plant extracts in the quick production of NPs [11].

3 Green Synthesis of Nanomaterials from Aquatic Plants

In both fundamental and applied research, scientists have recently shown great interest in the study and synthesis of inorganic crystalline NPs. There are numerous approaches for producing NPs, and creating low-cost, high-yield, non-toxic, and environmentally acceptable processes are becoming more crucial. The stability and synthesis rate of metal NPs made from aquatic plant-based sources are higher [4, 11, 21]. Green fabrication of diverse NPs employing various aquatic plant extracts is summarized in Table 1 and schematic layout depicted in Fig. 2.

Aquatic environments differ greatly from terrestrial ones and are a great reservoir of myriad bioactive substances. Aquatic plants have been shown to have a high amount of secondary metabolites, including alkaloids, flavonoids, polyphenols, and tannins, as well as antibacterial, antioxidant, and anticancer properties [75, 78]. However, few studies specifically focus on synthesizing NPs using aquatic plants. *Hydrilla verticilata* have been used in a successful and environmentally benevolent one-pot biogenic fabrication of AgNPs, as reported [62]. In a different study, Kaidi et al. [38] investigated the fabrication of AgNPs employing *Laminaria ochroleuca* extract.

For example, Santhoshkumar et al. [66] described the role of the *Nelumbo nucifera* in the green production of AgNPs. In their work, 12 mL of an aqueous extract of *Nelumbo nucifera* was mixed with 85 mL of a solution of silver nitrate (1 mM). The bio-reduction of Ag^+ into Ag^0 and the bio-fabrication, capping and stabilization of AgNPs are accounted for due to the active biomolecules of *Nelumbo nucifera.*

Using the aquatic seaweed *Sargassum polycystum*, Subramanian et al. [55] synthesized silver NPs by mixing 5 mL of an aqueous extract with 95 mL of 1 mM silver nitrate at 20 °C. The reaction mixture's color altered from colorless to brown, and the formation of AgNPs was verified by SPR spectra at 405 nm. The TEM spectra later demonstrated that the NPs were 28 nm in size and had a spherical morphology. Alkaloids, proteins, flavonoids, carbohydrates, phenols, and amino acids were all suggested to have a dual function as capping and reducing agents based on FTIR studies. Another aquatic plant, *Sargassum muticum*, has been found to demonstrate AgNP capping and biosynthesis [74]. An aqueous extract of the seaweed was mixed with a solution of silver nitrate (1 mM). The bio-reduction of Ag^+ during the fabrication of AgNPs caused the mixture's color to change to brown-yellow, and SPR spectra verified the synthesis of these particles at 400 nm.

The bio-fabrication and structural properties of AuNPs using a chloroauric acid solution and an aqueous extract of the *Pistia stratiotes* were reported by Anuradha et al. [3]. The UV–Vis spectra indicated the development of AuNPs. Their crystalline structure and diameter ranged from 20 to 155 nm for spherical, triangular, hexagonal, pentagonal, and truncated triangular particles. A brown sea alga called *Turbinaria conoides* was used to biosynthesize AuNPs [60]. Notably, the XRD data showed that the particles' sizes ranged from 6 to 10 nm and confirmed the produced NPs' crystalline structure.

Table 1 Aquatic plants-mediated NPs with physicochemical parameters

NPs type	Name of aquatic plant	Characterization	Size (nm)	Shape	Application	References
Ag	*Hydrilla verticilata*	UV–Vis, FTIR, Zeta potential, SEM	65.55	Spherical	–	[62]
	Laminaria ochroleuca	UV–Vis, EDX, XRD, TEM, FTIR, TGA	10–20	Spherical	Antibacterial activity	[38]
	Nelumbo nucifera	UV–Vis, FTIR, XRD, Particle sizer, zeta potential	50	Triangular, spherical, elongated and cubic or flower	Antimicrobial and photocatalytic study	[71]
	Nelumbo nucifera	UV–Vis, FTIR, XRD, TEM, SEM	45	Spherical, triangle, truncated triangles, and decahedral	Larvicidal activity	[66]
	Nymmphaea pubescens	UV–Vis, FTIR, AFM, EDAX	66–91	Spherical	Antimicrobial activity	[70]
	Nymphaea tetragona	UV–Vis, FTIR, SEM, EDAX, AFM	88	Spherical	Antibacterial activity	[58]
	Sargassum muticum	UV–Vis, XRD, FTIR, DLS, FETEM	50	Hexagonal	Antimicrobial and anticancer study	[72]
	Sargassum muticum	UV–Vis, FTIR, XRD, TEM	5–15	Spherical	–	[6]
	Sargassum polycystum	UV–Vis, FTIR, XRD, SEM, EDX, TEM,	28	Spherical	Antioxidant and anticancer study	[55]
	Sargassum muticum	UV–vis, XRD, TEM, FTIR, EDX, Zeta potential	20–54	Spherical	Larvicidal, antibacterial and photocatalytic activity	[74]
	Ulva armoricana	UV–vis, DLS, TEM, XRD, zeta potential	33	Spherical	Antimicrobial activity	[47]

(continued)

Table 1 (continued)

NPs type	Name of aquatic plant	Characterization	Size (nm)	Shape	Application	References
	Sargassum muticum	UV–vis, TEM, zeta potential, STEM, EDX, FTIR, ICP-OES	41.0 ± 5.7	Spherical	Antibacterial and antioxidant study	[25]
Ag/AgCl	*Ulva fasciata*	UV–vis, SEM, DLS, XRD, FTIR, TEM, Zeta potential	100	Spherical	Antifungal, antioxidant and cytotoxicity study	[18]
Ag-SiO_2	*Hydrilla verticillata*	SEM, EDX, FTIR, XRD	20.1–42.6	Spherical	Preservatives properties	[40]
Au	*Sargassum muticum*	UV–vis, zeta potential, FTIR, STEM, EDX, TEM, ICP-OES	10.4 ± 1.2	Spherical	Antibacterial and antioxidant study	[25]
	Pistia stratiotes	UV–vis, FTIR, SEM, EDAX, TEM, XRD	20–155	Spherical, triangular, hexagonal, pentagonal, and truncated triangular	–	[3]
	Sargassum muticum	UV–vis, TEM, STEM, EDAX, FTIR, Zeta potential	5–100	Spherical, rods, and triangles	–	[24]
	Turbinaria conoides	UV–vis, TEM, XRD, FTIR	6–10	Pseudo-spherical	–	[60]
	Ulva armoricana	UV–vis, ICP-OES, FESEM, TEM, XRD, XPS	10	Spherical	Catalytic reduction	[50]
Cu	*Sargassum vulgare*	UV–Vis, SEM, TEM, EDX, FTIR, XRD	40–50	Spherical	Antioxidant, antimicrobial and antibiofilm activity	[46]
Fe_3O_4	*Enteromorpha spp.*	UV–Vis, SEM, EDX, FTIR	78.83	Spherical	Cu^{2+} adsorption	[17]

(continued)

Table 1 (continued)

NPs type	Name of aquatic plant	Characterization	Size (nm)	Shape	Application	References
	Padina pavonica *Sargassum acinarium*	UV–Vis, TEM, EDX, FTIR	10–19.5 21.6–27.4	Spherical	Lead bioremediation	[15]
	Petalonia fascia *Colpomenia sinuosa* *Padina pavonica*	UV–Vis, TEM, EDX, FTIR	9.42 ± 2.84 17.78 ± 3.63 31.41 ± 2.3	Spherical Cubic Spherical	Reduction of metal ions	[16]
	Sargassum muticum	XRD, FESEM, XRF, VSM, FTIR, TEM	18 ± 4	Cubic	–	[45]
SrO	*Elodea canadensis*	FESEM, EDX, FETEM, AFM, XRD, XPS, FTIR, Zeta potential	28	Spherical	Antibacterial activity	[2]
ZnO	*Sargassum muticum*	–	–	–	Antiangiogenic and antiapoptotic activity	[65]
	Sargassum muticum	FTIR, FESEM, XRD, UV–Vis	3–57	Hexagonal	–	[5]
	Sargassum vulgare	UV–Vis, TEM, SEM, XRD, EDX, DLS, zeta potential	50–150	Spherical	Antifungal, antioxidant and photocatalytic study	[39]
Zn-Se	*Gracilaria corticata*	UV–Vis, FTIR, XRD, DLS, TEM, SEM, EDX, Zeta potential	50–250	Spherical	Antioxidant, antibacterial, antibiofilm and anticancer activity	[49]

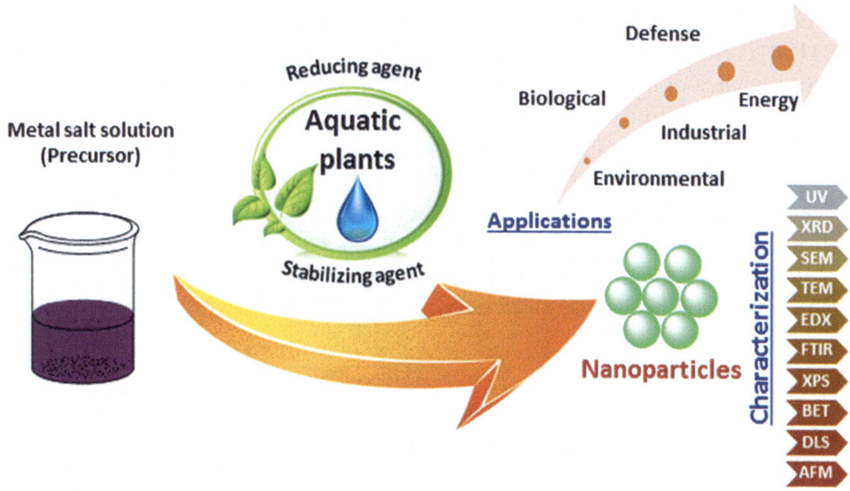

Fig. 2 Schematic layout for aquatic plant-mediated synthesis of NPs and their uses

Sargassum muticum, a marine brown alga, was used by Azizi et al. to study the formation of ZnO NPs. ZnO NPs were generated when the reaction's color altered from dark brown to pale white, and FTIR spectra suggested that the polysaccharide's hydroxyl and sulphate moieties may be the active biomolecules accountable for the bio-reduction and stability of the biogenically synthesized ZnO NPs [5].

4 Characterization of Nanomaterials

Size, morphology, and dispersity are the key characteristics used to describe synthetic NPs. Utilizing UV–visible spectroscopy, the produced NPs are initially characterized. Different metal and metal oxide NPs frequently use absorption between 200 and 800 nm. The surface charge and size distribution of the particles dispersed in a colloidal solution are characterized using the dynamic light scattering (DLS) analysis, and the stability of the particles is assessed using the Zeta Potential (ZP) analysis. Additionally, XRD analysis assesses if biologically generated and purified NPs are crystalline. In order to collect structural information, X-rays penetrated the NPs, and the ensuing diffraction pattern was compared with benchmarks. The capping behavior of active bio-molecules and potential organic functional groups (aldehyde, alcohol, amines, carbonyl, and hydroxyl groups, etc.) present in the aqueous extracts for the generation of NPs is confirmed by Fourier Transform Infrared (FTIR) spectroscopy in the range between 4000 and 400 cm^{-1}. Another often used technique for characterization is electron microscopy. Field Emission Scanning Electron Microscope (FESEM) is employed to characterize the topology of the produced NPs. Energy dispersive X-ray spectroscopy is applied to assess both the sample's quantitative and qualitative elemental makeup. High-Resolution Transmission Electron Microscopy (HRTEM) is employed to accurately measure the size and form of the NPs. The crystalline structures of the NPs and their crystalline lattice planes are obtained using a Selected Area Electron Diffraction (SAED) pattern. The size of the NPs acquired in the HRTEM picture is then measured using Image-J software, and the frequency distribution must be evaluated [21, 43].

Some aquatic plants are capable of producing metal NPs, according to earlier studies. The decrease of metal ions is caused by many functional groups found in aquatic plant extracts. As a result, it was discovered that synthesizing metal NPs using aquatic plants was successful, economical, and environmentally benign [70, 74, 75, 78].

5 Multifunctional Applications of NMs

5.1 Antimicrobial Properties

Worldwide, microbial infections are one of the worst causes of chronic diseases and are thought to be responsible for numerous fatalities. Due to their affordability and effectiveness, antibiotics have long been the treatment of choice for microbial illnesses [7]. However, overuse of antibiotics has resulted in the rise of multidrug-resistant microbial strains. There is an urgent need for innovative and successful approaches to treating microbial infections since multidrug-resistant bacteria have become a severe danger to world health [1]. As a result, numerous nanosystems with antimicrobial properties, such as biosynthesized metallic NPs, such as Ag, Cu, Au, TiO_2, ZnO and Fe_3O_4, have been thoroughly explored against multidrug-resistant pathogens alone or in synergy modules with the current and conventional antibiotics [11, 21, 30, 34, 48, 51, 64].

Numerous reports have been studied on the environmentally friendly synthesis of AgNPs with bactericidal properties using aquatic plants. For example, metallic AgNPs were greenly fabricated from the extract of *Sargassum muticum* and had strong bactericidal action against *Bacillus subtilis* (11.25 mm), *Klebsiella pneumoniae* (14.24 mm), *Escherichia coli* (13.35 mm), and *Salmonella typhi* (12.23 mm) [25]. In addition, AgNPs biosynthesized from aquatic plant extract; *Ulva armoricana* demonstrated bactericidal effect toward various types of Gram-positive and Gram-negative bacteria, including *Escherichia coli (ATCC 25,922), Pseudomonas aeruginosa* (ATCC 27,853), *Staphylococcus aureus* (ATCC 33,591), *and Staphylococcus epidermidis* (ATCC 35,984) [50].

However, due to their biocompatibility, inherent physicochemical features, chemical inertness, oxidation resistance, good scattering and absorption properties, AuNPs also possess antibacterial activities. Using the disc diffusion and microdilution method, González-Ballesteros et al. reported that the bactericidal effect of the aquatic plant *Sargassum muticum* mediated AuNPs on *Staphylococcus aureus*, *Staphylococcus saprophyticus*, and *Enterococcus faecal* [25].

Aquatic plants can environmentally synthesize other nanoparticles (NPs) beside Ag and Au, including Ag/AgCl, Ag-SiO_2, Cu, Fe_3O_4, SrO, ZnO, and Zn-Se NPs. These NPs have bactericidal performance against a variety of bacteria and fungi.

5.2 Antioxidant Performance

Antioxidants are chemicals that prevent the oxidation of natural components by scavenging reactive oxygen species (ROS) in the cells or neutralizing free radicals [79]. In this regard, antioxidants are one of the most cutting-edge frontiers in studying NMs, especially NPs. For instance, metallic oxide and metal NPs of organic materials and polymers have inherent bio-redox effects.

The DPPH assay was recently employed to assess the antioxidant activity of bio-extracted marine red algae via Ag and AuNPs. In this investigation, NPs are made to behave as electron donors, interacting with free radicals to make them more stable [25]. Similar investigations were done on the DPPH and H_2O_2 scavenging abilities of Cu NPs that were mediated by the marine seaweed *Sargassum vulgare*. For DPPH and H_2O_2, the absorbance peaks were found at 546 nm and 230 nm, respectively. The highest DPPH and H_2O_2 radicals inhibition activity is demonstrated by the Cu NPs, at 88% and 71%, respectively [46].

5.3 Anticancer Efficacy

After cardiovascular illnesses, cancer is the second biggest cause of mortality and a significant public health issue. However, successful treatment or diagnosis of cancer disorders has not yet been approved [42, 73]. Nanotheranostics refers to innovative therapeutic and diagnostic methods developed due to recent developments in nanotechnology, particularly the production and engineering of NPs. The management of cancer disorders, including drug administration, post-treatment response monitoring, imaging-guided focused therapy, drug-release monitoring, patient stratification, and tumor characterization, has recently benefited from cancer nanotheranostics.

Using the MTT assay, *Sargassum muticum*-mediated AgNPs were demonstrated to exhibit in vitro cytotoxic action against human breast cancer cells (MCF-7) AgNPs were given to MCF-7 cell lines at concentrations ranging from 3.12 to 50 ng/mL for 48 h, and NPs at a dosage of 12.5 μg were determined to be effective for 48 h [72]. Green synthesized AgNPs utilizing the aquatic plant *Sargassum polycystum* have also reportedly been reported as cytotoxic agents against human colon cancer (HT-29) cells at various doses, according to Subramanian et al. [55]. In this investigation, the IC_{50} value for AgNPs was 20 μg/mL, and as the quantity of AgNPs was increased from 5 to 50 μg/mL, the percentage of cell viability steadily dropped. Another study used the MTT assay to evaluate the anticancer properties of Zn-Se NPs mediated by *Gracilaria corticata* against the HTB-9 (ATCC 5637) and KB (ATCC CCL-17) cell lines [49]. Results showed that the HTB-9 and KB cell lines had IC_{50} values of 19.24 and 28.42 μg/mL, respectively.

5.4 Antibiofilm Study

According to several research reports, "nano-functionalized materials" that are coated, impregnated or embedded with NMs [53] prevent bacterial adherence and biofilm formation on surfaces. CuO, Ag, and ZnO, three inorganic NMs with antibacterial characteristics that prevent cell attachment and weaken the biofilm matrix, are potential sources [37]. Due to their huge surface area, which is inversely proportional to their particle size, nanostructured metal ions showed better efficiency. According to earlier studies [46, 49], surfaces of interest coated with ZnSe and Cu NPs from aquatic plants limit biofilm growth. However, the cellular processes underpinning the development of microbial biofilms are still poorly understood. The best method to manage issues brought on by the biofilm in marine environments is to target new receptors implicated in biofilm formation. Therefore, finding environmentally benign antibiofilm products from sources in the terrestrial and marine environments is urgently needed.

5.5 Environmental Remediation

In bioremediation's natural biotechnology process, live organisms break down hazardous waste and restore damaged habitats, including soil and water [52, 68]. The contamination brought on by various industrial pollutants can now be removed, recycled, and cleaned using nanotechnology. In this situation, Supraja et al. [71] conducted an experiment utilizing a reducing agent for AgNPs green synthesis that was an aqueous extract of *Nelumbo nucifera.* After 5 h of incubation, it was shown that the biosynthesized AgNPs were a functional reactant for methylene blue (MB) dye degradation. Similar research has shown that the brown seaweed *Sargassum muticum* can be used with stable AgNPs to bioremediate MB and methyl orange (MO) dye adsorption [74]. The ability of the synthesized AgNPs to remove MB and MO dye effectively suggests that they may be employed as an environmentally friendly option in wastewater treatment.

6 Concluding Remarks and Future Direction

The aquatic environment has recently received a lot of attention because it is rich in reducing and stabilizing agents like amino acids, polysaccharides, lipids, proteins, and peptides. These aquatic plants also contain valuable resources that have not yet been fully explored for the benefit of human life. An exceedingly promising field in "greener" nanotechnology, the bio-fabrication of metal NPs utilizing aquatic plants, has attracted more interest in recent years. Utilizing these numerous, freely accessible bio-resources and their bioactive components to create metal nanoparticles (NPs) with controlled size and shape is economical, capable of large-scale synthesis, and environmentally gracious. The current chapter also indicates that coating materials for aquaculture installations and other aquatic systems can be used to apply biologically manufactured, non-toxic, stable, clean, and ecologically acceptable NPs to avoid the development of fouling consortia. This can also be researched concerning pharmacological and biological uses. There is a growing desire to create environmentally friendly, greener NPs that use non-toxic compounds such as aquatic renewable resources, particularly seaweeds, to avoid negative consequences and impede future advancements in environmental and biomedical uses. Future studies on the specific features of NMs produced by aquatic plants will be crucial for biomedical sciences, pharmaceutical chemistry, and environmental sciences applications.

References

1. Ali J, Ali N, Wang L, Waseem H, Pan G (2019) Revisiting the mechanistic pathways for bacterial mediated synthesis of noble metal nanoparticles. J Microbiol Methods 159:18–25
2. Anbu P, Gopinath SCB, Salimi MN, Letchumanan I, Subramaniam S (2022) Green synthesized strontium oxide nanoparticles by Elodea canadensis extract and their antibacterial activity. J Nanostr Chem 12:365–373
3. Anuradha J, Abbasi T, Abbasi SA (2015) An eco-friendly method of synthesizing gold nanoparticles using an otherwise worthless weed pistia (Pistia stratiotes L.). J Adv Res 6:711–720
4. Arya AK, Durgapal M, Bachheti A, Deepti JKK, Gonfa YH, Bachheti RK, Husen A (2022) Ethnomedicinal use, phytochemistry, and other potential application of aquatic and semiaquatic medicinal plants. Evidence-Based Complement Alternative Med 2022:4931556. https://doi.org/10.1155/2022/4931556
5. Azizi S, Ahmad MB, Namvar F, Mohamad R (2014) Green biosynthesis and characterization of zinc oxide nanoparticles using brown marine macroalga Sargassum muticum aqueous extract. Mater Lett 116:275–277
6. Azizi S, Namvar F, Mahdavi M, Ahmad MB, Mohamad R (2013) 'Biosynthesis of silver nanoparticles using brown marine macroalga. Sargassum muticum aqueous extract. Materials 6:5942–5950
7. Balouiri M, Sadiki M, Ibnsouda SK (2016) Methods for in vitro evaluating antimicrobial activity: a review. J Pharmaceutical Anal 6:71–79
8. Barwant M, Ugale Y, Ghotekar S, Basnet P, Nguyen V-H, Pansambal S, Ananda Murthy HC, Sillanpaa M, Bilal M, Oza R (2022) Eco-friendly synthesis and characterizations of Ag/AgO/Ag2O nanoparticles using leaf extracts of Solanum elaeagnifolium for antioxidant, anticancer, and DNA cleavage activitie. Chem Papers, 1–13

9. Bekele ET, Sintayehu YD, Murthy HC, Shume MS, Ayanie GT, Turunesh DJ, Balachandran R, Tan KB, Chan K-Y, Ghotekar S (2022) Synthesis of ZnO nanoparticles mediated by natural products of Acanthus sennii leaf extract for electrochemical sensing and photocatalytic applications: a comparative study of volume ratios. Chem Pap 76:5967–5983
10. Chauhan A, Kumari S, Verma R, Dutta V, Ghotekar S, Kaur M, Kulshrestha S, Singh K, Andrew Lin K-Y, Kumar R (2022) Fabrication of copper oxide nanoparticles via microwave and green approaches and their antimicrobial potential. Chem Papers, 1–16
11. Cuong HN, Pansambal S, Ghotekar S, Oza R, Hai NTT, Viet NM, Nguyen V-H (2022) New frontiers in the plant extract mediated biosynthesis of copper oxide (CuO) nanoparticles and their potential applications: a review. Environ Res 203:111858
12. Dabhane H, Ghotekar SK, Tambade PJ, Shreyas Pansambal HC, Murthy A, Oza R, Medhane V (2021) Cow urine mediated green synthesis of nanomaterial and their applications: a state-of-the-art review. J Water Environ Nanotechnol 6:81–91
13. Dabhane H, Ghotekar S, Tambade P, Medhane V, Plant mediated green synthesis of lanthanum oxide (La2O3) nanoparticles: a review. Asian J Nanosci Mater 3:291–99
14. Dabhane H, Ghotekar S, Zate M, Kute S, Jadhav G, Medhane V (2022) Green synthesis of MgO nanoparticles using aqueous leaf extract of Ajwain (Trachyspermum ammi) and evaluation of their catalytic and biological activities. Inorg Chem Commun 138:109270
15. El-Kassas HY, Aly-Eldeen MA, Gharib SM (2016) Green synthesis of iron oxide (Fe3O4) nanoparticles using two selected brown seaweeds: characterization and application for lead bioremediation. Acta Oceanol Sin 35:89–98
16. El-Sheekh MM, El-Kassas HY, Shams NG, El-Din DI, Eissa, and Basma A El-Sherbiny. (2021) Green synthesis, characterization applications of iron oxide nanoparticles for antialgal and wastewater bioremediation using three brown algae. Int J Phytorem 23:1538–1552
17. Ercan G, Uzunoğlu D, Ergüt M, Ayla ÖZER (2019) Biosynthesis and characterization of iron oxide nanoparticles from Enteromorpha spp. extract: determination of adsorbent properties for copper (II) ions. Int Adv Res Eng J 3:65–74
18. Esmaeil L, Hamed MK, Alhameedawi AK, Marzban A (2021) Phyco-mediated synthesis of Ag/AgCl nanoparticles using ethanol extract of a marine green algae, ulva fasciata delile with biological activity. Biointerface Res App Chem
19. Ghotekar S (2019) Plant extract mediated biosynthesis of Al_2O_3 nanoparticles-a review on plant parts involved, characterization and applications. Nanochem Res 4:163–169
20. Ghotekar S, Pagar K, Pansambal S, Murthy HA, Oza R, A review on eco-friendly synthesis of BiVO4 nanoparticle and its eclectic applications
21. Ghotekar S, Pagar K, Pansambal S, Ananda Murthy HC, Oza R (2021) Biosynthesis of silver sulfide nanoparticle and its applications. In: Handbook of greener synthesis of nanomaterials and compounds (Elsevier)
22. Ghotekar S, Pansambal S, Bilal M, Pingale SS, Oza R (2021) Environmentally friendly synthesis of Cr_2O_3 nanoparticles: characterization, applications and future perspective—a review. Case Stud Chem Environ Eng 3:100089
23. Ghotekar S, Pansambal S, Andrew Lin K-Y, Pore D, Oza R (2022) Recent Advances in synthesis of CeVO4 nanoparticles and their potential scaffold for photocatalytic applications. Topics Catalysis, 1–15
24. González-Ballesteros N, Flórez-Fernández N, Torres MD, Domínguez H, Rodríguez-Argüelles MC (2021) Synthesis, process optimization and characterization of gold nanoparticles using crude fucoidan from the invasive brown seaweed Sargassum muticum. Algal Res 58:102377
25. González-Ballesteros N, Rodríguez-Argüelles MC, Lastra-Valdor M, González-Mediero G, Rey-Cao S, Grimaldi M, Cavazza A, Bigi F (2020) Synthesis of silver and gold nanoparticles by Sargassum muticum biomolecules and evaluation of their antioxidant activity and antibacterial properties. J Nanostr Chem 10:317–330
26. Hoseinpour V, Ghaemi N (2018) Green synthesis of manganese nanoparticles: applications and future perspective–a review. J Photochem Photobiol B 189:234–243
27. Huq MdA, Ashrafudoulla Md, Mizanur Rahman M, Balusamy SR, Akter S (2022) Green synthesis and potential antibacterial applications of bioactive silver nanoparticles: a review. Polymers, 14:742

28. Hussain I, Singh NB, Singh A, Singh H, Singh SC (2016) Green synthesis of nanoparticles and its potential application. Biotech Lett 38:545–560
29. Husen A (2022) Engineered nanomaterials for sustainable agricultural production, soil improvement and stress management. Elsevier Inc., 50 Hampshire St., 5th Floor, Cambridge, MA 02139, USA
30. Husen A, Iqbal M (2019b). Nanomaterials and plant potential. Springer International Publishing AG, Gewerbestrasse 11, 6330 Cham, Switzerland. https://doi.org/10.1007/978-3-030-05569-1
31. Husen A, Jawaid M (2020) Nanomaterials for agriculture and forestry applications. Elsevier Inc., 50 Hampshire St., 5th Floor, Cambridge, MA 02139, USA. https://doi.org/10.1016/C2018-0-02349-X
32. Husen A, Siddiqi KS (2022) Advances in smart nanomaterials and their applications. Elsevier Inc., 50 Hampshire St., 5th Floor, Cambridge, MA 02139, USA
33. Husen A (2017) Gold nanoparticles from plant system: synthesis, characterization and their application. In: Ghorbanpourn M, Manika K, Varma A (eds) Nanoscience and plant–soil systems, vol 48. Springer International Publishing AG, Gewerbestrasse 11, 6330 Cham pp 455–479
34. Husen A (2019) Natural product-based fabrication of zinc oxide nanoparticles and their application. In: Husen A, Iqbal M (eds) N***anomaterials*** and plant potential. Springer International Publishing AG, Gewerbestrasse 11, 6330 Cham, pp 193–291. (https://doi.org/10.1007/978-3-030-05569-1_7
35. Husen A (2020) Introduction and techniques in nanomaterials formulation. In: Husen, A, Jawaid M (eds) Nanomaterials for agriculture and forestry applications. Elsevier Inc. 50 Hampshire St., 5th Floor, Cambridge, MA 02139, USA, pp 1–14
36. Husen A, Rahman QI, Iqbal M, Yassin MO (2019) Plant-mediated fabrication of gold nanoparticles and their applications. In: Husen A, Iqbal M (eds) N***anomaterials*** and plant potential. Springer International Publishing AG, Gewerbestrasse 11, 6330 Cham, pp 71–110. https://doi.org/10.1007/978-3-030-05569-1_3
37. Ishwarya R, Vaseeharan B, Kalyani S, Banumathi B, Govindarajan M, Alharbi NS, Kadaikunnan S, Al-Anbr MN, Khaled JM, Benelli G (2018) Facile green synthesis of zinc oxide nanoparticles using Ulva lactuca seaweed extract and evaluation of their photocatalytic, antibiofilm and insecticidal activity. J Photochem Photobiol, B 178:249–258
38. Kaidi S, Belattmania Z, Bentiss F, Jama C, Reani A, Sabour B (2021) Synthesis and characterization of silver nanoparticles using alginate from the brown seaweed laminaria ochroleuca: structural features and antibacterial activity. Biointerface Res Appl Chem 12:6046–6057
39. Karkhane M, Lashgarian HE, Mirzaei SZ, Ghaffarizadeh A, Sepahvand A, Marzban A (2020) Antifungal, antioxidant and photocatalytic activities of zinc nanoparticles synthesized by Sargassum vulgare extract. Biocatal Agric Biotechnol 29:101791
40. Karthik C, Caroline DG, Dhanam Priya M, Pandi S, Prabha. (2021) Synthesis, characterization of Ag-SiO2 nanocomposite and its application in food packaging. J Inorg Organomet Polym Mater 31:2532–2541
41. Kashid Y, Ghotekar S, Bilal M, Pansambal S, Oza R, Varma RS, Nguyen V-H, Ananda Murthy HC, Mane D (2022) Bio-inspired sustainable synthesis of silver chloride nanoparticles and their prominent applications. J Indian Chem Soc, 100335
42. Khatami M, Sharifi I, Nobre MAL, Zafarnia N, Aflatoonian MR (2018) Waste-grass-mediated green synthesis of silver nanoparticles and evaluation of their anticancer, antifungal and antibacterial activity. Green Chem Lett Rev 11:125–134
43. Korde P, Ghotekar S, Pagar T, Pansambal S, Oza R, Mane D, Plant extract assisted eco-benevolent synthesis of selenium nanoparticles-a review on plant parts involved, characterization and their recent applications. J Chem Rev 2:157–68
44. Kumar R, Raizada P, Khan AAP, Nguyen V-H, Van Le Q, Ghotekar S, Selvasembian R, Gandhi V, Singh A, Singh P (2022) Recent progress in emerging BiPO4-based photocatalysts: Synthesis, properties, modification strategies, and photocatalytic applications. J Mater Sci Technol 108:208–225

45. Mahdavi M, Namvar F, Ahmad MB, Mohamad R (2013) Green biosynthesis and characterization of magnetic iron oxide (Fe3O4) nanoparticles using seaweed (Sargassum muticum) aqueous extract. Molecules 18:5954–5964
46. Marzban A, Mirzaei SZ, Karkhane M, Ghotekar SK, Danesh A (2022) Biogenesis of copper nanoparticles assisted with seaweed polysaccharide with antibacterial and antibiofilm properties against methicillin-resistant Staphylococcus aureus. J Drug Delivery Sci Technol 74:103499
47. Massironi A, Morelli A, Grassi L, Puppi D, Braccini S, Maisetta G, Esin S, Batoni G, Pina CD, Chiellini F (2019) Ulvan as novel reducing and stabilizing agent from renewable algal biomass: Application to green synthesis of silver nanoparticles. Carbohyd Polym 203:310–321
48. Mishra VK, Husen A, Rahman QI, Iqbal M, Sohrab SS, Yassin MO, Bachheti RK (2019) Plant-based fabrication of silver nanoparticles and their application. In: Husen A, Iqbal M (eds) N***anomaterials*** and Plant Potential. Springer International Publishing AG, Gewerbestrasse 11, 6330 Cham, p. 135–175. https://doi.org/10.1007/978-3-030-05569-1_5
49. Mirzaei SZ, Somaghian SA, Lashgarian HE, Karkhane M, Cheraghipour K, Marzban A (2021) Phyco-fabrication of bimetallic nanoparticles (zinc–selenium) using aqueous extract of Gracilaria corticata and its biological activity potentials. Ceram Int 47:5580–5586
50. Mukhoro OC, Roos WD, Jaffer M, Bolton JJ, Stillman MJ, Beukes DR, Antunes E (2018) Very green photosynthesis of gold nanoparticles by a living aquatic plant: photoreduction of AuIII by the seaweed Ulva armoricana. Chem Europ J 24:1657–66
51. Naikoo GA, Mustaqeem M, Hassan IU, Awan T, Arshad F, Salim H, Qurashi A (2021) Bioinspired and green synthesis of nanoparticles from plant extracts with antiviral and antimicrobial properties: a critical review. J Saudi Chem Soc 25:101304
52. Nasrollahzadeh M, Yek S-G, Motahharifar N, Gorab MG (2019) Recent developments in the plant-mediated green synthesis of Ag-based nanoparticles for environmental and catalytic applications. Chem Rec 19:2436–2479
53. Oliver S, Wagh H, Liang Y, Yang S, Boyer C (2018) Enhancing the antimicrobial and antibiofilm effectiveness of silver nanoparticles prepared by green synthesis. J Mater Chem B 6:4124–4138
54. Pagar K, Ghotekar S, Pardeshi O, Pansambal S, Pawar S, Machhi J, Pagar B (2022) Biogenic synthesis of lead-based nanoparticles and their recent applications. In: Plant and nanoparticles. Springer
55. Palanisamy S, Rajasekar P, Vijayaprasath G, Ravi G, Manikandan R, Prabhu NM (2017) A green route to synthesis silver nanoparticles using Sargassum polycystum and its antioxidant and cytotoxic effects: an in vitro analysis. Mater Lett 189:196–200
56. Pandit C, Roy A, Ghotekar S, Khusro A, Islam MN, Emran TB, Lam SE, Khandaker MU, Bradley DA (2022) Biological agents for synthesis of nanoparticles and their applications. J King Saud Univ Sci 34:101869
57. Pansambal S, Oza R, Borgave S, Chauhan A, Bardapurkar P, Vyas S, Ghotekar S (2022) Bioengineered cerium oxide (CeO2) nanoparticles and their diverse applications: a review. Appl Nanosc, 1–26
58. Prasad KS, Savithramma N (2016) Green Synthesis of Silver Bionanoparticles from Aquatic Resources to Control Bacterial Cell Proliferation. J Pharm Sci Res 8:721
59. Qureshi A, Blaisi NI, Abbas AAO, Khan NA, Rehman S (2021) Prospectus and development of microbes mediated synthesis of nanoparticles. In: Microbial nanotechnology: green synthesis and applications. Springer
60. Rajeshkumar S, Malarkodi C, Gnanajobitha G, Paulkumar K, Vanaja M, Kannan C, Annadurai G (2013) Seaweed-mediated synthesis of gold nanoparticles using Turbinaria conoides and its characterization. J Nanostr Chem 3:1–7
61. Rana A, Yadav K, Jagadevan S (2020) A comprehensive review on green synthesis of nature-inspired metal nanoparticles: Mechanism, application and toxicity. J Clean Prod 272:122880
62. Sable N, Gaikwad S, Bonde S, Gade A, Rai M (2012) Phytofabrication of silver nanoparticles by using aquatic plant Hydrilla verticillata. Nusantara Biosci, 4
63. Salem SS, Fouda A (2021) Green synthesis of metallic nanoparticles and their prospective biotechnological applications: an overview. Biol Trace Elem Res 199:344–370

64. Siddiqi KS, Husen A, Rao RAK (2018) A review on biosynthesis of silver nanoparticles and their biocidal properties. J Nanobiotechnol 16(14):1–28
65. Sanaeimehr Z, Javadi I, Namvar F (2018) Antiangiogenic and antiapoptotic effects of green-synthesized zinc oxide nanoparticles using Sargassum muticum algae extraction. Cancer nanotechnology 9:1–16
66. Santhoshkumar T, Rahuman AA, Rajakumar G, Marimuthu S, Bagavan A, Jayaseelan C, Zahir AA, Elango G, Kamaraj C (2011) Synthesis of silver nanoparticles using Nelumbo nucifera leaf extract and its larvicidal activity against malaria and filariasis vectors. Parasitol Res 108:693–702
67. Singh A, Gautam PK, Verma A, Singh V, Shivapriya PM, Shivalkar S, Sahoo AK, Samanta SK (2020) Green synthesis of metallic nanoparticles as effective alternatives to treat antibiotics resistant bacterial infections: a review. Biotechnol Reports 25:e00427
68. Singh J, Dutta T, Kim K-H, Rawat M, Samddar P, Kumar P (2018) 'Green'synthesis of metals and their oxide nanoparticles: applications for environmental remediation. J Nanobiotechnol 16:1–24
69. Singh KRB, Nayak V, Sarkar T, Singh RP (2020) Cerium oxide nanoparticles: properties, biosynthesis and biomedical application. RSC Adv 10:27194–27214
70. Siva Prasad K, Savithramma N (2015) 'Tapping of Aquatic Plant Nymphaea pubescens WILLD. For Synthesis of Silver Nanoparticles and Its Antimicrobial Evaluation. Int J Pharm Sci Rev Res 33:63–66
71. Supraja N, Avinash B, Prasad TNVKV (2017) Nelumbo nucifera extracts mediated synthesis of silver nanoparticles for the potential applications in medicine and environmental remediation. Adv Nano Res 5:373
72. Supraja N, Dhivya J, Prasad TNVKV, David E (2018) Synthesis, characterization and dose dependent antimicrobial and anticancerous efficacy of phycogenic (Sargassum muticum) silver nanoparticles against Breast Cancer Cells (MCF 7) cell line. Adv Nano Res 6:183
73. Thakur PK, Verma V (2021) A Review on green synthesis, characterization and anticancer application of metallic nanoparticles. Appl Biochem Biotechnol 193:2357–2378
74. Trivedi S, Alshehri MA, Panneerselvam C, Al-Aoh HA, Maggi F, Sut S, Dall'Acqua S (2021) Insecticidal, antibacterial and dye adsorbent properties of Sargassum muticum decorated nano-silver particles. S Afr J Bot 139:432–441
75. Vijayan SR, Santhiyagu P, Ramasamy R, Arivalagan P, Kumar G, Ethiraj K, Ramaswamy BR (2016) Seaweeds: a resource for marine bionanotechnology. Enzyme Microb Technol 95:45–57
76. Vu CM, Ghotekar S, Pham T-D, Viet NM, Oza R, Bilal M, Roy A (2022) Plant-mediated eco-friendly synthesis of platinum nanoparticles and their applications. In: Plant and Nanoparticles. Springer
77. Vu CM, Ghotekar S, Viet NM, Dabhane H, Oza R, Roy A (2022) Green synthesis of plant-assisted manganese-based nanoparticles and their various applications. In: Plant and nanoparticles. Springer
78. Yosri N, Khalifa SAM, Guo Z, Xu B, Zou X, El-Seedi HR (2021) Marine organisms: pioneer natural sources of polysaccharides/proteins for green synthesis of nanoparticles and their potential applications. Int J Biol Macromolecules 193:1767–98
79. Yousaf H, Mehmood A, Ahmad KS, Raffi M (2020) Green synthesis of silver nanoparticles and their applications as an alternative antibacterial and antioxidant agents. Mater Sci Eng, C 112:110901
80. Zhang D, Ma X-l, Gu Y, Huang H, Zhang G-W (2020) Green synthesis of metallic nanoparticles and their potential applications to treat cancer. Front Chem 8:799

Green and Cost-Effective Nanomaterials Synthesis from Desert Plants and Their Applications

Dalia G. Aseel, Said I. Behiry, and Ahmed Abdelkhalek

Abstract Nanotechnology is a growing area of biotechnology. In this connection, nanoparticles (NPs) synthesis is a significant step. Traditional ways of making NPs do not work as well as they could, so a greener way of doing it has been developed. Most of the time, plants and microorganisms are used in the green synthesis of metal NPs. Plants are increasingly being used to synthesize NPs, which have advantages over microorganisms because plants have a wide range of bio-molecular variability that can act as blocking/stabilizing agents and reducing agents, increasing the synthetic nanoparticle reduction rate and stability. Desert plants produces secondary metabolites for instance polyphenols (phenolic acids, flavonoids, and coumarins), alkaloids, steroids, saponins (spirostanol saponins, furostanol saponins, and open-chain steroidal saponins), and pregnane glycosides. Such phytochemicals derived from desert plants have many beneficial effects on human health, including antioxidants, anti-cancer, anti-diabetic, anti-inflammatory, antibacterial, hepatoprotective, and molluscicide. At the moment, very few papers discuss the use of desert plant extracts as a reducing agent in the biosynthesis of NPs. Therefore, this chapter discusses green nanomaterials' ability to be synthesized from desert plants and their applications.

Keywords Nanoparticles · Green synthesis · Desert plants · Biological applications

D. G. Aseel · A. Abdelkhalek (✉)
Plant Protection and Biomolecular Diagnosis Department, Arid Lands Cultivation Research Institute (ALCRI), City of Scientific Research and Technological Applications (SRTA-City), New Borg El Arab City 21934, Egypt
e-mail: abdelkhalek2@yahoo.com; aabdelkhalek@srtacity.sci.eg

S. I. Behiry
Agricultural Botany Department, Faculty of Agriculture (Saba Basha), Alexandria University, Alexandria 21531, Egypt

A. Husen (ed.), *Secondary Metabolites Based Green Synthesis of Nanomaterials and Their Applications*, Smart Nanomaterials Technology,
https://doi.org/10.1007/978-981-99-0927-8_16

1 Introduction

Nanotechnology is an emerging subject of inter-disciplinary study, particularly in biotechnology. Today, nanoparticle research is vital not only because of its significance but also because of the synthesis method [82]. The biosynthesis of nanoparticles (NPs), or the green synthesis of NPs, has garnered international attention in recent years [62] due to the biocompatibility, low toxicity, and eco-friendliness of the technique and NP products. Low energy requirements, moderate technology, and the absence of harmful chemicals are just a few benefits of synthesizing NP from biological agents for instance bacteria, yeasts, molds, microalgae, and plant extracts [3, 24, 68–75, 140, 141].

Additionally, it raised the desire for eco-friendly innovations through green synthesis and other biological methods. Occasionally, using diverse plant resources and their extracts to create NPs may be more advantageous than other biosynthetic processes that require extremely complex procedures for microbial culture maintenance [77]. This method employs phytochemicals in plant tissues as reducing, capping, chelating, and stabilizing agents. Polyphenols, flavonoids, ascorbic acid, alkaloids, terpenes, and reductase are all bioactive substances [98]. One more benefit is that explants can be taken from different parts of the plant, such as leaves, flowers, stems, roots, seeds, pods, and bark, while the plant's life cycle is still going on [89, 112]. Around 7,000 different plant species have been recognized for their potential therapeutic effects on human health, and this is all because of the chemical compounds they contain.

Secondary metabolites found in plants include tannins, terpenoids, alkaloids, and flavonoids, all of which have been shown to have anti-microbial, antioxidant, anticancer, and other biological effects [13, 63–72, 75]. For instance, emodin, a purgative resin containing quinone chemicals found in xerophyte plants, is responsible for silver nanoparticle formation [144]. Desert plants, also known as xerophytes, are the plants that grow in the desert. The plants that grow in one desert differ from those that grow in the other. However, the adaptation of xerophytes to thrive in extreme weather changes is a common feature of their lives. Desert plants, like desert animals, have adapted to extreme heat and aridity through physical and behavioral mechanisms. Such plants use various adaptation techniques to adapt to climatic changes. Other plants have adapted to the desert's complex, rocky, and dry soil. However, the desert is one of the most biodiverse places on Earth, with many different kinds of plants and animals living together. The vast majority of plants that can thrive in desert environments include medicinal qualities, and the native people who live there use these plants to treat a variety of illnesses [105]. There is evidence that plants native to hot deserts, such as *Leptadenia pyrotechnica*, have representatives of all of the major classes of phytochemicals, including alkaloids, flavonoids, saponins, and tannins [104]. This will open up new ways to use a plant that has been cut down as a reducing agent for biosynthesized NPs.

This chapter reviews the most up-to-date information available, offering an overview of the many techniques and desert plant materials utilized to synthesize

NPs, a discussion of the potential applications of these NPs, and an examination of the advantages and disadvantages of implementing this innovative methodology.

2 Nonmaterial Synthesis Methods

There are two primary approaches to the fabrication of NPs. First, the chemical synthesis of NPs include photochemical (utilize of photons) [133], physiochemical (c-radiation) [139], chemical reduction [61], microemulsion [60], electrochemical [45], microwave irradiation [123]. These techniques produce a large quantity of NPs but are limited by using hazardous substances and expensive, labor-intensive apparatus. The method that produces NPs most frequently among these is chemical reduction. Reducing agents in an aqueous medium include sodium borohydride, polyethylene glycol, sodium citrate, tollens, and elemental hydrogen transform metal ions (M+) into their metallic state, resulting in nanoparticle clusters [145]. Second, green biosynthesis is a valuable method that substitutes inexpensive hazardous chemicals with natural molecules like capping, stabilizing, and reducing agents. The hazardous chemicals are harmful to the environment, animals, and people. Therefore, most specialists refer to green synthesis as the bottom-up strategy, in which plant extracts (roots, leaves, fruit, and fruit peel) replace chemical compounds [3, 23, 38, 62, 67, 68, 109]. NPs have been continuously employed and changed over the years to allow their application in different industries, including agriculture and biomedicine. Due to the continued use of metallic NPs (MNPs) in various contemporary applications, the procedure for their creation must be safe to achieve greater control over the desired physicochemical and biofunctional features. For their synthesis, multiple techniques have been investigated [144]. Top-down and bottom-up strategies have been established for the fabrication of NPs. The top-down method decreases the bulk of attention to NPs, whereas the bottom-up method builds a lesser substance into the desired nanostructure [57]. The division of these tactics into physical, chemical, and biological techniques of NP production is shown in Fig. 1. Most physical techniques use a top-down approach, where bulk items are gradually cracked down to generate NPs [119]. Mechanical compression, electrical and radiation energy, melting, evaporation, and concentration are physical processes that can generate NPs. Instances contain inert gas concentration, vapor condensation, aerosol formation, laser excision, pyrolysis, and extra-energy ball grinding [119]. In the chemical processes for manufacturing NPs, reducing agents such as sodium citrate, elemental hydrogen, sodium borohydride, hydrazine, dimethylformamide, and ascorbate are utilized. The most common chemical techniques include sol–gel, hydrothermal, chemical vapor testimony, microemulsion, and polyol operations [119].

On the other hand, plant-based nanoparticle synthesis is quicker, safer, and lighter. It acts at low temperatures and only needs a few simple, safe for the environment ingredients [3, 58]. More people are interested in plant-based NPs because more people want to buy products that are good for the environment. Also, making NPs from plants has other benefits, such as using safer solvents, using less dangerous

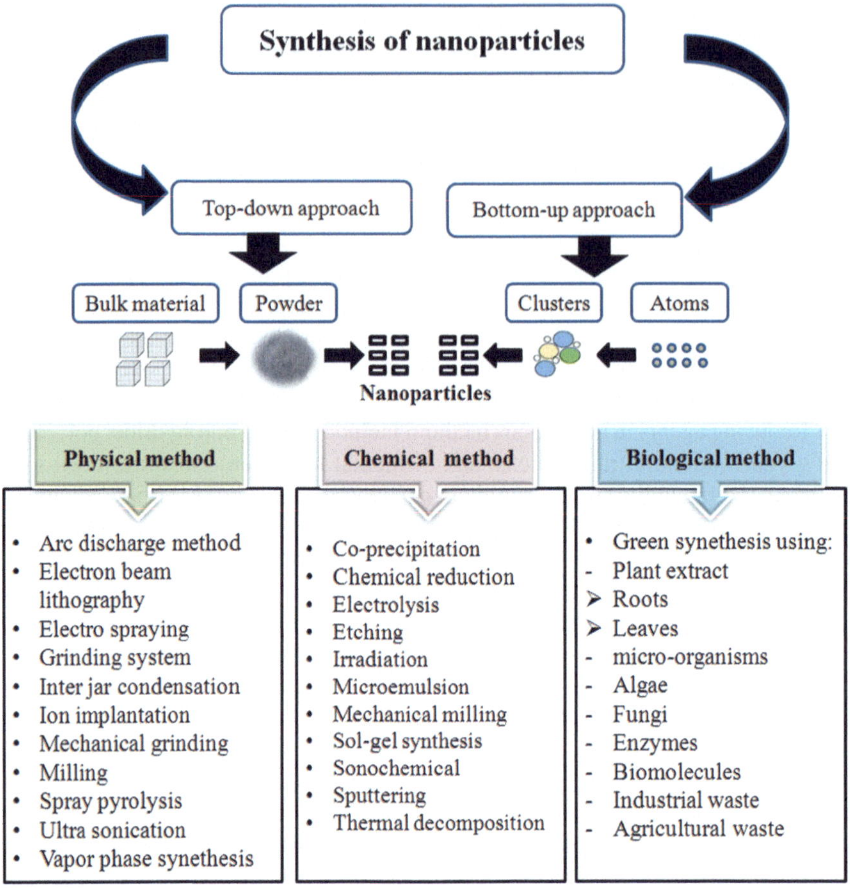

Fig. 1 The summary of the different methods used for nanoparticle synthesis, top-down and bottom-up

reagents, having milder reaction conditions, being possible, and being able to be used for medical, clinical, and pharmaceutical purposes [2].

3 Green Synthesis of NPs from Desert Plants

The green biosynthesis approach of NPs is a safe, inexpensive, and environmentally friendly alternative to conventional methods. Furthermore, using non-toxic extracts as reducing and stabilizing agents are an added benefit. Due to their composition (many bioactive compounds including polyphenols, enzymes, polysaccharides, esters, and terpenoids), plant leaf extracts can reduce and stabilize agents in nanoparticle manufacturing processes [9, 10, 62]. In this way, natural resources are protected, and opportunities for sustainable development are created at the same time. Several studies have shown that different plant species can be used to make bio NPs [5, 39, 42, 95]. Even though people are becoming more aware of plant-mediated NPs and their benefits, this area of research is still not well understood. Around the world, more than 80% of people utilize herbs and medicinal plants to treat conditions including diabetes, hypertension, cancer, and tuberculosis[94]. Also, 60% of all fabricated drugs utilized for medical purposes are made from plants which shows how important medicinal plants are [102].

In the last ten years, much research has been done on the idea of "green chemistry" for "sustainable development" [36]. Sustainable development is growth that fits the current demands devoid of hurting potential inventions' capability to serve their needs [129]. Sustainable improvement is important for many chemistry-based businesses because it cares about pollution's effects and the wasteful utilization of natural sources [114]. However, many different NPs have been made from plant parts like leaves, fruits, roots, stems, and seeds [108], Fig. 2. Plant extracts can make NPs with a specific size, shape, and chemical character. NPs created from plants are less likely than NPs made from chemicals to have negative side effects on people. They can also be employed in agriculture, food science and technology, bioengineering, cosmetics, or nanomedicine, and preserving human healthiness because they have a great biological potential (Fig. 2).

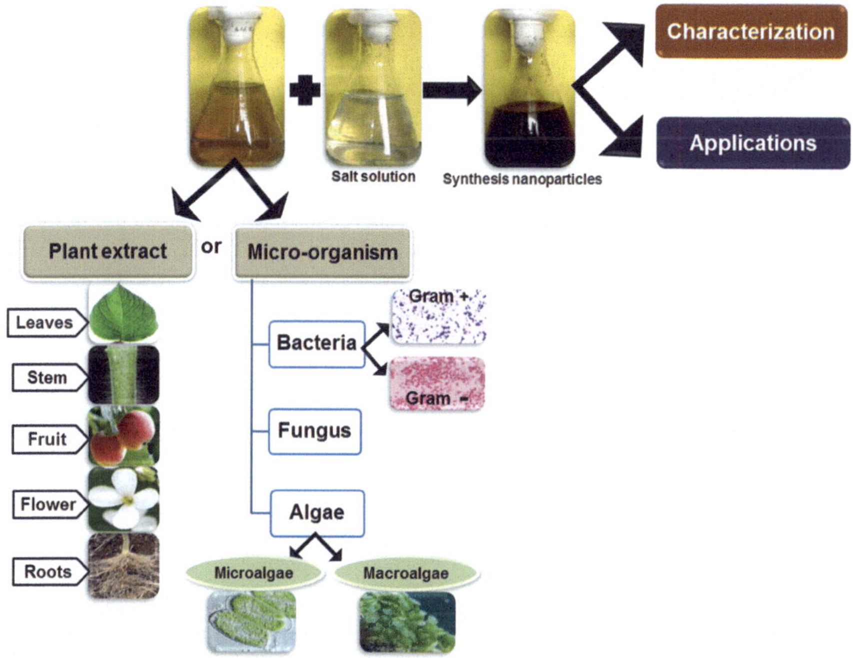

Fig. 2 Synopsis for biosynthesis, features, and functions of nanoparticles from a plant extract or produced by bacteria, fungi, and algae

The ratio of plant extracts to the chemical solution was the key factor influencing the size and stability of NPs. The majority of research supported the production of NPs from plant extracts. However, several difficulties and limits impede plant-based synthesis. Inadequacies in technicality and engineering methodology are among these constraints. In addition, economic restrictions were linked with the purity of plant extract, chemical solutions, the achievement of a reliable product with a good yield, the ratio of stoichiometric regents, and most crucially, characterization and applications. In addition, the absence of a life cycle evaluation, limits in process engineering, and operational stability are significant problems [122]. The plant-based green synthesis generates small particles suitable for different environmental and antimicrobial applications. The synthesis technique, application, and target organisms for additional studies are listed in Table 1.

Many researchers and scientists have been interested in NPs because of their unique properties. They have found that NPs have significant uses in many fields but that many nanoparticle materials are toxic at the nanoscale dimension [91]. MNPs were made from native South African plants like *Salvia africana-lutea* and *Sutherlandia frutescens* [39], *Galenia africana* [44], *Catharanthus roseus* [124], *Hypoxis hemerocallidea* [43]. Numerous indications exist that silver and gold NPs are formed in surviving plants [49, 50]. Some researchers have successfully produced MNPs using plant leaf extracts and explored their potential uses [138]. Moreover, they studied the bioreduction of chloroauric ions or silver ions by a broth of geranium (*Pelargonium graveolens*) or neem (*Azadirachta indica*) leaves [137]. In both Lemon grass (*Cymbopogon flexuosus*) and tamarind leaf extracts aggregations of NPs, such as Gold nanotriangles, have been discovered [116]. Scientists have recently proven the manufacture of silver NPs and gold nanotriangles from *Aloe vera* plant secretions. It was described that only proteins with a molecular weight of <3 kDa led to the creation of gold nanotriangles by reducing chloroaurate ions. Most NPs-plant extract experiments were utilized by boiling fresh plant leaves, while others used dried *Cinnamomum camphora* leaf mass. Treating silver ions with a *Capsicum annuum* extract expedites the synthesis of silver NPs [32].

3.1 Green Nanomaterials Characterization

To ensure uniformity in their manufacture, biological effects, and security, the NPs must be clearly and wholly featured. TEM, SEM, ATR, Raman spectroscopy, photoluminescence analysis, DLS, UV–visible spectroscopy, and FTIR are around the physicochemical techniques utilized to precisely characterize the synthesized NPs. Such methods are presented in Fig. 3.

Several studies have been done to determine what kind of metal is kept in plants. Some researchers have shown that the plants collect and store in their leaves and stems can be found as tiny pieces of natural gold metal [49]. X-ray absorption spectroscopy (XAS) and transmission electron microscopy (TEM) were used to look at *alfalfa sprouts* that had grown on agar with 320 mg/kg of gold chloride [49]. Individual NPs

Table 1 List of plant-based green synthesized nanoparticle materials (NMs) and their application

Plant species	NMs type	NMs size (nm)	Application	Key reference
Sutherlandia fructecens	ZnONPs	5–25	Antimicrobial	Aboyade et al. [4]
	AgNPs	15–20	Antibacterial and anti-cancer	[39]
Hypoxi shemerocallidea	AuNPs	9–27	Antibacterial and antiinflammatory	[43, 44]
Catharanthus roseus	AgNPs	49	Antimicrobial and wound healing	[14]
		35.55	Larvicidal	[124]
		6–25	Antimicrobial	Al-Yasiri et al. [15]
Galenia africana	AuNPs	9–27	Antibacterial	[44]
Aspalathus linearis	AuNPs	44	Antimicrobial	Joubert et al. [80]
	RhNPs	1.2		
Holoptelea integrifolia	$AgNO_3$	32–38	Anti-diabetic, Antibacterial, Anti-inflammatory	Kumar et al. [88]
Ricinus communis	Au	40–80	Antimicrobial, Anti-cancer	Ghramh et al. [55]
Moringa oleifera	Cu	35–95	Antifungal	Pagar et al. [117]
	FeO	15–21	Antibacterial	Aisida et al. [11]
Platanus orientalis	Fe_2O_3	38	Antifungal	Devi et al. [37]
Rhus punjabensis	$FeCl_3$	41–46	Disease control, Anti-cancer	Naz et al. [110]
Cucurbita seeds	ZnO_2	45–65	Larvicidal, Antibacterial, Antifungal	Velsankar et al. [153]
Solidago canadensis	Ag	100–300	Cytotoxicity	Botha et al. [27]
	Au			
	Ag-			
	Au			
Dracocephalum kotschyi	$AuCl_4$	5–21	Anti-cancer	Chahardoli et al. [31]
Aloe vera	Ag implePara>	15–15.6	Antimicrobial	Anju et al. [21]
Azadirachta indica	Ag, Au, Ag/Au	50–100	Antibacterial	Chinnasamy et al. [35]
Medicago sativa	ZnO	2–5.6	Antimicrobial	Król et al. [86]

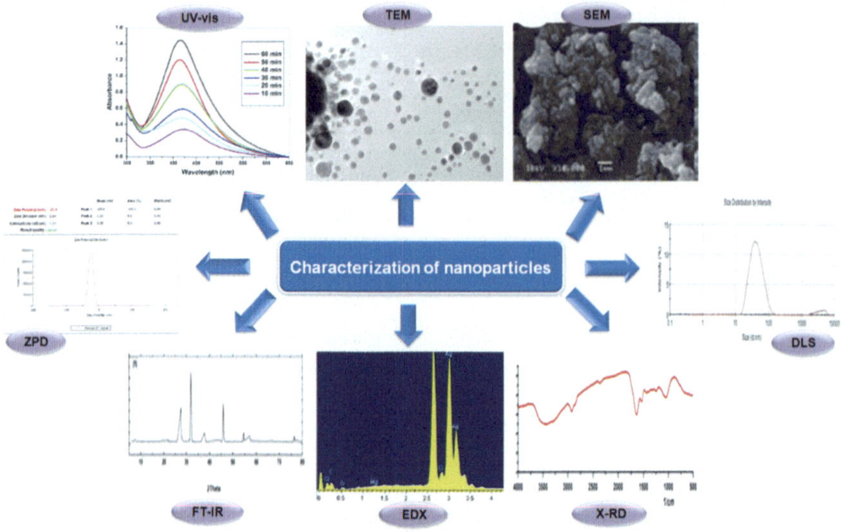

Fig. 3 Different methods of characterization of the synthesized nanoparticles

of 2–20 nm in diameter and particles that had joined together and were 20–40 nm in size were found in various parts of the plant tissue [49].

3.2 *The Advantages of Green Nanotechnology*

In recent years, the utilization of physiologically mediated synthetic procedures to create NPs has risen [155]. These protocols offer significant advantages, including (i) eco-friendly methods that do not employ harmful chemicals, (ii) being less expensive since they avoid high-pressure and high-energy costs, and (iii) capable of creating NPs of small size. Plant extracts may be used to greenly create NPs, which is advantageous since they are quick, simple, and ecologically safe. The potential for improving fertilizer, plant growth regulators, and pesticides in agriculture might be increased by nanomaterials made utilizing environmentally friendly and green technologies [53]. They also lower the number of harmful compounds that harm the environment. Because of its numerous applications in fields including medicine, the environment, and agriculture, nanotechnology has recently attracted a lot of interest [81]. As a result, it helps to reduce environmental toxins [65].

Regarding the green fabrication of NPs, the three most important things are to choose a green or ecologically approach able solvent (water, ethanol, and their mixtures are the most common), a good non-toxic reduction material, and a safe stabilizing ingredient. NPs have been made in many ways, but the most popular ones are physical, chemical, and biosynthetic. Chemical processes are usually too expensive and involve toxic and dangerous chemicals that pose several environmental risks [101]. On the other hand, green synthesis is a secure, biocompatible, and ecologically friendly way to make NPs for different uses, including biomedical ones [128]. Green chemistry can advance humanity's sustainability, security, and general safety [142]. It is a desirable method for creating, processing, and using chemicals that pose fewer risks to human health and the environment [136]. The plan calls for a thorough comprehension of the raw materials, particularly their synthesis into NPs and the resulting bioactivities that endanger people and the environment little to zero. Consequently, NPs can be fabricated using the natural resources at our disposal with the knowledge that potential hazards are reduced. Growing interest is being shown in using biological agents as reduction, capping, and stabilizing substances in the fabrication of MNPs [44]. The technique uses microorganisms and plant materials to reduce substances [25], creating benevolent and ecologically friendly circumstances that can boost the safety of people and animals [101]. The green nanomaterials have the potential to advance agriculture. Additionally, by using this technique, less dangerous compounds harm the environment [65].

3.3 The Disadvantages of Green Nanotechnology

Meanwhile, concerns have been raised concerning the possible negative consequences of NPs on biological systems and the environment, such as free radical toxicity, which causes lipid peroxidation and DNA damage. Other drawbacks include economic disruption and potential dangers to security, privacy, health, and the environment [6]. In this context, it is necessary to forecast the environmental impact of these NPs soon.

3.4 Applications of Plant-NMs

NPs are highly sought-after commercially because of their broad range of features in industries, electronics, the environment, energy, and mostly biomedical majors. NPs have been well studied in this area. They are remarkably interested in different biological applications (Fig. 4), including the most well-known Ag and Au NPs. Green NPs made from plants are less likely to cause major side effects than NPs made from chemicals. They have a wide variety of functions, involving but not restricted to those in agriculture, medicine, and the environment:

i. Nanomedicine and the protection of human health (antimicrobial, antiparasitic, antiproliferative, pro-apoptotic, pro-, or anti-oxidative varying on the perspective, anti-inflammatory effects, etc.) [18, 59, 106, 130] .

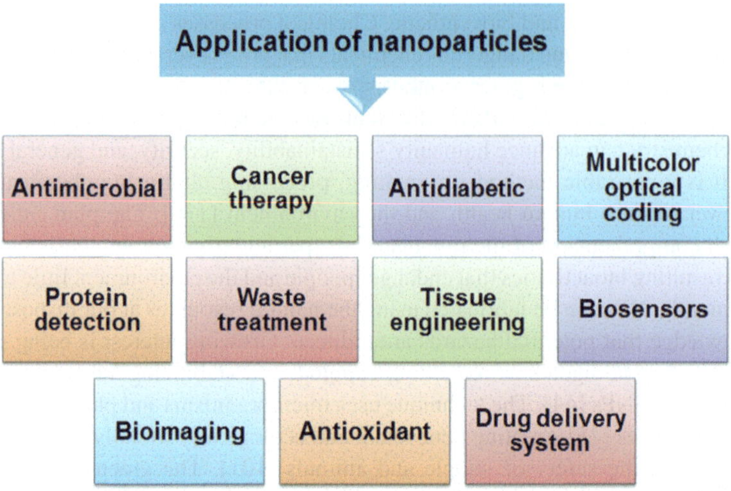

Fig. 4 Summary of the different nanoparticle applications

ii. Agriculture (accuracy farming with a controlled discharge of agrochemicals, target-certain delivery of biomolecules, more effective nutrients absorption, recognition, and control of plant infections, etc.) [17, 22].
iii. Bioengineering applications of food science and technology (processes for processing, storing, and packaging food) (biocatalysts, photocatalysts, biosensors, etc.) [135].
iv. Cosmetics (sunscreen, anti-aging, hair growing, bioactive components delivery, nano emulsion, etc.) [1].

3.4.1 Pre-clinical and Clinical Application of NMs

Nanomedicine is the application of nanotechnology to prevent, detect, and treat a wide range of diseases, including cancer. It also includes thorough processes and efficient methods for combating cancer through cancer detection and diagnosis, treatment, and possible personalized therapy [15–20]. Many plant-based NPs have some ability to inhibit cancer cells. ZnONPs created from a leaf extract of *Cassia auriculata* exhibited tumoricidal activity versus MCF-7 breast cancer cells while having no unfavorable effects on healthy MCF-12A human breast cells [121]. Compared to the bulk size NPS, the plants' silver/zinc oxide NPs are more effective as anti-cancerous compounds (Fig. 5). This makes sense given the greater accessibility of plant extracts that function as reduction and stabilizing substances in the initial phases of greener synthesis. Bimetallic NPs also slowed the proliferation of HepG-2. Because malignant cells differ from healthy cells, especially in terms of their metabolic needs, several types of cell toxicity are caused [125].

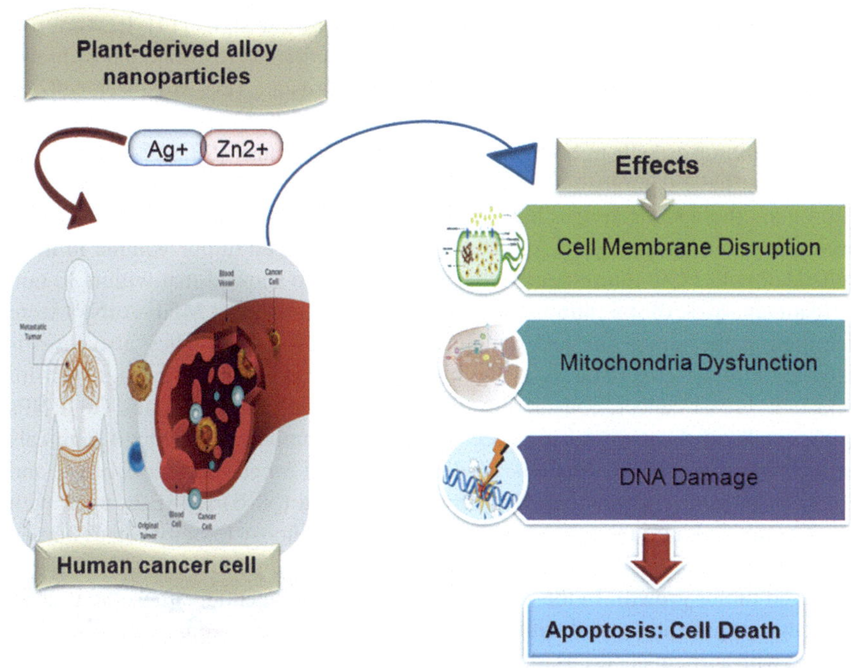

Fig. 5 impact of NPs on diseased cells managing cell death

Reactive oxygen species (ROS)-driven mortality was linked to the inhibition of HepG2 cancer cell growth by green AuNPs made from *Trachyspermum ammi* seed extract in a concentration-dependent manner [120]. According to a recent study, this technique may be linked to mitochondrial activity via ROS-induced Caspase-3 gene expression and enzyme action after plant-based NP-induced mitochondrial membrane potential disruption [22, 83]. Regardless of the emphasized ability of biosynthesized MNPs, only one of them has been rigorously assessed for in vivo applications [148, 152]. This may indicate that many MNP formulations have not fully satisfied the requirements of the pharmaceutical industry to support their use in clinical settings [16]. Nevertheless, up to this point, pre-clinical research has offered persuasive proof of the biogenic MNPs' potential for clinical use. Some biomedical applications include antibacterial, wound healing, drug delivery, and therapeutic agents [5, 14, 148]. These nanomaterials' benefits and long-standing application in treating medical conditions further justify their introduction into clinical settings [95, 150]. In fact, because of their biocompatibility, biogenic AgNPs can substitute MNPs made chemically. The biosynthesized MNPs can also destroy sick cells without attaching therapeutic or targeted moieties [95].

The capacity of biogenic *Catharanthus roseus* AgNPs to heal wounds in mice has been shown. The study also discovered that untreated mice showed indicators of microbial adulteration, pus growth, irritation, and bleeding. In contrast, mice cured with *C. roseus* AuNPs indicated no evidence of these signs [14]. Resveratrol-loaded AuNPs (RES-AuNPs) were shown to have anti-cancer action against breast (MDAMB-231), pancreatic (PANC-1) [158], and prostate (PC-3) cancer cells as well as against liver cancer in mice [96]. RES-AuNPs were more effective in inhibiting tumor growth in both studies than free resveratrol [96, 158]. Numerous experiments have demonstrated that MGF-AuNPs, either alone or in conjunction with chemotherapeutic drugs, have antitumor effects on cancer cells [5]. Within three weeks of therapy, MGF-AuNPs significantly reduced tumor volume in mice embedded with PC-3 tumor xenografts. This study also showed that within 30 min of treatment, 80% of the intratumorally dispensed radiolabeled MGF-AuNPs were identified at the tumor site and remained there for 24 h [84]. They also investigated the clinical effectiveness of the biogenic MNP articulation in patients with metastatic breast cancer, considering these positive pre-clinical results. Gold NPs synthesized with *Mangifera indica* peel extract were combined with phytochemicals from *Emblica officinalis, Mangiferaindica, Curcumin longa, Acacia nilotica,* and *Glycyrrhiza glabra* to generate Nano Swarna Bhasma (NSB) [148]. In mice with severe combined immunodeficiency, the NSB showed choosy toxicity versus human breast cancer (MDA-MB-23) and non-cancerous aortic endothelial (HAEC) cells and decreased tumor growth. In the research, significant therapeutic gains were seen in breast cancer patients treated with NSB, and these advantages were magnified when blended with anti-cancer drugs (doxorubicin and cyclophosphamide). Although NSB is the initial biogenic MNP to be utilized in human clinical studies [148], the Food and Drug Administration (FDA) has previously permitted the utilization of several other MNPs in clinical experiments. These comprise AuNPs permitted for the treatment of cancer (NCT00356980), diabetes (NCT02837094), skin (NCT02219074), and dental (NCT03669224); AgNPs approved for the treatment of skin infections (NCT03752424), and iron oxide NPs approved for the treatment of cancer (NCT02033447) and imagery (NCT02219074) (NCT04261777).

3.4.2 Agriculture Application of Nanomaterials

In the recent decade, nanotechnology in agriculture has acquired a significant pace due to the availability of community financing; yet, the current phase of improvement is satisfactory, even though various approaches have been categorized under agriculture. This could be linked to the open system aspect of agricultural production, in which energy and materials are freely transferred. Compared to industrial nanoproducts, the demand for input materials is often enormous, and unlike industrial nanoproducts, there is no control over the input of nanomaterials [103]. In addition to promising to decrease pesticide usage, nanotechnology delivers novel agrochemical substances and delivery systems to boost agricultural efficiency. Nanodevices for plant genetic engineering, plant infection diagnostics, animal healthiness, animal proliferation, poultry production, and postharvest administration are some of the functions of nanotechnology in agriculture. Other examples comprise (1) nano creations of agrochemicals for relating pesticides and fertilizers for crop development, (2) the function of nanosensors in crop defense for the documentation of illnesses and remains of chemicals, (3) nanodevices for the genetic engineering of plants, (4) plant disease diagnostics, and (5) plant disease diagnosis.

Sustainable development has advanced because of the deployment of green synthetic NMs in agricultural research. Plant propagation and genetics have also investigated nanoparticle tools. In addition, agricultural materials and/or byproducts may be utilized to create bio-nanocomposites. Using nanobiotechnology in the agro-based industry is one of the innovative techniques that can aid in the growth of a system via its growth and industrialization. The most recent developments in the manufacture of environmentally designed NPs for agricultural uses, such as gene-delivery in plant transformation and nano pesticides, nano fertilizers, and nano(bio)sensors, are shown in Fig. 6 [127].

Nano fungicides recent research has studied the use of nanosilver against the phytopathogen *Colletotrichum gloeosporioides*. In addition to their antibacterial capabilities, several NPs (Fe, Cu, Si, Al, Zn, ZnO, TiO_2, CeO_2, Al_2O3, and carbon nanotubes) were demonstrated to inhibit plant growth. Sometimes, NPs can hinder the development of helpful soil bacteria like *Pseudomonas putida* KT2440 [48]. Several research organizations have concentrated on the development of environmentally friendly pesticides. Like chemical pesticides, research is being conducted on nanoparticle-based pesticides and herbicides to apply antimicrobial compounds to shield crops from different diseases. Extensive research on nanoparticle-based technologies could reduce agriculture's requirement for pesticides. The antifungal

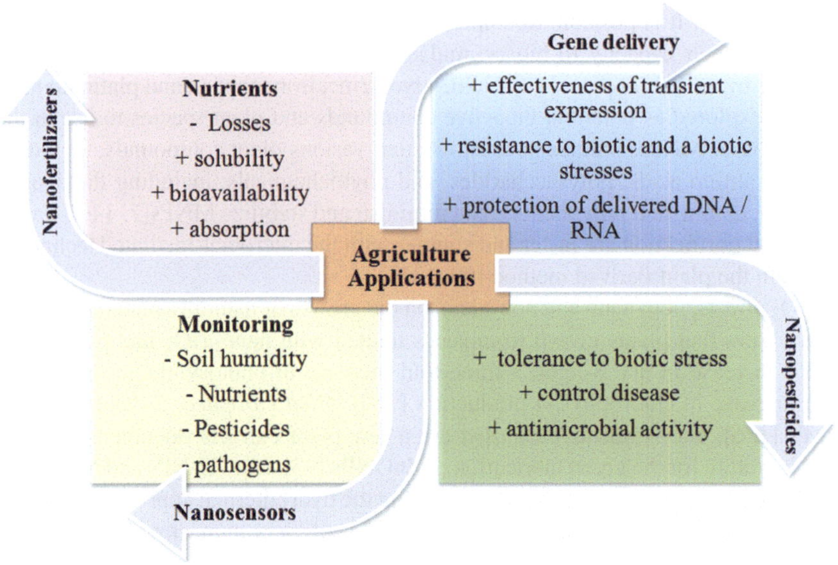

Fig. 6 Graphic interpretation of functions of nanotechnology in agriculture

capabilities of NPs can contribute to the creation of nanoparticle-based insecticides. Antimicrobial activity, like what was talked about in the last section, could be used to protect crops from agricultural pathogens. ZnONPs have shown their broad agricultural advantage by killing bacteria and fungi that harm plants. This was shown by the anti-phytopathogenic action of ZnONPs made from lemon fruit versus the soft rot bacteria pathogen *Dickeya dadantii* [64] and by the fungicidal effects of ZnONPs made from an extract of *Eucalyptus globules* against the major pathogens [8]. TiO_2NPs made from lemon fruit had the same antibacterial effect on *Dickeya dadantii* as ZnONPs [64].

Pesticide usage is expensive and not consistently successful, and the development and evolution of innovative pathogenic races is an ongoing issue. Nanomaterials have been investigated recently as an alternative method of managing plant pathogens [54]. Magnesium oxide (MgO) NPs were created in a greenhouse, and their impact on the green peach aphid (GPA) was evaluated [12]. Utilizing aqueous extracts of *Punica granatum* peels, *Olea europaea* leaves, and *Chamaemelum nobile* flowers, copper oxide (CuO), zinc oxide (ZnO), magnesium hydroxide (MgOH), and magnesium oxide (MgO) nanomaterials have been successfully created [53]. The proportion of green peach aphid mortality rose when generated bio NPs were screened. The accumulations of metal oxide NPs in the sweet green pepper fruits and leaves were investigated after the greenhouse testing. The findings revealed that there was no metal deposition in any of the plant fruits. An experiment on green pepper leaves utilizing foliar spray with produced MgOH NPs showed that spraying the leaves with 100–800 ppm metal NPs is helpful to plant growth and results in healthier plants with greener leaves and more excellent fruit quality than the control [51].

Agriculture and public health are threatened by insect pests' increased pesticide resistance [51]. It is possible to employ plants to transform metal ions into MNPs because of their capacity to bioaccumulate heavy metals. AgNPs were originally produced by an *Alfalfa* plant [99]. Gold, silver, zinc, iron, copper, and platinum have all been explored as potential bioactive compounds and plant species to use in the production of MNPs. Studies have shown that various plant compounds, including proteins, amino acids, polysaccharides, and phytochemicals, including flavonoids, alkaloids, tannin, and polyphenols, can diminish and stabilize MNPs[7, 149]. Fabrication and purification are faster and simpler with the microbial-mediated technique than with the plant-derived method [99, 131].

TiO_2 effects, both nano and non-nano, on the growth of naturally matured spinach seeds shows that plants grown from seeds treated with nano-TiO_2 had plants with a 73% increase in dry weight, a three-fold increase in photosynthetic rate, and a 45% increase in chlorophyll-a production [159]. *Olea europaea, Solanum nigrum* leaves [126], and *Azadirachta indica* are a few plant extracts documented in the open literature for the green fabrication of ZnONPs [93]. MgOH-NPs and MgO-NPs may be made using various methods, including the hydrothermal approach, water-in-oil microemulsion, and microwave reaction [157]. *Acacia* gum, *Brassica oleracea, Punica granatum* peels, Neem leaf extract, Citrus limon leaf extract, and these non-toxic, environmentally beneficial components were used to create MgOH [51, 52]. Insecticides and fertilizers based on nanotechnology that promote plant development and molecular farming using nano vectors that aim to replace viral vectors are just a few examples of how nanotechnology is being used in agriculture [34].

Spherical plant viruses are recognized as naturally occurring nanomaterials. The satellite tobacco necrosis virus has a diameter of 18 nm and is the tiniest plant virus yet discovered [76]. Plant viruses have genomes made up of RNA or DNA with one or both strands wrapped in a protein sheath. Their use in nanotechnology is necessary because of their capacity for infection, genome transfer to a precise location inside the host cell, reproduction, nucleic acid packaging, and orderly evacuation from the host cell. The utilization of plant viruses as bio prototypes for nanomaterials and their applications was recently thoroughly evaluated [156].

3.4.3 Food Science and Technology Application of Nanomaterials

The food industry dominates the manufacturing of nutrient-dense foods. High impermeable packaging nanomaterials, for example, are employed to protect food from UV radiation and provide extra strength to keep it safe from the environment. This increases the food's shelf life. Nanosensors are employed in food to identify chemicals, gases, and pathogens. Nanotechnology has recently made significant strides, revolutionizing several industries and scientific disciplines involving the food plant. In response to the increasing demand for nanoparticle applications in a variety of areas of food science and food microbiology, for instance as food processing, food packaging, functional food development, food safety, the recognition of food-borne pathogens, and the addition of food and/or food product shelf lives, as shown in Fig. 7, nanotechnology applications have emerged. This article discusses the potential uses of NPs in the food industry to give consumers clean, uncontaminated food and guarantee consumer acceptance of food with improved functional qualities. Additional information was offered on the challenges with safety and the regulatory issues with foods that have undergone nano-processing [143].

Today, food additives are transported in food items using nanocarriers as a delivery technique without changing their fundamental shape. Some cell lines may only successfully absorb submicron NPs, but not larger omicroparticles, indicating that particle size may immediately influence the distribution of any bioactive chemical to several locations throughout the body [46]. The capacity to transport the active ingredient precisely to the intended area, assures availability at a particular time and amount, and keeping active components at an acceptable level for an extended time are all desirable qualities for a delivery system (in storage condition). The creation of encapsulation, emulsions, biopolymer matrices, simple solutions, and association colloids using nanotechnology presents effective delivery methods with all the characteristics. Nano polymers are being used to try to restore traditional ingredients in food packaging. Nanosensors can identify mycotoxins, bacteria, and food contaminants [29]. Additionally, by examining how nanotechnology has improved food items in means of (i) texture, (ii) appearance, (iii) flavor, (iv) nutritional meaning, and (v) shelf life, it is possible to assess the relevance of nanotechnology in food processing.

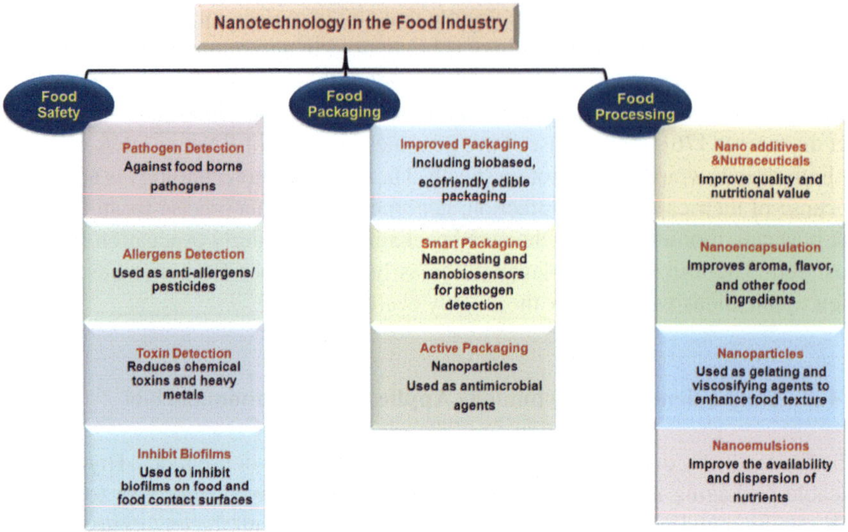

Fig. 7 Application of nanotechnology to food processing, packaging, and safety

Unexpectedly, nanotechnology significantly impacts the domains and has altered food items, giving them novel characteristics [143].

Biosensors are made with nanomaterials with elevated sensitivity levels and other unique qualities. In food microbiology, nanosensors or nano biosensors are utilized to monitor the availability of food components, identify infections in processing facilities or food material, and alert consumers and suppliers to food safety status [33, 63]. The nanosensor serves as a signal that reacts to environmental changes, such as changes in temperature or humidity in storing spaces, microbial contamination, or the deterioration of products [28]. Numerous nanostructures containing thin films, nanorods, NPs, and nanofibers, have been explored for utilization in biosensors [78]. Developing optical immunosensors based on thin films for detecting microbial elements or cells has led to quick and accurate detection methods. In these immunosensors, certain antibodies, antigens, or protein molecules are immobilized on nanofilms or sensor chips, which produce signs upon identifying objective molecules [146].

Several nanomaterials are utilized in food packaging as antibacterial agents, with silver NPs absorbing. This is due to its extensive use. Other NPs in use right now include titanium dioxide (TiO_2), zinc oxide (ZnO), silicon oxide (SiO), magnesium oxide (MgO), gold, and silver. Each has unique qualities and functions, such as zinc nanocrystal's antibacterial and antifungal properties [154]. NASA and the Russian Space Station used silver to sterilize and cleanse water [40], silver zeolite, and silver. Gold offers excellent antifungal and antibacterial activities against 150 distinct microorganisms [87, 92]as well as great temperature stability and minimal volatility. Due to silver's effectiveness in killing bacteria, the FDA approved its immediate use as a sanitizer in industrial water in 2009. Nanosilver particles covered with cellulose acetate phthalate displayed a comparable antibacterial activity versus *Escherichia coli, L. monocytogenes,* and *Staphylococcus aureus* [97, 111]. Several NPs have proven antifungal abilities. These fungi include yeast, *Aspergillus niger*, and *Candida albicans* [41, 147]. AgNPs are efficient versus methicillin-resistant *Staphylococcus aureus* as well [118]. Other NPs than silver have antibacterial properties, including titanium oxide (TiO_2). Its antibacterial action was evident in UV light. According to reports, zinc oxide has antibacterial properties in packaging materials [85]. Antibacterial activity of ZnO NPs was demonstrated using an aqueous extract of *Punicagranatum* peel against standard Gram-positive and Gram-negative like *S.aureus* and *E. coli* [56, 79].

3.4.4 Cosmetics Application of Nanomaterials

Unsurprisingly, the cosmetics sector quickly accepted nanotechnology in a marketplace where originality is typically the major reason behind business preferences. Its motivation came from its initial participation in biotechnology and bioscience. Since then, several nano-based products have been created using diverse nanomaterials with different compositions, shapes, and dimensions. These were principally selected for their capacity to conquer typical cosmetic constraints through increased penetration, improved ingredient stability, control of the release of active ingredients, and self-activation [100]. Additionally, these items' high surface-to-volume ratio and compact size enhanced the material dispersibility and textural quality [107].

The benefits of Nano cosmeceuticals are numerous. They specifically offer the measured release of active ingredients by regulating the drug release from carriers based on a diversity of variables, including the physical or chemical interaction between the compounds, the composition of the drug, polymer, and additives, as well as the proportion and preparation assay. They are utilized to cure hair loss and stop graying in hair care goods such as Identik Masque Floral Repair, Origem hair recycling shampoo, and Nirvel hair-loss control shampoo. The lifetime of scents like Chanel's Allure Parfum and Allure Eau de Parfum spray is increased using nano cosmeceuticals. These boost the potency of skin care goods and enhance sunscreens' UV safeguard, boosting their efficiency. Small particles enhance the surface area and enable active chemicals to be delivered to the skin. Occlusion facilitates increased skin hydration and enhances penetration. In addition to being more stable than traditional cosmetics, cosmetics feature strong sensory qualities and high entrapment efficiency. Most NPs are suited for medication delivery that is both lipophilic and hydrophilic. Nanomaterials are widely used to create anti-aging, hydrating, skin-bleaching, hair-restore shampoos, conditioners, and serums [113]. Figure 8 presented several advantages of nano cosmeceuticals.

Today, several cosmetic products, including the hair sunscreen Korres Hair Sun Protection Red Vine, the Nanocream® from Sinerga, and the Nano Gel from Kemira, produce translucent micellar emulsions with especially small size droplets of low

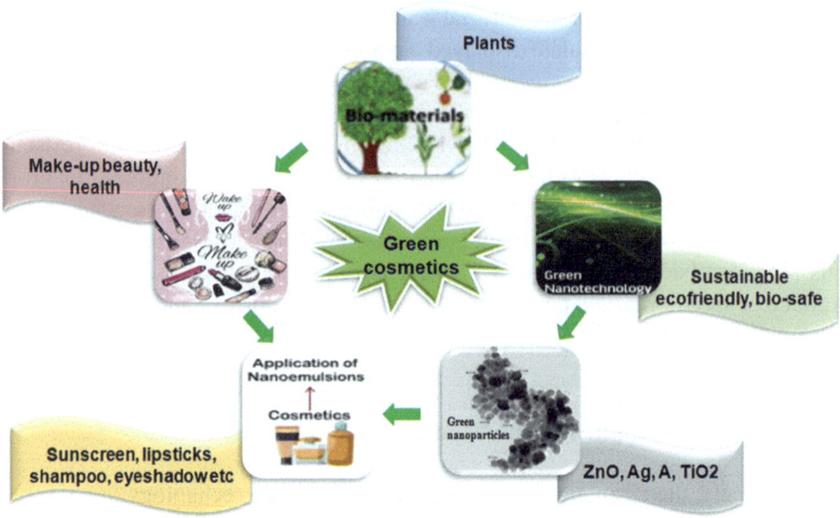

Fig. 8 Nonmaterial applications in cosmetics

viscosity and use nanoemulsions to encapsulate cosmeceuticals. This nanoformulation is well-tolerated by the customer due to its sensory features, for example, quick penetration, texture, and biophysical features, such as strong hydration capacity. Nanoemulsions are particularly helpful in moisturizing, anti-aging, and sun care products. Products like coenzyme Q10 encapsulated in proniosomes, which have stronger photoprotection than free coenzyme Q10, show the benefits of a class of nanomaterials above traditional forms available on the market. The initial indications of aging are brought on by genetic predisposition and environmental impacts on the skin [151]. Individuals age at various rates and experience aging differently. Genetic control, metabolism, way of life, and the creation of hazardous byproducts during metabolic processes all have a role in determining it. The primary factors that affect aging are broken down into four categories: (i) biological factors, which are genetically predetermined and unchangeable; (ii) environmental factors, which are influenced by lifestyle choices (such as time spent in the sun, smoking, etc.); (iii) mechanical aging, which is a result of muscle changes; and (iv) other factors, such as diet, sleeping patterns [47] (Table 2).

Currently, NPs are utilized in cosmetics; however, how many nanomaterials are used depends on how these materials are classified. Given that some nanomaterials may be more bioavailable or harmful than others, it is important to consider whether these NPs might be absorbed via the skin before making a safety judgment. A substance's stability may also play a role in safety questions, as unstable NPs are less likely to be absorbed. The FDA's post-market authority over cosmetics in the United States governs NPs used in cosmetic products [90].

The following is a list of the solid particles that are now used in skincare products:

Table 2 Examples of cosmetic products, including nanomaterials, are presently accessible on the market

Nanomaterial	Active components and functions	Brand/producer
Nanosomes™	Gingko biloba, for example, is a plant extract. Almond and lavender oils are natural substances (caffeine, amino acids, and polyphenols). High cationic charge density promotes functional, active component and fragrance deposition at the target spot and lessens rinse-off. such as Revita® shampoo and Viterol. A (Cream for Wrinkles and Expression Lines) (Anti-hair loss, hair follicle stimulus)	DS Laboratories
Liposomes	Green tea extract, vitamins A, C, and E, and coenzyme Q10. With improved UV protection and antioxidant effects from the active components, encapsulation utilizing RovisomeTM. Rovisome ACE Plus, RovisomeDefence, and Rovisome Q10, for instance	Rovi Cosmetics International GmbH
	Q10 Coenzyme increases collagen and elastin formation in connective tissue and improves antioxidant, anti-wrinkle, and penetration capabilities. e.g., Ageless Anti-Wrinkle Serum with Ageless Face Firming Serum	I-Wen Naturals
	Oils and natural extracts (*Chondrus crispus, Prunus armeniaca, Aloe barbadensis, Symphytum officinale, Rosa damascena*). Higher antioxidant effects and skin hydration are brought about by liposome encapsulation, which enhances natural product delivery to the target location for instance, Liposome Concentrate	Russell Organics

(continued)

Table 2 (continued)

Nanomaterial	Active components and functions	Brand/producer
	Sunflower oil, superoxide dismutase (SOD), vitamin E, and hyaluronic acid. Hyaluronic acid, SOD, and encapsulated vitamin E improve skin protection from oxidative stress and UV damage. Enhancement of skin renewal, e.g., Liposome Neck and Face Cream	Clinicians Complex
	Ginseng-infused liposomes preparation to reduce skin aging symptoms. Anti-Age Response Cream, for instance, Simply Man Match	Nouvelle, HSA® cosmetics
SLN and NLC	Plant oils and extracts have anti-inflammatory and anti-scaling properties that aid skin renewal; for instance, Phyto NLC Active Cell Repair	Sirehemas™
	Oil, Q10, macadamia and hemp oils, and Edelweiss extract. The encapsulation enhanced skin regeneration and suppleness. Example: Xcelent Uplift Q10	Dr. Rimpler
Nanoemulsion	Vitamin B3, provitamin B5, and E. increased absorption of the active substances that fight pollutants and promote skin renewal. For instance, Bepanthol-Protect Facial Cream Ultra	Bayer HealthCare
	UVA and UVB filters are included in clear cationic Nanoemulsions to protect hair. such as NanolipobelleTM	Mibelle Biochemistry, Switzerland
Nanocomplex with TiO2 and ZnO	Include zinc oxide particles, UV radiation blockers to protect against the sun's rays. For instance, Sheer Zinc Dry-Touch Broad Spectrum SPF 50	Neutrogena®

(continued)

Table 2 (continued)

Nanomaterial	Active components and functions	Brand/producer
	Contains vitamins, plant extracts, titanium dioxide, iron oxide, and more. They have bronze mineral pigments that provide the appearance of a tan from the sun. Example: body butter with instant bronze	Mineral Fusion®
Gold nanoparticles	With botanical (extracts and oils) and plant stem cell extracts bonded to 24 karat gold and silk, this product promotes antioxidant, smoothing, clarity, and brightness capabilities. Contains wrinkle-, wrinkle-, and redness-reducing characteristics. Skincare items (e.g., Nano Gold Energizing Cream and Nano Gold Energizing Eye Serum)	Chantecaille
	Gold powder, plant extracts (coffee seed (Coffee arabica), *aloe barbadensis*, and pumpkin seed), vitamins (A, C, and E), and hyaluronic acid are among the active substances. Product for skin care with anti-aging qualities. Examples include Nano Gold Anti-Aging Lifting Serum	Nuvoderm
Nanospheres	Vitamin E, hyaluronic acid, and oils from plants (olive, Canadian WillowherbTM, green tea) provide anti-inflammatory and anti-wrinkle properties and encourage the production of collagen and fibroblasts. e.g., Tender Eyes	Kara Vita

(continued)

1. Owing to their exceptional stability, high loading capability, and ease of morphological feature modification, gold NPs have a wide variety of functions, including colorant, anti-aging, preservative, and carrier [30].
2. Broad-spectrum antibacterial and antifungal substances, pigments, and aesthetic uses of silver NPs have all been thoroughly explored [132].

Table 2 (continued)

Nanomaterial	Active components and functions	Brand/producer
	Vitamin E, natural ingredients (soy firming agent), and UVA/UVB filters are present. The active components are released gradually via nanospheres. The mixture supports skin restoration on damaged tissues, anti-inflammatory and moisturizing properties, and UV radiation defense. The photo-reflecting substance gives the skin brightness. E.g., Cream DNA Filler Intense	CellAct Switzerland

3. Using the sol–gel and precipitation techniques, titanium oxide NPs and zinc oxide NPs are produced. Due to their capability to signify and scatter UV-A and UV-B light, UV inorganic filters [134].
4. According to the European Commission (2019), amorphous silica makes up most of the items that fall under silica NPs. They are used in rinse-off and leave-on cosmetics to enhance the product's texture and/or to offer the area where the product is applied a matte (i.e., opaque) finish.
5. Due to copper's crucial function in developing skin extracellular matrix and its biocidal effects, copper-based nanomaterials have been explored for anti-aging and biocidal aims [26]. Their purpose is still up for debate, and cosmetics seldom use them.
6. While aluminum oxide NPs are on the list of permitted nanomaterials by the EU, they are utilized in cosmetics rather than skincare products (such as mineral foundations, concealers, and foundations) [115]. The first two nanotubes have been utilized in skincare. However, the third nanotube-along with other non-harmful and less expensive ones (such as naturally occurring halloysite)—is frequently employed as a pigment carrier in haircare [133].

4 Conclusion and Future Perspective

Green synthesis technology offers a clean, non-toxic, and environmentally acceptable method for the synthesis of NPs. This technology is of significant interest due to the economic prospects and the fact that it is feasible. Desert plants have an incredible capacity for optimizing their production of proteins, enzymes, and other biomolecules, which are necessary for biosynthesis, stabilizing NPs, and producing a larger quantity of these substances. We believe that using them as sustainable routes for synthesis and creation of various types of NPs is the future approach to improving metal nanoparticle production by adopting the eco-friendly green synthesis approach for more advanced applications. This will make it possible for us to make more NPs that are better and in larger quantities. However, additional protocol modifications are required to make these technologies competitive with conventional approaches to the mass manufacturing of NPs in terms of cost and efficiency. Moreover, most of these approaches are in their formative stages, so there are still obstacles to overcome. These include controlling crystal formation, shape, and size and preventing NPs from aggregating. Further investigation into nanoparticle purification and isolation is also crucial. In conclusion, researchers and people in charge of making decisions should think about the direct and indirect ways greener technologies could improve nanoparticle production, performance, and competitiveness.

The generation of NPs using green synthesis has just lately become widespread. Prior to the widespread application of green NPs generated using several techniques, fine-tuning the final characteristics is undoubtedly the most common and difficult challenge to resolve. Indeed, the possibilities for the future need to be discussed in light of:

- Analysis of the potential of desert plants from various landmasses and ecosystems to provide compounds that can be used in the synthesis.
- Research into the reduction, nucleation, growth, and stability of organ, tissue, cell, or organelle specificity.
- Consideration of intraspecific, physiological, seasonal, and geographical variation in formation profile.
- Assessment of biotic and abiotic influences on plant species employed in the generation process.
- Analyzing how a population's genetic history influences its propensity for effective offspring development.
- Desert plant metabolic pathways must be modulated to regulate species' physical, chemical, and structural properties.
- The synergy of molecular components from various species during synthesis and its possible evaluation.
- Molecular characterization of constituents related to chemical reduction, nucleation, growth, and stabilization to be identified.
- Use of high-throughput and combinatorial methods to determine which compounds should be used.
- Characterization and biological assay methods should be consistent for meaningful comparisons to be made.
- Adsorption and absorption by various surfaces and materials have been studied, along with their kinetics, dynamics, and reactivity.
- Screening for effects and mechanisms of action in a variety of species using cutting-edge methods.
- Green synthesis vs. traditional chemical/physical synthesis: an assessment of biological consequences.

- Research into the effects on a wide variety of plant and animal species to serve as examples of potential future use.
- Analysis of unusual uses in fields including forensics, electronics, and other niche areas.

References

1. Abbasi BH, Fazal H, Ahmad N, et al (2020) Nanomaterials for cosmeceuticals: nanomaterials-induced advancement in cosmetics, challenges, and opportunities. Nanocosmetics 79–108
2. Abdelhalim MAK, Jarrar BM (2011) Gold nanoparticles administration induced prominent inflammatory, central vein intima disruption, fatty change and Kupffer cells hyperplasia. Lipids Health Dis 10:133. https://doi.org/10.1186/1476-511X-10-133
3. Abdelkhalek A, Al-Askar AA (2020) Green synthesized ZnO nanoparticles mediated by mentha spicata extract induce plant systemic resistance against tobacco mosaic virus. Appl Sci 10:5054
4. Aboyade OM, Styger G, Gibson D, Hughes G (2014) Sutherlandia frutescens: the meeting of science and traditional knowledge. J Altern Complement Med 20:71–76
5. Aboyewa JA, Sibuyi NRS, Meyer M, Oguntibeju OO (2021) Gold nanoparticles synthesized using extracts of cyclopia intermedia, commonly known as honeybush, amplify the cytotoxic effects of doxorubicin. Nanomaterials 11:132
6. Agrawal S, Rathore P (2014) Nanotechnology pros and cons to agriculture: a review. Int J Curr Microbiol App Sci 3:43–55
7. Ahmad T, Bustam MA, Irfan M et al (2019) Mechanistic investigation of phytochemicals involved in green synthesis of gold nanoparticles using aqueous Elaeis guineensis leaves extract: Role of phenolic compounds and flavonoids. Biotechnol Appl Biochem 66:698–708
8. Ahmad H, Venugopal K, Rajagopal K et al (2020) Green synthesis and characterization of zinc oxide nanoparticles using Eucalyptus globules and their fungicidal ability against pathogenic fungi of apple orchards. Biomolecules 10:425
9. Ahmed S, Saifullah AM et al (2016) Green synthesis of silver nanoparticles using Azadirachta indica aqueous leaf extract. J Radiat Res Appl Sci 9:1–7
10. Ahmeda A, Zangeneh A, Zangeneh MM (2020) Green synthesis and chemical characterization of gold nanoparticle synthesized using Camellia sinensis leaf aqueous extract for the treatment of acute myeloid leukemia in comparison to daunorubicin in a leukemic mouse model. Appl Organomet Chem 34:e5290
11. Aisida SO, Madubuonu N, Alnasir MH et al (2020) Biogenic synthesis of iron oxide nanorods using Moringa oleifera leaf extract for antibacterial applications. Appl Nanosci 10:305–315
12. Al-Antary T, Al-Dabbas M, Shaderma A (2014) Effect of UV-radiation on methomyl, oxamyl, and carbosulfan residues in tomato juice. Fresen Environ Bull 23:924–928
13. Al-Saleem MS, Awaad AS, Alothman MR, Alqasoumi SI (2018) Phytochemical standardization and biological activities of certain desert plants growing in Saudi Arabia. Saudi Pharm J 26:198–204
14. Al-Shmgani HSA, Mohammed WH, Sulaiman GM, Saadoon AH (2017) Biosynthesis of silver nanoparticles from Catharanthus roseus leaf extract and assessing their antioxidant, antimicrobial, and wound-healing activities. Artif Cells Nanomedi Biotechnol 45:1234–1240
15. Al-Yasiri AY, Khoobchandani M, Cutler CS et al (2017) Mangiferin functionalized radioactive gold nanoparticles (MGF-198 AuNPs) in prostate tumor therapy: green nanotechnology for production, in vivo tumor retention and evaluation of therapeutic efficacy. Dalt Trans 46:14561–14571
16. Alphandéry E (2020) Natural metallic nanoparticles for application in nano-oncology. Int J Mol Sci 21:4412

17. Anjum S, Anjum I, Hano C, Kousar S (2019) Advances in nanomaterials as novel elicitors of pharmacologically active plant specialized metabolites: Current status and future outlooks. RSC Adv 9:40404–40423
18. Anjum S, Hashim M, Malik SA et al (2021a) Recent advances in zinc oxide nanoparticles (Zno nps) for cancer diagnosis, target drug delivery, and treatment. Cancers (Basel) 13:4570
19. Anjum S, Ishaque S, Fatima H et al (2021b) Emerging applications of nanotechnology in healthcare systems: Grand challenges and perspectives. Pharmaceuticals 14:707
20. Anjum S, Khan AK, Qamar A et al (2021c) Light Tailoring: Impact of UV-C Irradiation on Biosynthesis, Physiognomies, and Clinical Activities of Morus macroura-Mediated Monometallic (Ag and ZnO) and Bimetallic (Ag–ZnO) Nanoparticles. Int J Mol Sci 22:11294
21. Anjum S, Komal A, Abbasi BH, Hano C (2021) Nanoparticles as elicitors of biologically active ingredients in plants. Nanotechnol Plant Growth Promot Prot Recent Adv Impacts 170–202
22. Anjum S, Komal A, Abbasi BH, Hano C (2021d) Nanoparticles as elicitors of biologically active ingredients in plants. Nanotechnol Plant Growth Promot Prot Recent Adv Impacts 170–202
23. Arumugam A, Karthikeyan C, Hameed ASH et al (2015) Synthesis of cerium oxide nanoparticles using Gloriosa superba L. leaf extract and their structural, optical and antibacterial properties. Mater Sci Eng C 49:408–415
24. Bachheti RK, Abate L, Deepti, Bachheti A, Madhusudhan A, Husen H (2021) Algae-, fungi-, and yeast-mediated biological synthesis of nanoparticles and their various biomedical applications. In: Kharisov B, Kharissova O (Eds) Handbook of Greener Synthesis of Nanomaterials and Compounds, Volume 1: Fundamental Principles and Methods. Elsevier Inc. 50 Hampshire St., 5th Floor, Cambridge, MA 02139, USA, pp. 701–734 https://doi.org/10.1016/B978-0-12-821938-6.00022-0
25. Bar H, Bhui DK, Sahoo GP et al (2009) Green synthesis of silver nanoparticles using latex of Jatropha curcas. Colloids surfaces A Physicochem Eng Asp 339:134–139
26. Borkow G (2014) Using copper to improve the well-being of the skin. Curr Chem Biol 8:89–102
27. Botha TL, Elemike EE, Horn S et al (2019) Cytotoxicity of Ag, Au and Ag-Au bimetallic nanoparticles prepared using golden rod (Solidago canadensis) plant extract. Sci Rep 9:1–8
28. Bouwmeester H, Dekkers S, Noordam MY et al (2009) Review of health safety aspects of nanotechnologies in food production. Regul Toxicol Pharmacol 53:52–62
29. Bratovčić A, Odobašić A, Ćatić S, Šestan I (2015) Application of polymer nanocomposite materials in food packaging. Croat J food Sci Technol 7:86–94
30. Cao M, Li J, Tang J et al (2016) Gold nanomaterials in consumer cosmetics nanoproducts: Analyses, characterization, and dermal safety assessment. Small 12:5488–5496
31. Chahardoli A, Karimi N, Fattahi A, Salimikia I (2019) Biological applications of phytosynthesized gold nanoparticles using leaf extract of Dracocephalum kotschyi. J Biomed Mater Res Part A 107:621–630
32. Chandran SP, Chaudhary M, Pasricha R et al (2006) Synthesis of gold nanotriangles and silver nanoparticles using Aloevera plant extract. Biotechnol Prog 22:577–583
33. Cheng Q, Li C, Pavlinek V et al (2006) Surface-modified antibacterial TiO2/Ag+ nanoparticles: Preparation and properties. Appl Surf Sci 252:4154–4160
34. Chinnamuthu CR, Boopathi PM (2009) Nanotechnology and agroecosystem. Madras Agric J 96:17–31
35. Chinnasamy G, Chandrasekharan S, Koh TW, Bhatnagar S (2021) Synthesis, characterization, antibacterial and wound healing efficacy of silver nanoparticles from Azadirachta indica. Front Microbiol 12:611560
36. Clark JH, Macquarrie DJ (2008) Handbook of green chemistry and technology. John Wiley & Sons
37. Devi HS, Boda MA, Shah MA et al (2019) Green synthesis of iron oxide nanoparticles using Platanus orientalis leaf extract for antifungal activity. Green Process Synth 8:38–45

38. Dhillon GS, Brar SK, Kaur S, Verma M (2012) Green approach for nanoparticle biosynthesis by fungi: current trends and applications. Crit Rev Biotechnol 32:49–73
39. Dube P, Meyer S, Madiehe A, Meyer M (2020) Antibacterial activity of biogenic silver and gold nanoparticles synthesized from Salvia africana-lutea and Sutherlandia frutescens. Nanotechnology 31:505607
40. Duncan TV (2011) Applications of nanotechnology in food packaging and food safety: barrier materials, antimicrobials and sensors. J Colloid Interface Sci 363:1–24
41. Egger S, Lehmann RP, Height MJ et al (2009) Antimicrobial properties of a novel silver-silica nanocomposite material. Appl Environ Microbiol 75:2973–2976
42. Elbagory AM, Cupido CN, Meyer M, Hussein AA (2016) Large scale screening of southern African plant extracts for the green synthesis of gold nanoparticles using microtitre-plate method. Molecules 21:1498
43. Elbagory AM, Hussein AA, Meyer M (2019) The in vitro immunomodulatory effects of gold nanoparticles synthesized from Hypoxis hemerocallidea aqueous extract and hypoxoside on macrophage and natural killer cells. Int J Nanomedicine 14:9007
44. Elbagory AM, Meyer M, Cupido CN, Hussein AA (2017) Inhibition of bacteria associated with wound infection by biocompatible green synthesized gold nanoparticles from South African plant extracts. Nanomaterials 7:417
45. Elemike EE, Onwudiwe DC, Fayemi OE, Botha TL (2019) Green synthesis and electrochemistry of Ag, Au, and Ag–Au bimetallic nanoparticles using golden rod (Solidago canadensis) leaf extract. Appl Phys A 125:1–12
46. Ezhilarasi PN, Karthik P, Chhanwal N, Anandharamakrishnan C (2013) Nanoencapsulation techniques for food bioactive components: a review. Food Bioprocess Technol 6:628–647
47. Farage MA, Miller KW, Elsner P, Maibach HI (2007) Structural characteristics of the aging skin: a review. Cutan Ocul Toxicol 26:343–357
48. Gajjar P, Pettee B, Britt DW et al (2009) Antimicrobial activities of commercial nanoparticles against an environmental soil microbe, Pseudomonas putida KT2440. J Biol Eng 3:1–13
49. Gardea-Torresdey JL, Gomez E, Peralta-Videa JR et al (2003) Alfalfa sprouts: a natural source for the synthesis of silver nanoparticles. Langmuir 19:1357–1361
50. Gardea-Torresdey JL, Rodriguez E, Parsons JG et al (2005) Use of ICP and XAS to determine the enhancement of gold phytoextraction by Chilopsis linearis using thiocyanate as a complexing agent. Anal Bioanal Chem 382:347–352
51. Ghidan AY, Al-Antary TM, Awwad AM (2016) Green synthesis of copper oxide nanoparticles using Punica granatum peels extract: Effect on green peach Aphid. Environ Nanotechnol Monit Manag 6:95–98
52. Ghidan AY, Al-Antary TM, Awwad AM, Akash MW (2017a) Aphidicidal potential of green synthesized magnesium hydroxide nanoparticles using Olea europaea leaves extract. ARPN J Agric Biol Sci 12:293–301
53. Ghidan AY, Al-Antary TM, Salem NM, Awwad AM (2017b) Facile green synthetic route to the zinc oxide (ZnONPs) nanoparticles: Effect on green peach aphid and antibacterial activity. J Agric Sci 9:131–138
54. Ghidan AY, Al-Antary TM, Awwad AM, Ayad JY (2018) Physiological effect of some nanomaterials on pepper (Capsicum annuum L.) plants. Fresenius Environ Bull 27:7872–7878
55. Ghramh HA, Khan KA, Ibrahim EH, Setzer WN (2019) Synthesis of gold nanoparticles (AuNPs) using Ricinus communis leaf ethanol extract, their characterization, and biological applications. Nanomaterials 9:765
56. Gold K, Slay B, Knackstedt M, Gaharwar AK (2018) Antimicrobial activity of metal and metal-oxide based nanoparticles. Adv Ther 1:1700033
57. Gomes JF, Garcia AC, Ferreira EB et al (2015) New insights into the formation mechanism of Ag, Au and AgAu nanoparticles in aqueous alkaline media: alkoxides from alcohols, aldehydes and ketones as universal reducing agents. Phys Chem Chem Phys 17:21683–21693
58. Goodsell DS (2004) Bionanotechnology: lessons from nature. John Wiley & Sons
59. Gul R, Jan H, Lalay G et al (2021) Medicinal plants and biogenic metal oxide nanoparticles: a paradigm shift to treat Alzheimer's disease. Coatings 11:717

60. Guo Q, Li Y, Lou Y et al (2019) Bacillus amyloliquefaciens Ba13 induces plant systemic resistance and improves rhizosphere microecology against tomato yellow leaf curl virus disease. Appl Soil Ecol 137:154–166
61. Guzmán MG, Dille J, Godet S (2009) Synthesis of silver nanoparticles by chemical reduction method and their antibacterial activity. Int J Chem Biomol Eng 2:104–111
62. Heflish AA, Hanfy AE, Ansari MJ et al (2021) Green biosynthesized silver nanoparticles using Acalypha wilkesiana extract control root-knot nematode. J King Saud Univ 33:101516
63. Helmke BP, Minerick AR (2006) Designing a nano-interface in a microfluidic chip to probe living cells: challenges and perspectives. Proc Natl Acad Sci 103:6419–6424
64. Hossain A, Abdallah Y, Ali MA et al (2019) Lemon-fruit-based green synthesis of zinc oxide nanoparticles and titanium dioxide nanoparticles against soft rot bacterial pathogen Dickeya dadantii. Biomolecules 9:863
65. Huang S, Wang L, Liu L et al (2015) Nanotechnology in agriculture, livestock, and aquaculture in China. A review. Agron Sustain Dev 35:369–400
66. Husen A, Siddiqi KS (2014) Phytosynthesis of nanoparticles: Concept, controversy and application. Nanoscale Res Lett 9(229):1–24
67. Husen A (2017) Gold nanoparticles from plant system: synthesis, characterization and their application, In: Nanoscience and plant–soil systems Vol.–48 (Eds. Ghorbanpourn M, Manika K, Varma A) Springer International Publishing AG, Gewerbestrasse 11, 6330 Cham pp. 455–479
68. Husen A (2019) Natural product-based fabrication of zinc oxide nanoparticles and their application, *In*: N*anomaterials* and Plant Potential (Eds. Husen A, Iqbal M) Springer International Publishing AG, Gewerbestrasse 11, 6330 Cham pp. 193–291 (https://doi.org/10.1007/978-3-030-05569-1_7
69. Husen A (2021) Traditional Herbal Therapy for The Human Immune System. Taylor & Francis Group, LLC, 6000 Broken Sound Pkwy. NW
70. Husen A (2022a) Exploring Poisonous Plants: Medicinal Values, Toxicity Response and Other Potential Uses. Taylor & Francis Group, LLC, 6000 Broken Sound Pkwy. NW
71. Husen A (2022b) Environmental Pollution and Medicinal Plants. Taylor & Francis Group, LLC, 6000 Broken Sound Pkwy. NW
72. Husen A (2022c) Herbs, Shrubs and Trees Of Potential Medicinal Benefits. Taylor & Francis Group, LLC, 6000 Broken Sound Pkwy. NW
73. Husen A, Jawaid M (2020) Nanomaterials for Agriculture and Forestry Applications. Elsevier Inc., 50 Hampshire St., 5th Floor, Cambridge, MA 02139, USA https://doi.org/10.1016/C2018-0-02349-X
74. Husen A, Siddiqi KS (2022) Advances in Smart Nanomaterials and their Applications. Elsevier Inc., 50 Hampshire St., 5th Floor, Cambridge, MA 02139, USA
75. Husen A, Bachheti RK, Bachheti A (2021) Non-Timber Forest Products (*Food, Healthcare and Industrial Applications*). Springer Nature Switzerland AG, Gewerbestrasse 11, 6330 Cham, Switzerland
76. Höglund S (1968) Some electron microscopic studies on the satellite tobacco necrosis virus and its IgG-antibody. J Gen Virol 2:427–436
77. Iqbal Z, Sarkhosh A, Balal RM et al (2021) Silicon Nanoparticles Mitigate Hypoxia-Induced Oxidative Damage by Improving Antioxidants Activities and Concentration of Osmolytes in Southern Highbush Blueberry Plants. Agronomy 11:2143
78. Jianrong C, Yuqing M, Nongyue H et al (2004) Nanotechnology and biosensors. Biotechnol Adv 22:505–518
79. Jones N, Ray B, Ranjit KT, Manna AC (2008) Antibacterial activity of ZnO nanoparticle suspensions on a broad spectrum of microorganisms. FEMS Microbiol Lett 279:71–76
80. Joubert E, Gelderblom WCA, Louw A, De Beer D (2008) South African herbal teas: aspalathus linearis, Cyclopia spp. and Athrixia phylicoides—A review. J Ethnopharmacol 119:376–412
81. Kah M, Hofmann T (2014) Nanopesticide research: current trends and future priorities. Environ Int 63:224–235

82. Khan N, Ali S, Latif S, Mehmood A (2022) Biological Synthesis of Nanoparticles and Their Applications in Sustainable Agriculture Production. Nat Sci 14:226–234
83. Khan R, Ghazali FM, Mahyudin NA, Samsudin NIP (2021) Aflatoxin Biosynthesis, Genetic Regulation, Toxicity, and Control Strategies: A Review. J Fungi 7:606
84. Khoobchandani M, Katti KK, Karikachery AR et al (2020) New approaches in breast cancer therapy through green nanotechnology and nano-ayurvedic medicine–pre-clinical and pilot human clinical investigations. Int J Nanomedicine 15:181
85. Kim B, Kim D, Cho D, Cho S (2003) Bactericidal effect of TiO2 photocatalyst on selected food-borne pathogenic bacteria. Chemosphere 52:277–281
86. Król A, Railean-Plugaru V, Pomastowski P, Buszewski B (2019) Phytochemical investigation of Medicago sativa L. extract and its potential as a safe source for the synthesis of ZnO nanoparticles: the proposed mechanism of formation and antimicrobial activity. Phytochem Lett 31:170–180
87. Kumar R, Münstedt H (2005) Silver ion release from antimicrobial polyamide/silver composites. Biomaterials 26:2081–2088
88. Kumar PV, Kala SMJ, Prakash KS (2019) Green synthesis of gold nanoparticles using Croton Caudatus Geisel leaf extract and their biological studies. Mater Lett 236:19–22
89. Kuppusamy P, Yusoff MM, Maniam GP, Govindan N (2016) Biosynthesis of metallic nanoparticles using plant derivatives and their new avenues in pharmacological applications–An updated report. Saudi Pharm J 24:473–484
90. Kätzel U, Vorbau M, Stintz M et al (2008) Dynamic light scattering for the characterization of polydisperse fractal systems: II. Relation between structure and DLS results. Part Part Syst Charact 25:19–30
91. Lateef A, Ojo SA, Elegbede JA (2016) The emerging roles of arthropods and their metabolites in the green synthesis of metallic nanoparticles. Nanotechnol Rev 5:601–622
92. Liau SY, Read DC, Pugh WJ et al (1997) Interaction of silver nitrate with readily identifiable groups: relationship to the antibacterialaction of silver ions. Lett Appl Microbiol 25:279–283
93. Madanat HM, Al Antary TM, Zarqa M (2016) Toxicity of six ethanol plant extracts against the green peach aphid Myzus persicae Sulzer (Homoptera: Aphididae). Fresen Environ Bull 25:706–718
94. Mahomoodally MF (2013) Traditional medicines in Africa: an appraisal of ten potent African medicinal plants. Evidence-Based Complement Altern Med 2013:
95. Majoumouo MS, Sharma JR, Sibuyi NRS et al (2020) Synthesis of biogenic gold nanoparticles from Terminalia mantaly extracts and the evaluation of their in vitro cytotoxic effects in cancer cells. Molecules 25:4469
96. Manisha DR, Alwala J, Kudle KR, Rudra MPP (2014) Biosynthesis of silver nanoparticles using flower extracts of Catharanthus roseus and evaluation of its antibacterial efficacy. World J Pharm Pharm Sci 3:877–885
97. Marambio-Jones C, Hoek E (2010) A review of the antibacterial effects of silver nanomaterials and potential implications for human health and the environment. J nanoparticle Res 12:1531–1551
98. Marimuthu S, Antonisamy AJ, Malayandi S et al (2020) Silver nanoparticles in dye effluent treatment: A review on synthesis, treatment methods, mechanisms, photocatalytic degradation, toxic effects and mitigation of toxicity. J Photochem Photobiol B Biol 205:111823
99. Martínez Espinosa JC, Carrera Cerritos R, Ramírez Morales MA et al (2020) Characterization of silver nanoparticles obtained by a green route and their evaluation in the bacterium of pseudomonas aeruginosa. Curr Comput-Aided Drug Des 10:395
100. Mihranyan A, Ferraz N, Strømme M (2012) Current status and future prospects of nanotechnology in cosmetics. Prog Mater Sci 57:875–910
101. Mittal AK, Chisti Y, Banerjee UC (2013) Synthesis of metallic nanoparticles using plant extracts. Biotechnol Adv 31:346–356
102. Mordeniz C (2019) Introductory chapter: traditional and complementary medicine. Traditional and Complementary Medicine. Cengiz Mordeniz, IntechOpen

103. Mukhopadhyay SS (2014) Nanotechnology in agriculture: prospects and constraints. Nanotechnol Sci Appl 7:63
104. Munazir M, Qureshi R, Munir M (2015) Preliminary phytochemical screening of roots and aerial parts of Leptadenia pyrotechnica. Pak J Bot 47:659–664
105. Mustafa G, Ahmed S, Ahmed N, Jamil A (2016) Phytochemical and antibacterial activity of some unexplored medicinal plants of Cholistan desert. Pakistan J Bot 48:2057–2062
106. Nadeem M, Khan R, Afridi K et al (2020) Green synthesis of cerium oxide nanoparticles (CeO2 NPs) and their antimicrobial applications: a review. Int J Nanomedicine 15:5951
107. Nanda S, Nanda A, Lohan S, et al (2016) Nanocosmetics: performance enhancement and safety assurance. In: Nanobiomaterials in galenic formulations and cosmetics. Elsevier, pp 47–67
108. Narayanan KB, Sakthivel N (2011) Green synthesis of biogenic metal nanoparticles by terrestrial and aquatic phototrophic and heterotrophic eukaryotes and biocompatible agents. Adv Colloid Interface Sci 169:59–79
109. Nasseri MA, Shahabi M, Allahresani A, Kazemnejadi M (2019) Eco-friendly biosynthesis of silver nanoparticles using aqueous solution of Spartium junceum flower extract. Asian J Green Chem 3:382–390
110. Naz S, Islam M, Tabassum S et al (2019) Green synthesis of hematite (α-Fe2O3) nanoparticles using Rhus punjabensis extract and their biomedical prospect in pathogenic diseases and cancer. J Mol Struct 1185:1–7
111. Necula AM, Dunca S, Stoica I et al (2010) Morphological properties and antibacterial activity of nano-silver-containing cellulose acetate phthalate films. Int J Polym Anal Charact 15:341–350
112. Nguyen NTT, Nguyen LM, Nguyen TTT, et al (2022) Recent advances on botanical biosynthesis of nanoparticles for catalytic, water treatment and agricultural applications: A review. Sci Total Environ 154160
113. Nohynek GJ, Lademann J, Ribaud C, Roberts MS (2007) Grey goo on the skin? Nanotechnology, cosmetic and sunscreen safety. Crit Rev Toxicol 37:251–277
114. Omer AM (2008) Energy, environment and sustainable development. Renew Sustain energy Rev 12:2265–2300
115. Otulak K, Kozieł E, Garbaczewska G, et al (2014) Seeing is believing. The use of light, fluorescent and transmission electron microscopy in the observation of pathological changes during different plant—Virus interactions. Microsc Adv Sci Res Educ 6th ed; Mendez-Vilas, A, Ed 43:367–376
116. Paciotti GF, Myer L, Weinreich D et al (2004) Colloidal gold: a novel nanoparticle vector for tumor directed drug delivery. Drug Deliv 11:169–183
117. Pagar K, Ghotekar S, Pagar T et al (2020) Antifungal activity of biosynthesized CuO nanoparticles using leaves extract of Moringa oleifera and their structural characterizations. Asian J Nanosci Mater 3:15–23
118. Panáček A, Kvitek L, Prucek R et al (2006) Silver colloid nanoparticles: synthesis, characterization, and their antibacterial activity. J Phys Chem B 110:16248–16253
119. Patra JK, Baek K-H (2014) Green nanobiotechnology: factors affecting synthesis and characterization techniques. J Nanomater 2014:
120. Perveen K, Husain FM, Qais FA et al (2021) Microwave-assisted rapid green Synthesis of gold nanoparticles using seed extract of Trachyspermum ammi: ROS mediated biofilm inhibition and anticancer activity. Biomolecules 11:197
121. Prasad KS, Prasad SK, Ansari MA et al (2020) Tumoricidal and bactericidal properties of ZnONPs synthesized using Cassia auriculata leaf extract. Biomolecules 10:982
122. Qamar SUR, Ahmad JN (2021) Nanoparticles: Mechanism of biosynthesis using plant extracts, bacteria, fungi, and their applications. J Mol Liq 334:116040
123. Rafique M, Rafique MS, Kalsoom U et al (2019) Laser ablation synthesis of silver nanoparticles in water and dependence on laser nature. Opt Quantum Electron 51:1–11
124. Rajagopal T, Jemimah IAA, Ponmanickam P, Ayyanar M (2015) Synthesis of silver nanoparticles using Catharanthus roseus root extract and its larvicidal effects. J Environ Biol 36:1283

125. Rajendran K, Sen S, Suja G et al (2017) Evaluation of cytotoxicity of hematite nanoparticles in bacteria and human cell lines. Colloids Surfaces B Biointerfaces 157:101–109
126. Ramesh M, Anbuvannan M, Viruthagiri G (2015) Green synthesis of ZnO nanoparticles using Solanum nigrum leaf extract and their antibacterial activity. Spectrochim Acta Part A Mol Biomol Spectrosc 136:864–870
127. Ramkumar S, Baskar V, Skymoon R, et al (2022) Green synthesis of nanoparticles and their uses in agriculture. In: Nano-enabled Agrochemicals in Agriculture. Elsevier, pp 247–271
128. Razavi M, Salahinejad E, Fahmy M, et al (2015) Green chemical and biological synthesis of nanoparticles and their biomedical applications. Green Process Nanotechnol 207–235
129. Robert KW, Parris TM, Leiserowitz AA (2005) What is sustainable development? Goals, indicators, values, and practice. Environ Sci policy Sustain Dev 47:8–21
130. Saleem K, Khursheed Z, Hano C et al (2019) Applications of nanomaterials in leishmaniasis: a focus on recent advances and challenges. Nanomaterials 9:1749
131. Salem SS, Fouda A (2021) Green synthesis of metallic nanoparticles and their prospective biotechnological applications: an overview. Biol Trace Elem Res 199:344–370
132. Salvioni L, Galbiati E, Collico V et al (2017) Negatively charged silver nanoparticles with potent antibacterial activity and reduced toxicity for pharmaceutical preparations. Int J Nanomedicine 12:2517
133. dos Santos MA, Paterno LG, Moreira SGC, Sales MJA (2019) Original photochemical synthesis of Ag nanoparticles mediated by potato starch. SN Appl Sci 1:1–13
134. Schneider SL, Lim HW (2019) A review of inorganic UV filters zinc oxide and titanium dioxide. Photodermatol Photoimmunol Photomed 35:442–446
135. Shafiq M, Anjum S, Hano C et al (2020) An overview of the applications of nanomaterials and nanodevices in the food industry. Foods 9:148
136. Shah M, Fawcett D, Sharma S et al (2015) Green synthesis of metallic nanoparticles via biological entities. Materials (Basel) 8:7278–7308
137. Shankar SS, Ahmad A, Pasricha R, Sastry M (2003) Bioreduction of chloroaurate ions by geranium leaves and its endophytic fungus yields gold nanoparticles of different shapes. J Mater Chem 13:1822–1826
138. Shankar SS, Ahmad A, Sastry M (2003) Geranium leaf assisted biosynthesis of silver nanoparticles. Biotechnol Prog 19:1627–1631
139. Sharma AK, Kaith BS, Shanker U, Gupta B (2020) γ-radiation induced synthesis of antibacterial silver nanocomposite scaffolds derived from natural gum Boswellia serrata. J Drug Deliv Sci Technol 56:101550
140. Siddiqi KS, Husen A (2016) Fabrication of metal and metal oxide nanoparticles by algae and their toxic effects. Nanoscale Res Lett 11(363):1–11
141. Siddiqi KS, Husen A (2016) Fabrication of metal nanoparticles from fungi and metal salts: scope and application. Nanoscale Research Letters 11(98), 1–15Singh P, Kim Y-J, Zhang D, Yang D-C (2016) Biological synthesis of nanoparticles from plants and microorganisms. Trends Biotechnol 34:588–599
142. da Silva BF, Pérez S, Gardinalli P et al (2011) Analytical chemistry of metallic nanoparticles in natural environments. TrAC Trends Anal Chem 30:528–540
143. Singh T, Shukla S, Kumar P et al (2017) Application of nanotechnology in food science: perception and overview. Front Microbiol 8:1501
144. Singh P, Kim Y-J, Zhang D, Yang D-C (2016) Biological synthesis of nanoparticles from plants and microorganisms Trends Biotechnol 34 588 599
145. De Souza CD, Nogueira BR, Rostelato MECM (2019) Review of the methodologies used in the synthesis gold nanoparticles by chemical reduction. J Alloys Compd 798:714–740
146. Subramanian A, Irudayaraj J, Ryan T (2006) A mixed self-assembled monolayer-based surface plasmon immunosensor for detection of E. coli O157: H7. Biosens Bioelectron 21:998–1006
147. Sánchez-Valdes S, Ortega-Ortiz H, Ramos-de Valle LF et al (2009) Mechanical and antimicrobial properties of multilayer films with a polyethylene/silver nanocomposite layer. J Appl Polym Sci 111:953–962

148. Tavakoli J, Miar S, Zadehzare MM, Akbari H (2012) Evaluation of effectiveness of herbal medication in cancer care: a review study. Iran J cancer Prev 5:144
149. Thakkar KN, Mhatre SS, Parikh RY (2010) Biological synthesis of metallic nanoparticles. Nanomedicine nanotechnology, Biol Med 6:257–262
150. Tyavambiza C, Elbagory AM, Madiehe AM et al (2021) The antimicrobial and anti-inflammatory effects of silver nanoparticles synthesised from Cotyledon orbiculata aqueous extract. Nanomaterials 11:1343
151. Ünlü E, Erdem C (2010) Deri yaşlanmasında korunma ve tedavi yöntemleri. Dermatoz 1:23–31
152. Velidandi A, Dahariya S, Pabbathi NPP et al (2020) A review on synthesis, applications, toxicity, risk assessment and limitations of plant extracts synthesized silver nanoparticles. NanoWorld J 6:35–60
153. Velsankar K, Sudhahar S, Maheshwaran G (2019) Effect of biosynthesis of ZnO nanoparticles via Cucurbita seed extract on Culex tritaeniorhynchus mosquito larvae with its biological applications. J Photochem Photobiol B Biol 200:111650
154. Vermeiren L, Devlieghere F, Debevere J (2002) Effectiveness of some recent antimicrobial packaging concepts. Food Addit Contam 19:163–171
155. Villaseñor-Basulto DL, Pedavoah M-M, Bandala ER (2019) Plant materials for the synthesis of nanomaterials: Greener Sources. Handb Ecomater by Martínez, L, Kharissova, O, Kharisov, B(eds) Springer, Cham 105–121
156. Young M, Debbie W, Uchida M, Douglas T (2008) Plant viruses as biotemplates for materials and their use in nanotechnology. Annu Rev Phytopathol 46:361–384
157. Yousefi M (2014) A fast method for synthesis magnesium hydroxide nanoparticles, thermal stable and flame retardant poly vinyl alcohol nanocomposite. J Nanostructures 4:383–388
158. Zhang D, Zhang J, Zeng J et al (2019) Nano-gold loaded with resveratrol enhance the anti-hepatoma effect of resveratrol in vitro and in vivo. J Biomed Nanotechnol 15:288–300
159. Zheng L, Hong F, Lu S, Liu C (2005) Effect of nano-TiO2 on strength of naturally aged seeds and growth of spinach. Biol Trace Elem Res 104:83–91

The manufacturer's authorised representative in the EU is Springer Nature Customer Service Centre GmbH, Europaplatz 3, 69115 Heidelberg, Germany. If you have any concerns regarding our products, please contact ProductSafety@springernature.com

Printed and bound by CPI Group (UK) Ltd, Croydon, CR0 4YY
01/10/2026
02225892-0012